Ornamental Horticulture

Ornamental Horticulture

Gary L. McDaniel

Reston Publishing Company
A Prentice-Hall Company
Reston, Virginia

Library of Congress Cataloging in Publication Data
McDaniel, Gary L
    Ornamental horticulture.

    Includes index.
    1. Ornamental horticulture. 2. Ornamental
horticulture—Vocational guidance. I. Title.
SB404.9.M33   635.9'023   78-25589
ISBN 0-8359-5346-7

10   9   8   7   6   5   4   3   2   1

Printed in the United States of America

To my father,
Lester L. McDaniel

Contents

Preface, xiii

Part One: Ornamental Horticulture: A Field with Promise

1 **The History of American Ornamental Horticulture, 3**
It Began at Jamestown, 4
Noted Early American Horticulturists, 6
Postrevolutionary Horticulture, 7
The Florist Industry, 10
America's Laws Were an Important Influence, 11
The Future for Ornamental Horticulture, 16
Variety or Cultivar? 17
Terms to Know, 17
Study Questions, 18
Suggested Activities, 18

2 **Career Opportunities in Ornamental Horticulture, 19**
Occupational Descriptions and Qualifications, 23
Floriculture Industry, 23
Nursery Industry, 30
Turf Grass Industry, 35
Ornamental Horticulture: A Field with Promise for Men and
    Women, 37
Selected References, 38

Teaching Aids, 40
Terms to Know, 41
Study Questions, 41
Suggested Activities, 41

3   Judging Flowers and Ornamental Plants, 42
Contest Eligibility, 42
Selection of State Judging Teams, 44
Contest Rules, 44
Individual and Team Scoring, 45
Sections of the Horticulture Contest, 46
Selected References, 88
Terms to Know, 89
Study Questions, 89
Suggested Activities, 90

Part Two: Greenhouse Production and Marketing

4   The Greenhouse, 93
Greenhouse Structures, 94
Construction Materials for Greenhouses, 98
Comparison of Construction Styles, 99
Greenhouse Benches, 101
Greenhouse Environmental Controls, 107
Hotbeds and Coldframes, 115
Shadehouses, 116
Greenhouse Employees, 116
Selected References, 117
Terms to Know, 117
Study Questions, 117
Suggested Activities, 118

5   Plant Production in the Greenhouse, 119
Plant Propagation, 119
Greenhouse Growing Media, 127
Pots and Containers for Florists' Crops, 128
Fertilizing and Watering Greenhouse Plants, 130
Greenhouse Pest Control, 138
Proper Handling of Pesticides, 143
Summary, 147
Selected References, 148
Terms to Know, 150
Study Questions, 150
Suggested Activities, 151

6   Specialized Crop Production, 152
Chrysanthemums, 152
Poinsettia, 160
Florists' Forcing Azaleas, 165
Potted Easter Lily, 168
Growing Carnations, 173
Cut Rose Production, 180
Selected References, 185
Terms to Know, 186
Study Questions, 186
Suggested Activities, 186

7   The Foliage Plant Business, 187
Terrarium Construction, 188
Planning the Terrarium, 191
Planting the Terrarium, 195
Planters and Dish Gardens, 201
Hanging Baskets, 208
The Plant Shop Employee, 216
Selected References, 216
Terms to Know, 217
Study Questions, 218
Suggested Activities, 218

8   House Plant Care, 219
Caring for Plants in the Shop, 220
House Plants in the Home, 223
Diagnosing House Plant Problems, 228
Diseases and Insect Pests on House Plants, 232
Plants in Interior Design, 238
Review of the Diagnosis of Plant Disorders, 239
Selected List of Plant Requirements for Interior Plantings, 240
Selected References, 242
Terms to Know, 243
Study Questions, 243
Suggested Activities, 244

Part Three: The Florist Business

9   The Retail Flower Shop, 247
Types of Flower Shops, 247
Employees of a Flower Shop, 249
Handling Cut Flowers and Potted Plants, 255
Wholesale Florists, 259

Selected References, 262
Terms to Know, 263
Study Questions, 263
Suggested Activities, 263

10   **Floral Designs — Arrangements, 264**
History of Design, 264
Principles of Design, 265
Design Elements, 268
Selection of Floral Containers, 271
Selection of Flowers and Foliage, 273
Mechanical Aids for Arrangements, 274
Floral Arrangements, 279
Selected References, 287
Terms to Know, 288
Study Questions, 289
Suggested Activities, 289

11   **Floral Designs — Corsages and Specialty Pieces, 290**
Corsage Construction, 290
Designing Corsages, 299
Wedding Designs, 311
Funeral Designs, 312
Wreaths, 317
Selected References, 319
Terms to Know, 319
Study Questions, 319
Suggested Activities, 320

12   **Customer Relations and Sales, 321**
Personal Requirements for Selling, 322
Personal Selling Techniques, 326
Retail Sales Techniques, 329
Making Telephone Sales, 333
Selling Through Visual Merchandising, 336
Becoming Proficient in Selling, 339
Selected References, 340
Terms to Know, 341
Study Questions, 341
Suggested Activities, 341

Part Four: The Nursery Industry

13   **Production of Nursery Stock, 345**
The Wholesale Nursery, 346

Nursery Stock Propagation, 347
Propagation by Vegetative Plant Parts, 348
Propagation of Nursery Plants from Seeds, 356
Field Production of Nursery Stock, 359
Container Production of Nursery Stock, 368
Standard Grades for Nursery Stock, 375
Selected References, 376
Terms to Know, 377
Study Questions, 377
Suggested Activities, 377

14   Landscape Design, 379
Plan Before You Plant, 379
Adding Design to the Landscape, 381
Design for Family Personality, 383
Landscaping Principles, 383
Groundcover Plants for Landscaping, 402
Perennial Vines in the Landscape, 406
Rock Gardens in the Landscape, 408
Selected References, 412
Terms to Know, 413
Study Questions, 413
Suggested Activities, 413

15   Turf Grass Maintenance, 415
Establishing and Maintaining Lawns, 416
Establishing Turf Grass Lawns from Seed, 425
Establishing Lawns by Vegetative Methods, 426
Maintaining an Established Lawn, 429
Renovation of Lawns, 435
Lawn Pest Control, 438
Golf Course Maintenance, 441
Selected References, 444
Terms to Know, 445
Study Questions, 446
Suggested Activities, 446

16   Landscape Maintenance, 447
Planting Ornamental Trees and Shrubs, 448
Fertilizing Established Trees and Shrub Plantings, 465
Pruning Landscape Plants, 467
Pruning Ornamental and Shade Trees, 480
Pruning and Care of Roses, 485
Selected References, 488
Terms to Know, 489

Study Questions, 489
Suggested Activities, 490

Glossary, 491
Appendices, 504
Index, 519

Preface

The ornamental horticulture industry in the United States comprises a complex grouping of businesses involved in the production and sales of nursery stock and florists' crops, design of interior and exterior landscapes, and development of recreational areas for our enjoyment. The student of horticulture who wishes to find employment in one of these businesses will benefit from the combined training received from classroom instruction and practical experience.

The book was written to provide a general text for persons interested in ornamental horticulture as a vocation. The author has attempted to provide the student of ornamental horticulture with an overview of the various types of businesses in the industry. A description of the job requirements and duties performed is presented for various vocational fields. The material presented is intended for use in introductory horticulture courses at the secondary and vocational-technical school or college levels. It would also serve as a reference text for those presenting information to students trying to select a vocation or for those persons entering various fields of ornamental horticulture who have received little previous training.

Part I of the text offers an overview of the fields of ornamental horticulture by highlighting the influencing factors in the progress of American horticulture. Career opportunities are considered from a realistic viewpoint which will help the reader select the vocation and level of training best suited to him. The judging of flowers and ornamental plants is considered for the vocational student who will be entering competitive judging events as part of their horticulture training program.

The production and marketing of greenhouse plants is covered thoroughly in Part II. A detailed discussion of greenhouse structures and crop production methods is presented. The specific growth requirements of selected floriculture crops are outlined to assist the reader in understanding how they are grown in commercial greenhouses. Special attention is also given to the foliage plant industry, with chapters devoted to the discussion of plant uses and care.

The florist business and floral design methods are presented in Part III. The techniques used by florists in the creation of designs are thoroughly examined, with sections devoted to arrangements and specialty pieces. Selling techniques used in all types of retail shops are explained in detail in the final chapter of this unit. The serious student of horticulture will refer to this unit long after the course is completed.

Part IV provides a detailed explanation of the production, uses, design, and maintenance of nursery and landscaping materials. The production of nursery stock is presented in detail for the student interested in a career in this field. Landscape design principles and methods are explained sufficiently to provide a working knowledge of plant material functions in a well-designed landscape planting. The maintenance of landscape plantings is presented from the standpoint of both turf grass and landscape plantings. Pruning, planting, and general care of landscape plants are explained. The planting and care of lawn grasses for ornamental and recreational purposes are given special attention in this unit.

Finally, a glossary of terms unique to the field of horticulture is provided to assist students. The appendices then outline step-by-step procedures for the production of specific crops and present other important aides for the teacher and student of ornamental horticulture.

Gary L. McDaniel

PART ONE

Ornamental Horticulture:
A Field with Promise

CHAPTER ONE

The History of American Ornamental Horticulture

American history has been influenced greatly by the horticultural sciences as it has shaped the development of this country. The earliest settlers of North America depended largely upon the growth of crops that could be used for food and survival. As these early pioneers in the new lands established their homes and communities, more time could be spent in the pursuance of aesthetic arts and pleasures. Early horticulture was based more on a haphazard art, which relied heavily on luck. Little information was available concerning the growth of plants and development of hardier varieties. Modern horticulturists require little guesswork in the production of crops, since the field of horticulture has evolved into a science.

Ornamental horticulture evolved originally as a hobby pursued by the rich and influential. Persons who were accredited with early horticultural discoveries were often physicians, lawyers, or political figures. These individuals collected interesting and rare plants for display in their gardens. The love of new and rare plants is not restricted to modern society. It is interesting to note that the first recorded plant-hunting expedition was organized more than 3000 years ago (about 1570 B.C.) by Queen Hatshepset of Egypt. Like other rulers since those early days of history, unusual plants were used to decorate the royal gardens and were added to collections to demonstrate wealth and power.

Religion has also influenced the development of ornamental horticulture since the early history of mankind. Flowers and plants have been used since the first record of history for decorating tables at feasts or other religious festivities. The greatest influence was evolved from the work of monks living in monestaries of Medieval Europe. These monks collected the many wild plants known at the time to select those which were most useful for medicinal purposes. When the art of printing was developed, these monks recorded the descriptions of the various plant types known at the period. They also documented the then-common methods for plant culture. If it had not been for these early records, much knowledge concerning the ancient plants and plant culture would be lost to modern society.

By the time North America was being explored and colonized, the European gardens had become quite popular. The English manor and formal botanical gardens were highly fashionable, even among the commoners of western European countries. Although these gardens were often small, they were highly formal and attended by nearly every household. These gardens contained collections of wild flowers and shrubbery which were largely nondecorative and of a short blooming period. There existed at that time a basic lack of knowledge concerning the culture, propagation, and breeding of plants. For this reason, the plants were sheared into many unusual shapes so that they might appear more interesting.

IT BEGAN AT JAMESTOWN

The first settlers of the North American continent brought their ideas of gardening from western Europe during the seventeenth century. The first permanent settlement in America was in Jamestown, Virginia, in 1607. These settlers had brought the seeds and cuttings of plants they knew best from their homes in England. Among the plants were various types of herbs for medicinal purposes and vegetables. If it had not been for these familiar plants, the settlement could not have survived.

These early colonists obtained a vast knowledge about the uses of North American plants from the Indians they found in the new country. These Indians had obtained an extensive botanical knowledge. They taught these colonists which plants could be used for such things as food, medicine, dyes, deodorant, hair tonic, fiber, and utensils. Many Indian tribes cultivated vegetables for food and introduced the use of maize (corn) to the colonists. Wildflowers were grown by some Indians for decoration at ceremonial events. The colonists quickly adapted the use of these many new plants and began trading and exchanging them with European plant collectors. Ship captains found it a challenge to keep these plants alive on the long voyages between the continents. Consequently, seeds and cuttings were often shipped.

After their need for survival was met, the pilgrims planted flower gardens for their personal pleasures. They were interested in continuing the fashions of Europe and planted front yard gardens. Herbs were also used extensively in the ornamental gardens, where they added interesting texture, color tones, and aroma. The herbs were also practical for their medicinal qualities.

Topiary art.   During the period of colonization, topiary art was prevalant. The act of pruning and shaping plants into elaborate geometric forms was not new. This formal garden technique, which can be seen reconstructed at Williamsburg, Virginia, dates back to the Roman Empire (Figure 1–1). It was used in early France and became fashionable in England during the sixteenth and seventeenth centuries. Boxwood and yew were generally used for topiary designs and trees were pruned into fantastic shapes. Maze gardens were also popular during this period. These shrub barriers bordered paths which guided people through the formal gardens. A relaxing atomosphere was created by these designs. The wealthy citizens and politicians used this type of garden design exclusively, because such gardens required continous upkeep and maintenance.

FIGURE 1–1.   Topiary art and formal gardens reconstructed at Williamsburg, Virginia.

Greens.   Also popular in colonial America were "greens." A forerunner to parks as we know them today, these open areas of turf were used for public and private recreation. They were social gathering places where the colonists listened to village band presentations and played small group games. The greens served as a marketplace as well as military drill areas.

Colonial plant culture revolved around the belief that all living things were correlated to the moon. The colonists planted, pruned, weeded, and harvested according to the moon signs. They religiously followed the *almanac,* a publication stating astronomical and meteorological data. It also contains a varied assortment of other data, such as statements of philosophy, facts, and statistics.

Beginning of the nursery industry. With a growing demand for plants, especially as food, Americans began to exchange plant stock. Fruit and nut trees were the most desired plants. The first record of what might be called a nursery business was in 1648, when Henry Walcott, Jr. of Connecticut filled orders for fruit trees. In that one year, he received 32 orders from 25 persons.

The colonists were dissatisfied with the plants that were transported from England. Many arrived dead, and the seeds would not germinate after the long voyage between continents. Consequently, the nursery business began to flourish in America.

NOTED EARLY AMERICAN HORTICULTURISTS

John Bartram. John Bartram is recognized as a great early American horticulturist during the late part of the seventeenth century. This physician, who was known to use many herbal remedies, established one of the most well-known American botanical gardens. He gathered many of the plants from plant-hunting expeditions. As a leader among early American plant explorers, he brought many species into cultivation from the wilds of America.

Benjamin Franklin. Benjamin Franklin, a Pennsylvanian, was one of the first Americans to suggest formal education in horticulture—as early as 1750. In his *Proposals Relating to the Education of Youth in Pennsylvania,* he wrote:

> Might not a little gardening, planting, grafting, and inoculating be taught . . . The improvement of agriculture being useful to us all, and skill in it no disparagement to any.

Americans then and now have recognized the truth and wisdom in such thinking and have incorporated agricultural courses into the curriculum of our country's best school systems. Benjamin Franklin is also recognized for his bits of wisdom published in *Poor Richard's Almanac.* As noted earlier, almanacs played an important role in the lives of early Americans. The dates for plant cultivation and care were reverently adhered to by most persons who believed in following the moon signs.

Martha Logan. The first horticulture book written in America was authored by Martha Logan when she was 72 years old.

Mrs. Logan's book was titled *The Gardener's Kalender*. This early treatise on horticulture was apparently published several times around 1774–1779. No copy is known to be in existence today.

Her influence on gardening was great, especially in the Charles Town, South Carolina, area, where she was known as a community member of high standing. Her father was Robert Daniel, a proprietory governor of South Carolina. Mrs. Logan was also widely known for exchanging plants and seeds with noted horticulturists of the period, such as with John Bartram.

POSTREVOLUTIONARY HORTICULTURE

A gradual, but important change in gardening took place immediately following the revolution. The formal, English-style gardens were being replaced by a naturalistic style of gardening. Topiaries, geometrical designs, and avenues of shaped trees were being destroyed. This seems to be related to the expression of freedom which was the general attitude of Americans following the acquisition of their freedom from England. During this time, as in the earlier colonial days, the homes of the rich were adorned with elaborate gardens made up of plant collections, while the homes of the poor were devoid of ornamental plantings.

Fences were an important part of the colonial and post-revolutionary period. The little cottage with a white picket fence, often spoken of in literature, was the dream of the lower class. Actually, these fences served an important function. Livestock was allowed to roam free and fences were a necessity to prevent the animals from ruining the gardens. In the 1700s, wrought-iron gates became fashionable. Many types of fences were used, but rails and stone walls were especially prevalent because of readily available materials.

As the naturalistic garden came more into its own, *ha ha's* became popular. They were also called "ha haw" or "aha ha", because these terms were often used when visitors came close to stumbling into them. Actually, they were ditches, which sometimes incorporated mounds of earth on one or both sides. They prevented animals from entering the garden while allowing an unobstructed view of the surrounding landscape. Postrevolutionary gardens have been recreated at Mount Vernon and Monticello, the homes of Washington and Jefferson.

Thomas Jefferson. Thomas Jefferson made several important contributions to horticulture. He had studied landscape gardening in France. In 1803 he obtained funds in Congress for the Lewis and Clark Expedition up the Missouri to the Pacific. An important purpose of this project was to learn more about the plants of the West. So important was this endeavor that Lewis was sent to Philadelphia for nine months before the expedition to study botany from William Bartram, a noted professor and gifted son of John Bartram. In addition to the acquisition of much new botanical knowledge, many plants and seeds were brought back from this important exploration of the new world.

The Shakers.   The Shakers, a religious group, were important in a new development to horticultural marketing—packaging. They grew seeds of high quality and were the first to sell the seeds in packets. Herbs were also grown and packaged for sale outside the colony. Although it was against their religion to grow flowers for adornment and beauty, roses were cultivated for the production of rosewater, which was sold as flavoring and to bathe sick persons. Poppies were also grown to produce crude opium, which was sold for medicinal purposes. They also pioneered in *commission marketing*. Seeds were placed in stores to be sold for a commission. What was not sold was picked up at the end of the season. Living in celibacy, this religious sect could grow only by converting new members. Consequently, it did not grow and flourish.

The Prince Nursery.   Although several persons sold and exchanged small amounts of nursery stock, the Prince Nursery is America's first major commercial nursery. Nursery catalogs are available from 1771 through 1850 from this famous nursery. Four generations of the Prince family operated this nursery business. They led other nurseries in size and number of species and varieties of stock offered for sale. They are said to have advertised: *"Send your orders from anywhere in the world addressed to Prince—America"* and they get there. Their catalogs are good literature, representing horticulture of the early nineteenth century.

Johnny Appleseed.   One of the most eccentric characters in horticultural history is Johnny Appleseed (1777–1847). His name by birth was John Chapman. This pioneer was a preacher who was a friend to all. He acquired apple seeds from the cider mills and gave the seeds to any settler who would agree to plant them and promise to take care of the young seedlings. Sometimes where he found rich, loamy soil, he would plant the area and cover it with brush to prevent the animals from destroying the young plants. He would then tell the farmers to help themselves to the seedlings.

Wearing merely a heavy sack with holes cut for his neck, arms, and legs and using a cooking pot for a hat, he traversed barefooted across much of the frontier (especially Ohio and Indiana), spreading his philosophy and planting apples. His philosophy was not to injure any living thing. He believed that wickedness was practiced through pruning and grafting; therefore, thousands of acres of seedling apple trees were left which were only able to produce apples fit for cider. None of his seedlings are now in cultivation.

Bernard McMahon.   Bernard McMahon (1775–1818) was an Irish immigrant who arrived in Philadelphia in 1796. He started a lucrative seed and nursery business. As a writer, he published an extensive nursery/seed catalog and *The American Gardener's Calendar* in 1804. This calendar was the standard authority on plant materials and gardening throughout a 50-year period. Eleven editions of this publication were printed. McMahon was also recognized for the marketing of plants which had been discovered during the Lewis and Clark Expedition.

Oriental influence on horticulture.   Many of America's garden and greenhouse plants originated in the Orient. Missionaries and plant explorers located many fine flowering plants, such as the camellia (mid-1500s); the florist chrysanthemum and the *Rhododendron indicum* (early 1800s); oriental arbor vitae and the China aster (1740–1756); the speciosum lily, China rose, and *Rosa banksiae* (1780–1817); and the *Lilium auratum* (1860).

Gregor Mendel.   Gregor Mendel discovered the role of dominant and recessive characters in plant breeding in 1865. Modern genetics is based on his theories; however, at the time he published his findings, people were reluctant to recognize his concepts as being accurate. It was not until 1900 that his theories were actually put into practice in breeding.

Before Mendel, plant breeders had already learned some important concepts concerning genetic crossing of plants. One of these concerned the "spontaneous degeneration of horticultural varieties." This refers to the genetic situation where seeds from an exceptional parent plant will produce plants more related to the average of past generations.

The relationship of pollen and breeding had been discovered in 1676 by Nehemiah Grew. The world's first recorded artificial hybrid had been produced by Thomas Fairchild, a commercial grower, in 1717. By 1721, it was recognized that pollination of some plants was done by insects. But Mendel's theory clearly was a breakthrough in the understanding of genetics. Geneticists and plant breeders began an extensive breeding program aimed at the improvement of food and ornamental crops in the early years of the present century. Their efforts have resulted in plants that are disease- and nematode-resistant, more nutritious, and mature more rapidly with less attention by the grower. During the past half-century, plant exploration was reduced to a minimum, with the genetic approach to crop improvement being used as the primary method for development of new varieties. The recent boom in foliage plant popularity has again spurred an active search for new and different plants from exotic regions of the world for adaptation to American markets.

THE FLORIST INDUSTRY

Plant breeding was especially important to the florist industry. Flowers could now be produced with the qualities that were more desirable for use in homes and ceremonies. They could be produced in the most desirable colors and would last considerably longer than flowers from the garden or woods. Florist plants began to be more in demand once the public was made aware of these desirable qualities.

In 1780, there were less than 100 florists in America, and their combined glass structures could have covered 50,000 square feet. Not only was plant breeding a significant factor in the increase of the florist industry, but new progress in greenhouse structures was a major reason for the growth of the industry. In the 1880s and 1890s, florist crops were grown in pots and tubs. The flowers were small and short stemmed, had to be wired, and were best used when placed in closely packed nosegays.

By 1929, greenhouses were reported to have covered over 170,000,000 square feet of space (3400 times as much as in 1880). Eighty-nine percent of the business consisted of cut flowers and flowering plants, with 11 percent of the receipts from sale of vegetables. At this peak of production, only 10 percent of the buying public purchased flowers, leaving a great potential for expansion of the industry.

Greenhouses.

Plant-forcing houses have been in existence since the time of the Romans. They can be traced back as far as about 500 B.C. In Roman times, the walls were built of translucent mica sheets and heating was done through wall ducts. Later the Dutch led in the use of greenhouses. These crude greenhouses used charcoal braziers for heat. The English pitched their glass roofs at a 45-degree angle to the sun and heated them with wood-burning stoves. Heating frames were also used in the early 1700s, with heat provided by fermenting manure.

The first greenhouse in America is documented to have been constructed in Boston around 1710. This greenhouse was built by Andrew Faneuil on the estate of Governor Bellingham. Another early greenhouse (1730s) was located in Virginia and was owned by Colonel Byrd.

He had stocked it primarily with orange trees for his own use.

These early greenhouses were usually small, oblong houses with small glass panes used in windows along the sides. They possessed shingled roofs. Lean-to houses were also constructed for use as greenhouses. The walls were hollowed out to allow the passage of smoke from a fire for heating. As early as 1830, hot-air flues were used for heating. These were long heating tubes which carried heat from one end of the greenhouse to the opposite end. With these advancements in heating of greenhouses, the use of many exotic plants became fashionable, with orchids starting to be grown for sale.

Glass for these houses was imported to America from France and Belgium because American glass was waxy and spotted with bubbles. This caused plants to be burned by concentrated sun rays resulting from the defective glass. Hot-water heating for greenhouses was introduced in 1830, which created a boom in the building of greenhouses. In addition to hot-water heating, the concept of incorporating steel and glass into the structure was a major improvement. This innovation was the idea of J. C. Loudon, a writer/horticulturist of this period. Wide acceptance for these glass greenhouses became prevalent from the popularity of the famous Crystal Palace, which was built in England in 1851. J. C. Paxton, gardener for the Duke of Devonshire, designed and built this magnificent structure, which utilized the new techniques of both Loudon and himself. This fabulous structure inspired the building of many other botanical greenhouses and conservatories.

A craze for blue glass existed around the 1850s. It was placed in homes and greenhouses because many believed that if the sun's rays were filtered through blue glass, the resulting light would be healthful to human beings and plants. This theory would appear to be based on sound scientific information, since the photosynthetic reactions in plants are most efficient in the blue and red wavelengths of light. However, this blue glass reduced the total light transmission to a point where plants were unable to grow well.

Currently, flower, and to some extent foliage plant, production still dominates the use of greenhouses. Ninety percent of the production is in florist's crops, with 10 percent of American greenhouse space being used for vegetable crops.

AMERICA'S LAWS WERE AN IMPORTANT INFLUENCE

Horticulture in America began to expand with the westward movement of families. It was aided by the progress of the railroads, which made it easier to transport people and materials.

The Morrill Land-Grant Act. The *Morrill Land-Grant Act* (1862) was important legislation for the people of the United States. It made federal lands available for sale to endow colleges whose aim was to promote "liberal and practical education—in the several pursuits and professions of life." Colleges throughout the United States opened to the people. Horticulture was an important subject taught in these schools. Today, every state in the nation maintains at least one of these land-grant universities for the purpose of teaching any high school graduate the science and trades associated with home economics and the various agricultural sciences, including ornamental horticulture.

The Homestead Act. The *Homestead Act* (1862) of Congress encouraged families to move farther west. Therefore, the nursery business began to flourish, with an increased need for fruit and windbreak trees. After the Civil War, money was plentiful enough that nursery stock could be sold for what was considered a good price for the time. Pear trees sold for 35 cents each and apple trees sold for 18 to 20 cents each. Around 1872, nurserymen and farmers experienced severe problems with winter-kill of their crops. Times were hard and the prices for fruit trees nearly doubled. Many western nurserymen were put out of business by these hardships.

The Timber Culture Act. Following the *Timber Culture Act* (1873), homesteaders were required to have a tree-planting program before title to their new land in the western frontier could be claimed. Forty acres of timber trees were required on each homestead. These trees could be planted no more than 12 feet apart. This meant that approximately 7500 trees were required for each homestead. Production of deciduous tree seedlings became important, especially throughout the prairie states, where trees were not an abundant natural resource. By 1878, unscrupulous tree agents were plentiful. They traveled throughout the homesteaded regions, swindling nurserymen as well as the general public.

In 1876, an official trade organization was created. It adopted the name of the American Association of Nurserymen, Florists, and Seedsmen. Salesmen and dealers in horticultural supplies and implements were also included in the membership. Some of their purposes included: (1) quicker and better transportation of nursery products at more reasonable rates (this problem still persists today, even with modern transporation facilities); (2) avoidance of dishonest agents; and (3) prevention of taxation of nursery stock growing in the fields. This organization has since divided into more individualized associations, such as the Society for American Florists (SAF) in 1884, but the development of one large group was a major step in the coordination of efforts within this vast industry.

The Trade Mark Law.   The *Trade Mark Law* (1881) required the United States Department of Agriculture (USDA) to maintain an American Plant Register of Horticulture. All names of cultivated plants were to be listed, and new varieties were to be entered upon introduction. The plant originators were entitled to "exclusive propagation and sales rights" for a stated period of time. This law created disputes in some cases when more than one person claimed to be the plant originator. Finally, by 1930, the patent laws were amended to include plants. Plant patenting was then handled through the Department of Commerce. The first plant patent was issued in 1931 to H. F. Bosenberg, who was with the New Jersey Agricultural Experiment Station. It was for an everblooming rose produced from the Van Fleet climbing rose and named "New Dawn."

Communications by florists and nurserymen were done by telegraph prior to the general use of the telephone. Florists developed their own wire service, the Florists' Telegraph Delivery (FTD) in 1910, and nurserymen used a specialized telegraph code developed by W. F. Heikes of Huntsville, Alabama. This code used cipher words to describe messages in short form.

In the 1900s, mail-order firms were becoming a common method for marketing seeds and nursery stock. They distributed nursery stock and seeds of all types through their catalogs. Competition was created by the mail-order establishments. During the early years of the present century, fruit production was especially important. Railroads provided better accessibility to markets for horticultural products. Cut flowers and other perishable nursery stock could be shipped from coast to coast once the refrigerated railroad cars were in prominent use. By the early 1890s, iced boxcars were in heavy use by the horticulture industry for the transportation of their products. Freight classification and rates still posed problems for those who relied on railroads to transport their products to market.

In the 1920s, florists, seedsmen, and nurserymen were busy promoting and marketing their products (Figure 1–2). The insignia of the American Association of Nurserymen was created by Norman Rockwell and read "It's not a home until it's planted." This was such an effective promotion that it was continued for many years. These were the boom years; there was an enormous overproduction of horticultural products until the stock market crash of 1929.

In the early 1930s, a worldwide depression occurred. It was especially difficult for nurserymen; however, the depression did not hurt the flower industry as much as it did many other industries. Flowers for specific occasions were not considered to be a luxury, and the florist industry continued to grow through these economically hard times. Many of these greenhouse operations turned to production of

FIGURE 1–2. St. Louis World's Fair (1904). Nursery stock production boomed in the early 1900s, with fruit trees and cut flowers being widely distributed.

vegetable crops to provide food products for their families and customers through the cold winter months.

The War Years (1941–1945). All of America was seriously affected by World War II. Crucial controls were placed on materials and supplies to be used for the war effort. The War Production Board also placed heavy controls on manpower. Plant materials were needed for concealment and protective camouflage of vital industries, airfields, and emplacements. Nurserymen used their designing expertise and their largest plants to help in this effort. Every available ornamental plant was used for these camouflage purposes.

The Victory Garden Program was one in which all available space was converted to food production. Victory Garden Harvest Shows were conducted to help encourage the program. The American Association of Nurserymen was extensively involved in helping to promote and carry out this program for the war effort.

Education and training programs, which were conducted as a result of the war, helped to train persons in the field of horticulture. The War Manpower Commission conducted an Apprentice Training Service. These apprenticeship programs offered 2 years of on-the-job training for veterans of the war. The government paid a portion of the trainee's wages. These programs were especially popular with nurserymen, who required many field laborers and much technical assistance.

Vocational Education Act. As a result of the *Vocational Education Act* (1963), vocational courses in horticulture are included as acceptable courses in vocational programs at high school, post-high school, and technical institutes. The superior training by these educational institutions since the passage of this act has enabled ornamental horticulture students to find supervisory positions in nurseries, greenhouses, and other plant businesses.

Plant Variety Protection Act.   During the late 1960s, several proposals were made by the Commission on the Patent System to abolish the existing plant patent regulations. The dropping of these regulations would have created some serious problems for the horticulture industry, and a lack of inducement for developing new varieties would have resulted. During the following years, much debate was expressed by the various horticultural associations concerning these proposals. Finally, in 1970 a *Plant Variety Protection Act* was passed to end these debates. This act essentially encourages the development of new plant varieties which are seed-propagated (Figure 1–3). The plant patent laws as they were originally established are still valid. However, the plant patent regulatory duties are now administered by the U. S. Department of Agriculture rather than by the Patent Office of the Department of Commerce.

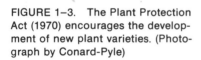

FIGURE 1–3.   The Plant Protection Act (1970) encourages the development of new plant varieties. (Photograph by Conard-Pyle)

Grades and standards.   Both the floriculture and nursery industries have attempted many times since the late nineteenth century to establish a uniform set of standards for their products. During the early years of development of the nursery industry, each grower was forced to use his own judgment concerning the quality of the goods he was buying. This became especially difficult when a landscape contractor or architect specified "first-class" plants, only to receive poorly developed or half-dead specimens. The industry badly needed some guidelines for establishing uniformity between nurseries or florists in price and size or quality of products.

Standard grades for nursery stock were the first to be recognized by the industry. Fruit tree classifications were developed, followed by those for shade and ornamental trees. These were classified by a system which considered both height of the tree and diameter of the trunk (caliper). Canned nursery stock was not included in these early standards until the 1960s. Once this new innovation had found industry-wide acceptance, grades for container stock were quickly adopted. These were based primarily on the size of container and the rate of growth of the plant species.

The Colorado Carnation Growers Association pioneered in the development of grades and standards for floriculture crops. Carnation grades were established using a color-coded tag. These were classified according to quality and length of stem. Grades have since been established for many other florists' crops, based on the same criteria.

THE FUTURE FOR ORNAMENTAL HORTICULTURE

The ornamental horticulture industry will continue to grow at a rapid rate; however, certain changes will take place which will alter the methods for growing and marketing plants as we know them today. Greenhouse production areas have steadily declined since World War II, succumbing to the encroachment of the cities onto their valuable lands. Although many large greenhouse ranges will continue to operate in the northern regions of the country, many more commercial florists' crops will be grown in the southern and western states. This will be largely a result of the dwindling and expensive fuel supplies available. Although not available for use in commercial greenhouses at this time, solar fuel systems will eventually assist in reducing the heating costs of production.

Greenhouses to be constructed for commercial production will generally be made of plastic rather than of glass glazing surfaces. Those elegant glass and steel structures that were once so prominent in every large community were expensive to heat and maintain. Today's modern greenhouses may extend over as much as 15 acres or more, all under polyethylene film. These structures may be constructed at a fraction of the cost of a similar glass range and may be heated for nearly half the cost. Such innovations must be adapted to hold the cost of production down, while maintaining high standards for the crops being produced.

The nursery industry will be challenged by the steady growth of new housing developments requiring landscaping. Some of the more prevalent changes now taking place include (1) urbanization at a rate that exceeds the production of landscape plants, (2) a gradual movement of the public into planned community areas and condominiums, (3) a shortage of skilled employees to fill positions in the industry, and (4) an increased willingness and awareness by the public to beautify their homes, businesses, and general environment. These factors provide a tremendous opportunity for those horticulture students who are willing to obtain further education and experience and to dedicate themselves to hard work. A person who possesses these qualities will find a rewarding career in the horticulture industry.

VARIETY OR CULTIVAR?

Throughout this textbook references to particular plants will be listed as either cultivars or as varieties. These terms are often confused and may be improperly interchanged by students unfamiliar with their true meaning. For many years industry members have attempted to formulate guidelines for the naming of horticultural plants. In a broad sense, use of the word *variety* applies only to those plants which are closely related within the same botanical species.

According to the International Code of Nomenclature for Cultivated Plants (1961), a *cultivar* may be: "vegetatively propagated as a clone and originally derived from a single plant (which includes such mutations as chimeras) or those plants which are seed propagated and retain the characters of the parents (a true variety will not)." An example of the two proper methods for writing cultivar names would be either (1) *Petunia hybrida* cv. Orchid Lace or (2) *Petunia hybrida* 'Orchid Lace.' Although the term "variety" will continue to be used among horticulturists as a matter of convenience, the proper use of the term "cultivar" is becoming more widely accepted.

TERMS TO KNOW

Aesthetic	Commission Marketing	Ha Ha's
Almanac	Crystal Palace	Herbs
Apprentice	Cultivar	Nomenclature
Botanical Garden	Foliage Plants	Shakers
Caliper	Genetics	Topiary

STUDY QUESTIONS

1. Describe the importance of horticulture in Colonial America. Discuss the needs that it fulfilled for the colonists.

2. What difficulties were encountered by the colonists in obtaining plants from Europe and England?

3. How did the following persons or groups influence early American horticulture?
   (a) John Bartram
   (b) Benjamin Franklin
   (c) Martha Logan
   (d) Thomas Jefferson
   (e) The Shakers
   (f) The Prince Family
   (g) Johnny Appleseed
   (h) Bernard McMahon

4. How is plant exploration and plant breeding beneficial to modern horticulture?

5. Describe the types of greenhouses that have been used throughout history.

6. What is the Morrill Land-Grant Act and why is it one of the important factors in the development of modern horticultural science?

7. State why it is important to have established grades and standards in the floriculture and nursery industries.

SUGGESTED ACTIVITIES

1. Research Mendel's theory of genetics. Write a paper about his theory and present your findings to the class.

2. Locate an almanac and compare its suggestions about planting and cultivation to scientific publications that can be obtained from your county extension service.

3. Review pictures of gardens in Colonial Williamsburg or sixteenth- and seventeenth-century England and show some of the characteristic designs to the class.

4. Collect information about Mount Vernon and Monticello (the homes of Washington and Jefferson). Write an essay describing these gardens.

5. Have a class debate about the importance of plant patents and note the financial implications for requiring them.

CHAPTER TWO

Career Opportunities in Ornamental Horticulture

The related fields of ornamental horticulture are part of a rapidly expanding industry. This relatively recent growth in popularity has come about from the aesthetic and recreational interests of the American public. As our rural areas are becoming more urbanized, fewer people have a part in the traditional role of farming. This rapid growth of cities, suburbs, and small towns has created a great demand on the horticultural industries for goods and services.

In addition to the rapid movement of the population away from the rural farm, Americans today are enjoying a greater economic standard, shorter workdays and workweeks, and have more time and money for the pursuit of leisure-time activities. Americans spend millions of dollars annually on such sports as tennis, golf, football, and other athletic events which are all to some extent dependent upon ornamental horticulturists for their design or maintenance.

The American public is moving into new housing areas at such a high rate that the landscape and nursery businesses cannot expand fast enough to meet the needs for new landscape plantings. Sales of landscape maintenance equipment has expanded to the point where millions of lawn mowers are sold annually from many different types of businesses. Many homeowners are not interested in maintaining their own home grounds and can afford the services of landscape maintenance firms to do this for them. The need for aesthetic beautification of our highways, public buildings, and recreational parks has prompted the development of highly trained maintenance horticulturists by various federal, state, and local agencies. These agencies are responsible for millions of acres of public use areas.

Every state in the nation has a thriving ornamental horticulture industry. Specific states are known in particular for their individual industries: California for its flowers, Florida and Texas for foliage plants, South Carolina and Alabama for liner nursery stock, and Texas for commercial rose stock production. Although these states have made these industries famous, every state provides an economic opportunity for retailers, wholesalers, and growers of the various plant products associated with ornamental horticulture.

The student of ornamental horticulture may look forward to a challenging and rewarding career in any of these industries. A high school graduate having some training in ornamental horticulture may enter one of these career areas in various ways. Most of the practical education which is beneficial to the beginning horticulturist will be obtained through a combination of classroom education and practical work experience. The first jobs available to the horticulture student will involve a great deal of manual labor, without much opportunity for decision making. During this period of training, the student will be learning the intricate details of the particular business. Such classroom subjects as sales, production, management, and human relations will be observed firsthand. Although the wages earned during this early phase of training are low, the student will be gaining very valuable experience for later employment.

Following graduation from a vocational high school program, the horticulture student may pursue further education at a technical junior college or a four-year college or university which offers a degree program in horticulture. At the two-year technical college, the student will receive more advanced training in ornamental horticulture than was offered in the high school program. A major portion of the educational program of a technical school will be directed at individual development of the skills required for more advanced employment positions within the industry. Upon graduation, these students will receive an Associate degree in agriculture (horticulture).

Students who are qualified may elect to go on to advanced study of ornamental horticulture from a state college or university that offers a degree program in this discipline. In these four-year schools, the student will receive a diversified education, consisting of courses in the various areas of horticulture production, marketing, and science. In addition, these students will receive instruction in the scientific fields of chemistry, mathematics, English, botony, economics, and agriculture. A student may also be able to minor in such related areas as business administration, interior design, botany, or agronomy. A Bachelor of Science degree is awarded to those students who complete an approved program of study from these educational institutions.

A college degree is not required for all employees in any ornamental horticulture business. Many of the more successful managers and owners of nurseries, greenhouses, or other horticultural businesses have received their training and education by working for other businesses first. They gained their training through experience while they earned a living. Some professions within the field of horticulture will require an advanced degree for employment. These individuals are trained as specialists in their chosen field and will receive higher wages for their training and experience. Even the graduates of vocational junior colleges and four-year universities will benefit from working for a successful horticulture business for several years before attempting to start their own business.

The employment opportunities in the various ornamental horticulture occupations are extremely good, with many jobs available to students who have received training from a vocational horticulture program and are willing to work. Most of these jobs allow the employees to work outdoors with plants and to be involved with construction or mechanical equipment. The various occupations available for the student of ornamental horticulture are as follows:

1. *Floriculture Industry*
   (a) *Greenhouse Production*
      Greenhouse Manager or Owner
      Crop Manager–Grower
      Crop Production Foreman
      Sales Manager
      Greenhouse Maintenance Foreman
      Greenhouse Worker
   (b) *Flower Shop or Plant Store*
      Flower Shop Owner/Manager
      Floral Designer
      Salesperson
      Delivery Personnel
      Interior Landscape and Maintenance Employee
   (c) *Wholesale Florist/Broker*
      Owner or Manager
      Salesperson
      Delivery Personnel
      Stock Buyer and Control Manager
2. *Nursery Industry*
   (a) *Nursery Production*
      Nursery Manager or Owner
      Crop Production Manager–Grower
      Crop Production Foreman
      Plant Propagator

Plant Breeder
Sales Manager
Nursery Worker
(b) *Garden Center*
Garden Center Manager or Owner
Landscape Designer/Consultant
Landscape Installation and Maintenance Supervisor
Salesperson
Delivery Personnel
(c) *Landscaping*
Landscape Architect
Landscape Designer/Consultant
Landscape Contractor
Draftsman
Landscape Installation Worker
Landscape Maintenance Worker
(d) *Arboriculture*
Tree Pruning and Repair Employee
Utility Company Lineman and Arborist
Arboretum and Botanical Garden Arborist

3. *Turf Grass Industry*
Golf Course Superintendent
Greenskeeper
Golf Course Employee
Park Superintendent
Groundskeeper
Athletic Field Maintenance Employee
Sod Production Manager and Employees

4. *Related Areas in Ornamental Horticulture*
Writer or Editor for Horticultural Publications
Chemical Company Representative
Horticultural Equipment and Supply Broker
Arboretum and Botanical Garden Maintenance
Equipment Mechanic

5. *Some Technical Professions Requiring Post-Secondary Education*
Teacher (High School, Technical School, College)
Research Technician
Extension Service Specialist
Agronomist
Landscape Architect
Landscape Designer
Plant Breeder
Plant Quarantine Inspector

OCCUPATIONAL DESCRIPTIONS AND QUALIFICATIONS

Many of the career opportunities within these fields of ornamental horticulture require either extensive training through work experience or a college education in horticulture. The more common career positions are discussed, listing the type of work done by these professionals and the qualifications needed by a person wishing to enter these positions. Most of the professional positions listed can be found represented in various businesses in larger cities and even in some small communities. The student of ornamental horticulture should seek a position within such a business which matches his qualifications. Through the experience gained at the job-entry level, the student can gain the knowledge and expereince necessary to advance to the more technical professional positions within the chosen field of interest.

FLORICULTURE INDUSTRY

Greenhouse Production

Greenhouse worker. The greenhouse worker may expect to be involved with all aspects of the growing of crops in a commercial greenhouse (Figure 2–1). The employee will learn to propagate plants from cuttings and seed, handle soil mixes, grow potted plants (and cut flowers, in some greenhouses), and assist with maintenance of the structures. While the employee may not make decisions on crop production and rotations, he may observe these practices and aid in the implementation of their production.

FIGURE 2–1. Greenhouse workers are involved with the many aspects of growing floriculture crops. Two employees are shading a crop of chrysanthemums. (Courtesy DuPage Horticulture School)

After several years of experience as a greenhouse worker, the employee will have had the opportunity to observe the production of many types of crops; will have learned how to replace or maintain greenhouse equipment and structures; learned to prepare soil mixes, fertilizers, and pesticides for use on the various crops; will be able to identify crop production problems before they become serious; and will have learned how the crops are handled and marketed. Often a greenhouse is associated with a nursery operation, so the greenhouse worker will also be involved with the many aspects of nursery crop production as well as with the greenhouse maintenance. Once the greenhouse worker has gained adequate experience and training, he will be qualified to advance to a more technical position within the business.

The student who wishes to enter the field of floriculture as a greenhouse worker should be seriously interested in working with greenhouse plants. A high school education, with training in ornamental horticulture from a vocational program, would be beneficial. The work is largely year-round, with greater demands during holiday crop production seasons. In addition to the technical knowledge of horticulture, the greenhouse worker will benefit from some training in such areas as electrical wiring and appliance repair, small engine repair, plumbing, and construction trades.

Crop manager/Grower.   The crop manager is responsible for the production practices required to grow a specific crop or manage a range of greenhouse production benches. The grower is a specialist who has spent many years learning to produce the highest-quality cut flowers or potted plants. He may have received formal education from a technical college or a university, or he may have been trained by another crop manager. The grower will often have one or more assistants whom he is training in the intricate details of production. Together, the growers are responsible for all the physical labor required in the production of the crops they are in charge of. In a large greenhouse range, several growers will be employed to produce specific crops. In smaller greenhouse businesses, only one grower is required.

Most successful growers have received some training above the high school level. They may be graduates of either a two-year post-high school program or a four-year college program in horticulture. At least two years of practical greenhouse experience would be expected for a person wishing to become a grower. Grower's assistants may be high school students who are enrolled in a vocational horticulture program or recent high school graduates. The major requirement for persons wishing to become crop managers is a strong interest in the production of florists' crops. These employees must be dedicated and acutely aware of the status of the plants during production. The greenhouse manager or company executives depend upon the grower for the profits to be earned by the business.

Crop production and maintenance foreman. The crop production foreman is employed by large greenhouse businesses to coordinate the efforts of the various growers or individual crop managers. He is responsible for maintaining the proper crop scheduling and producing quotas for the business. He must see to it that the quality of the flowers and plants remains as high as possible.

The greenhouse maintenance foreman is in charge of the complex operation of the greenhouse facilities. Together, the two foremen must maintain the greenhouses in the best possible condition for the growth of the plants. The maintenance foreman will maintain a crew of greenhouse workers who assist him in performing the maintenance duties.

The greenhouse crop production foreman will often have received a B.S. degree in horticulture from a state university. More often the foreman has received a degree from a two-year technical horticulture program and has spent several years as a grower himself. He must be skilled in all areas of crop production and management and be able to supervise several employees. The maintenance foreman should have similar qualifications but should also be capable of making the various mechanical repairs and installations necessary for the profitable operation of the greenhouse ranges.

Sales Manager. The greenhouse crop sales manager works closely with the crop production foreman to determine which crops are to be grown and the timing of their harvest. He is responsible for selling these floral products to prospective buyers, which generally is done through wholesale florist firms. The sales manager must see that each crop is stored properly and shipped immediately whenever sold. The sales manager may supervise a large staff of sales assistants and clerical workers in larger wholesale production greenhouses. His rank is considered to be quite high in the business because he has such an important role in the acquisition of profits for the greenhouse business. The sales manager is placed under the immediate supervision of the greenhouse manager and often holds a position on the board of directors of greenhouses operated as a corporation.

The sales manager requires a strong background in business administration or sales. He may have received a B.S. degree from a college or university in agricultural business or business administration. The sales manager may also reach this position after receiving an Associate degree from a business college and several years of experience with a greenhouse production business or a wholesale florist firm. Since most of this work is largely conducted over the telephone and the salaries are generally high, this position is highly sought after by most junior employees.

Greenhouse manager. The greenhouse manager is the ultimate supervisor of all employees and activities conducted within the greenhouse business. The manager may also own the business, but more often he is a principal member of the corporate board of directors. In large greenhouse operations, the greenhouse manager supervises most directly the activities of the production foreman, the maintenance foreman, and the sales manager. In smaller greenhouse businesses, the greenhouse manager may have to do one or more of these duties in addition to his own. His main responsibility is to coordinate the efforts of all the departments to ensure that an efficient and profitable business is maintained.

Among the qualifications required by a successful greenhouse manager are a thorough knowledge of all aspects of the greenhouse production business, an ability to sell his products, a solid background in basic business practices, and an ability to supervise and counsel a large staff of employees. The greenhouse manager should hold a B.S. degree in horticulture from a college or university. In addition to having received the major education in horticulture and the sciences, additional courses in business administration are quite beneficial. Although a four-year education is most desirable, graduates of junior colleges in horticulture may also find employment as greenhouse managers after several years of training in a production greenhouse business. In either case, from three to five years of practical experience in a successful greenhouse business is necessary following graduation from a post-high school vocational program before one should consider owning a greenhouse business.

Flower Shop or Plant Store

Floral designer.   The floral designers in a florist shop must be able to create arrangements from available floral materials to the satisfaction of the customers. The designers must be able to arrange a wide array of designs, such as corsages, wedding designs, funeral pieces, centerpieces, bouquets, and various party pieces. Experienced floral designers are able to create floral arrangements of their own design and often aid customers in planning the arrangements for special or solemn occasions. In a larger shop, the head floral designer may supervise several assistant designers who are being trained in the basic skills of commercial floral design. The head designer is responsible for coordinating all the employees and the work load within the design room of the shop.

An experienced and highly skilled floral designer may expect to receive a comfortable salary from a successful shop. Although no special college training is necessary to become a floral designer, attendence at a good floral design school will be highly beneficial. All prospective floral designers should work in a successful flower shop for several years before expecting to become a proficient floral designer. Although many floral designers in the United States are women, some of the most successful designers are men. A successful florist must become skilled in the art of floral design before considering ownership of a flower shop.

Salesperson. Most persons who enter a career in the florist shop or plant shop businesses begin as sales clerks. Here they will learn the business from active participation in the sales of plant products and designs (Figure 2–2). This is a very important position in any shop, since the sales personnel are in constant contact with the buying public. While acting as a sales person, the employee may observe the floral designers or interior landscape designers as they perform their tasks. As sales clerks become more confident and skilled in the art of selling, they will be called upon to assist these designers with their work. After a short period of experience, they might be advanced to a designer's assistant position.

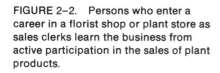

FIGURE 2–2.   Persons who enter a career in a florist shop or plant store as sales clerks learn the business from active participation in the sales of plant products.

Whether associated with a flower shop, plant store, garden center, or a mass-market chain store, the sales personnel should have a basic knowledge of the plants and plant products being sold. This includes correct identification of plants, pronunciation of their names, and a working knowledge of plant care and maintenance. In addition, the sales staff will be required to give advice on remedies for plant disorders and make suggestions as to specific chemical pesticides. These employees must become very quickly familiar with all types of plant problems and be capable of providing assistance to customers.

Although a sales position does not require much formal education in horticulture, sales personnel play a very important role in the field of floriculture. The beginning sales clerk may be currently enrolled in a vocational horticulture program while working in a store on a part-time basis. During the first few months of employment, the fundamental sales techniques may be learned from more experienced sales personnel. If a student is planning to obtain more horticultural education from a post-high school, college, or university, this early experience in sales will prove highly beneficial.

Interior landscape and maintenance employee.   Any person who has gained adequate experience and training in the use of plants and in plant care in interior plantings may become an interior landscape designer. The duties generally include consultation with prospective customers concerning design aspects of building interiors for suitable plant growth and proper display. Once a design is agreed upon, the interior landscape designer will draw a plan and recommend plants that are best suited to the specific locations. Following the installation of an interior landscape, a maintenance crew of qualified employees is then often contracted to maintain the health and appearance of these planters.

A person who wishes to become an interior landscape designer should enroll in the horticulture program offered by a junior or four-year college. In addition to the basic courses offered in plant identification and care, landscape design, and basic sciences, the student should obtain further education in the related areas of salesmanship, business administration, art, and interior design.

Wholesale Florist/Broker

Sales personnel.   Personnel who work on the sales staff of a wholesale florist business spend most of their working hours taking telephone orders and preparing deliveries for their retail customers. These employees play a very important role in the wholesale brokerage of flowers and floral products because they must be in constant contact with the retail shop owners who purchase these products. These salespersons generally do not require a post-high school degree to enter the wholesale florist business. Some experience in greenhouse operations, crop production, or flower shop experience, in addition to the training received from a vocational high school horticulture program, are also very beneficial to the sales personnel of these wholesale floral businesses.

Stock buyer and control personnel. The stock buyer at a whole-sale florist business spends a large part of each day telephoning various commercial greenhouse sales managers from whom they customarily purchase flowers. The buyers are responsible for maintaining a steady supply of cut flowers, greenery, and hardware items for their own customers. In order to do this, they must locate and purchase the best-quality merchandise for the lowest possible prices from many floral product producers throughout the country. These buyers must also inspect each shipment when it is received to determine that the quality and type of flowers or hardware items are the same as those purchased. In a large wholesale flower business, the inventory over which the buyer is in charge may include many thousands of items. To help the buyer with such a large task, computer systems may be used to record every purchase and sales transaction of the business.

The stock buyer will require several years of experience in a wholesale florist business before being capable of handling this position. Some training or experience following graduation from a high school vocational horticulture program is necessary. Often, these employees have worked as growers in production greenhouses for several years prior to employment with a wholesale florist firm. In addition to having a sound background in floriculture, a successful buyer will also benefit from additional training in business and marketing. A post-high school degree program should include studies in horticulture, business administration, marketing, and computer programming.

Delivery personnel. All delivery personnel regardless of the type of business for which they are employed, must have similiar qualifications. The primary qualification is that the employee be a qualified driver who holds a current and valid driver's license for the type of vehicle to be operated. In most large production greenhouse operations and wholesale florist firms, the delivery personnel will be required to occasionally operate such trucks as step vans, large refrigerated vans, semitractors with trailers, in addition to the more common panel trucks. A chauffeur's license is required by most states to operate a motor vehicle for a salary.

The delivery personnel will not require any specific formal post-high school training. The position is generally a place for the beginning employee to find a job while still in high school or immediately following graduation. While the employee performs the delivery duties, he will be able to observe and assist in the other functions of the particular ornamental horticulture business. During this time he will also gain the necessary personality characteristics that will be used later in sales and general business management.

NURSERY INDUSTRY

Nursery Production

Crop production manager–Grower.   The crop production manager for a commercial nursery performs the same basic functions as his counterpart in the greenhouse industry. The grower is generally a specialist who is responsible for the production of a specific crop (such as azaleas) or a block of related nursery stock in a field. The grower will often have spent several years as a nursery worker and assistant grower before becoming a production manager.

The crop production manager should be thoroughly trained in the art of producing the specific crops he is in charge of. He may obtain this training from a technical junior college or four-year college horticulture program. Many growers learn their profession firsthand from more experienced growers while assisting them.

Plant propagator.   A plant propagator is a specialist in the art of increasing plant varieties or species from seed or by various vegetative parts. The propagator is a highly skilled horticulturist who has spent many years training with other experienced nurserymen. The plant propagator plays a very important role in the production of nursery stock (and some florists' crops), since these wholesale nurseries often maintain very large facilities for starting the crops. The propagation facilities will often include greenhouses for seed propagation, refrigerated storage for seed ripening, cold frames for cutting and grafting stock, and various other structures for liner production. Plant propagators supervise a large staff of assistants and nursery workers to accomplish this important job.

A person who wishes to become a plant propagator should plan on enrolling in a suitable horticulture program in a technical junior college or four-year university following graduation from high school. Upon completion of a college program, the student will then require several more years as an assistant propagator or nursery worker before being capable of supervising this area of the nursery business.

Plant breeder.   A plant breeder is employed by large nurseries, greenhouses, or seed companies for the purpose of improving and maintaining named varieties of plant material. The plant breeder may also spend considerable time traveling to foreign countries in search of different plants from which to select for breeding purposes (Figure 2–3).

A plant breeder must be highly skilled in the art of plant breeding and propagation. Plant breeding can only be learned by enrolling in a suitable program in a four-year college or university. A student should plan on majoring in horticulture, plant breeding, or genetics, with special emphasis placed on genetics. Once the student has graduated from a college program, he may expect to work with more experienced plant breeders for several years before taking a position as a head plant breeder for a company.

FIGURE 2–3.   A plant researcher investigating production problems. These highly skilled employees require advanced college degrees.

Nursery worker.   The nursery worker may be involved with many of the operations conducted at a nursery. The nursery worker will do the general labor required to produce, store, and sell the crops. Most of the labor is outside work, but is not necessarily heavy or tiresome. Most employees enter the nursery business by first becoming nursery workers. During this period of training the employee may observe and assist the more experienced specialists in the performance of their duties.

The nursery worker may seek employment with a nursery business while still enrolled in a vocational high school horticulture program on a part-time basis. The seasonal nature of the nursery business requires additional personnel during the spring and autumn months. This is an excellent opportunity for the student to determine whether he will enjoy a profession in the nursery industry. Upon graduation from high school, the student may wish to enroll in a horticulture program at a junior college or four-year college to learn the more advanced techniques required for the more highly skilled professional positions in the nursery business.

Garden Center

Salesperson.   The sales personnel at a garden center have an opportunity to work with and learn about the plants used for landscaping homes (Figure 2–4). The sales staff must acquire a basic knowledge of plant care, identification, and uses in the landscape to be effective in selling these products. They must also learn the fundamentals of selling and handling customer requests. These fine points of selling are best learned while on the job and will be important learning activities for later professional life.

FIGURE 2–4. Garden center employees have the opportunity to work with plants as they learn the cultural requirements and uses of landscape plant material.

Most persons begin a career in the garden center business by first becoming a sales clerk. Since they are in constant contact with the buying public, they must possess the qualifications of a knowledgeable horticulturist and have sales ability. A student currently enrolled in a vocational high school horticulture program may become a part-time employee at a garden center. Following graduation, the student may wish to obtain further landscaping or nursery training from a junior or four-year college to become qualified for a higher-paying position in a garden center business.

Garden center manager/owner. A manager of a garden center generally has received training from at least a junior college or university horticulture program and has also been employed by a garden center or nursery for several years following graduation. The manager may be employed by a garden center chain company or he may own his business. To become qualified as a store manager, he must have received considerable training and experience in the following areas: business management, general horticulture, nursery production and plant care, plant identification, landscape design, marketing, and sales. The manager supervises all the activities of the garden center. This is the position that most garden center employees eventually hope to obtain, since being able to operate one's own business also means a greater opportunity for a better salary.

The garden center manager must also be willing to work long, hard hours and enjoy working with plants and the public. Garden center work is highly seasonal, corresponding to the planting seasons of spring and autumn. He must be able to generate work for his employees and manage the company finances during the periods between these peak periods of business. An enterprising garden center manager may become a successful member of the community and obtain a comfortable living from his profession.

Landscaping

Landscape architect.   A landscape architect is a highly trained individual who has received a specialized degree in that field. Often a landscape architect must attend a special college degree program for at least five years following graduation from high school. This degree program includes studies in such areas as engineering, architecture, design, plant material uses and care, and residential and city planning.

A landscape architect's work is highly demanding but is also a very prestigious position in the community. A landscape architect may be self-employed or associated with a retail nursery, garden center, or engineering firm. The major responsibilities are preparing plans, bidding on landscaping contracts, and supervising the installation of the landscape work. Since this position requires a highly trained and educated individual, the salaries are also quite high in relation to other horticultural professionals.

Landscape designer.   A landscape designer is a highly trained and educated individual but has not completed a certified landscape architecture degree program. A landscape architect must hold a degree in landscape architecture before using this title professionally or he must pass a rigorous examination before he can be certified and licensed to practice. A landscape designer is generally educated in horticulture but has received extensive education in plant material identification and use of plants in design. Both of these professionals must possess a high degree of skill in the design aspects of plant material and drafting. Most such persons are highly artistic and enjoy the creation of landscapes.

An individual who wishes to become a successful landscape designer should plan to attend a post-high school training program in horticulture. Special emphasis should be placed on design, drafting, plant material identification and uses, and business management courses. The landscape designer may be self-employed as a landscape contractor or associated with a retail nursery or garden center.

Landscape contractor. A landscape contractor may also be a qualified landscape designer, but in addition must be qualified to perform the various construction tasks necessary to install landscapes. This person generally owns his own business, including many pieces of machinery required to complete an extensive landscape installation. He may employ several crews of landscape workers who are responsible for handling the plant material, planting the nursery stock, following the landscape plan, and operating the heavy machinery.

An individual who desires to become a landscape contractor should obtain a post-high school education from a junior college or four-year college which offers a degree in horticulture and specialized training in landscape design, landscape construction, business management, and contracting.

Draftsman. Most two-year technical schools and junior colleges offer post-high school training in drafting and engineering technology. Graduates from these programs are capable of reading construction and design plans and also are skilled in their preparation. A person who has some artistic ability and enjoys preparing professional design plans from sketches made by a landscape designer or architect will enjoy this type of work. While working as a draftsman, an individual will be able to participate in the functions of landscape design and may eventually be qualified to make design decisions. Many draftsmen eventually attend a four-year college to obtain a degree in horticulture with a specialization in landscape design or landscape architecture.

Landscape installation/maintenance personnel. The employees who do the actual planting, bed preparation, installation of landscape materials, and maintain the landscape plantings are under the direct supervision of a landscaping foreman. They may be employed by a garden center, landscape contractor, nursery, or a landscape maintenance firm. Landscape installation work is seasonal, with the greatest amount of work done in the spring and autumn months. Maintenance work can extend throughout most of the warmer months of the year. The work is largely done outdoors during sunny weather.

Students enrolled in a vocational horticulture program may have the opportunity to work part-time as landscape workers until they graduate. During this period of training the students will gain the experience and knowledge that will help them to become competent landscape designers later. Although no post-high school education is required for entering the business as a landscape worker, the motivated student will desire more formal training and advancement in their career. After attending a junior or four-year college in a horticulture curriculum, the student may be qualified for one of the other more skilled professional positions in the landscape industry.

Arboriculture

Arborist. An arborist is a person trained in the science of pruning and repairing large trees in the correct manner. The work is highly demanding and very dangerous, so considerable training is required before an individual is allowed to begin work. Although some colleges and universities still teach courses in the science of arboriculture, most individuals gain this training from other professional arborists, through extensive training provided by the various utility companies or private tree service companies, or from one of the various post-high school training programs. This training instructs the tree worker on the proper methods for scaling a tree (climbing), securing themselves and large branches before cutting, safety, removal of large trees, and repairing tree damage or diseased portions.

The arborist may find employment from various areas of the ornamental horticulture industry: public parks and private industry, arboreta and botanical gardens, public utility companies, and private tree service companies. Most arborists must begin their careers by assisting the professionals from the ground. After a suitable period of training, they will have received the proper training and experience to perform the tree work themselves.

TURF GRASS INDUSTRY

Golf course worker/greenskeeper. Golf course employees and greenskeepers are primarily responsible for the maintenance of the greens, tees, and fairways in suitable playing condition. These employees must be qualified to operate the specialized equipment used for maintaining turf grass surfaces and to perform the various duties required to provide the best possible maintenance programs. The greenskeeper supervises the work of the golf course employees and is directly responsible to the golf course superintendent.

A golf course greenskeeper may have received his education and training from a junior or four-year college that offers a curriculum in turf grass management. However, most greenskeepers begin their careers as golf course laborers and receive their training from a more experienced greenskeeper. If a student wishes to become a greenskeeper, he may begin work at a golf course during summers between school years or immediately following graduation. In the most southern states, the work may be available for most months of the year. However, in more northern climates, the work for a greens-keeper is reduced to only the warmer months of the year.

Golf course superintendent. A golf course superintendent supervises the activities of the golf course. His responsibilities not only include the maintenance of good turf grass playing conditions, but may also include the operation of the clubhouse, restaurant, and pro shop; arranging for tournaments; and operation of the lounge and all other physical facilities of the course. These responsibilities demand a great deal of time from the superintendent, so some of these duties may be allocated to assistants. A qualified superintendent may develop the maintenance schedules to be used by the greenskeeper and his crews.

A golf course superintendent generally will hold a college degree from a field of study such as horticulture (turf grass management), business administration, or restaurant management. Many of the more successful superintendents have gained some experience as golf course greenskeepers so that they might better understand the problems encountered in golf course turf grass maintenance. Although an interest and skill in the art of playing golf is not a primary prerequisite to becoming a golf course superintendent, this ability often makes the position attractive to prospective employees and is a benefit in the performance of these duties.

Park superintendent. The various public and private parks are maintained by skilled horticulturists. The overall supervision of the various maintenance crews is the responsibility of the park superintendent. He is also responsible for the various other activities conducted at the park. A park superintendent often holds a college degree in horticulture, wildlife management, or forestry. However, he may also have acquired this professional position after years of experience as a park employee.

Groundskeeper. A professional groundskeeper may be employed by a public park, private industry, school, or a public agency (Figure 2–5). The specific responsibilities of a groundskeeper will vary somewhat depending on the type of agency in which he is employed. The more common responsibilities include maintenance of the lawns, ball fields, and other physical facilities used by park visitors; maintenance of the landscape trees and shrubs; planting and maintenance of ornamental flower beds; and control of insect and disease pests.

A student may seek employment as a groundskeeper while still enrolled in a vocational horticulture program in high school. Following graduation, more advanced training may be obtained from a horticulture program at a post-high school vocational school or junior college. Specific skills that a groundskeeper will require include knowledge of plant growth and maintenance, pruning, mowing, operation of various types of maintenance equipment, and fertilization and pest control.

FIGURE 2–5. Groundskeepers at Colonial Williamsburg, Virginia. These employees must keep the formal gardens appearing as they might have looked during the Colonial period.

Sod production employee. Turf sod is produced at specialized nurseries for use in landscaping. The employees of such nurseries are trained in the specific science of sod production and management. They must be capable of establishing fields of turf grasses from seed, stolons, sprigs, or plugs. These nursery fields must then be intensively maintained to exclude weeds and promote a healthy, rapidly growing turf sod. When the sod is ready for sale, the nursery bed is cut with a sod cutter and the sod rolls must be loaded onto trucks. The employees of the sod nursery may also be responsible for laying the sod at the new landscaping site.

These employees must be capable of handling the equipment used at a sod nursery. They should be able to recognize disease or insect pests and their symptoms on the crop, as well as being able to manage the fertilization and watering practices of the nursery fields. A student currently enrolled in a high school horticulture program may seek employment as a sod nursery worker during the summer months. Following graduation from high school, further education in turf grass management may be obtained from a junior or four-year college that offers a curriculum in this field.

ORNAMENTAL HORTICULTURE: A FIELD WITH PROMISE FOR MEN AND WOMEN

The many professional areas within the field of ornamental horticulture offer a multitude of career opportunities for the student of horticulture. These jobs are available for both men and women, with many jobs going unfilled each year. The need for more individuals who are qualified to take these positions in the various industries will continue to grow (Figure 2–6).

Within the ornamental horticulture field are career opportunities for individuals who prefer outdoor work, greenhouse crop production, designing landscapes or floral arrangements, management, and sales. The unskilled worker who enters the horticulture industry will receive somewhat lower salaries than the national average; however, additional post-high school education and experience in the field of interest will eventually be rewarded in high salaries. The ability to work with plants and to eventually own one's own business are highly motivating. Those students who graduate from a post-secondary school training program in horticulture will enjoy a rewarding professional career.

FIGURE 2–6. The highly motivated horticulture student may obtain an advanced college degree and become a skilled laboratory researcher.

SELECTED REFERENCES

American Association of Nurserymen. 1971. *Career Opportunities in the Nursery Industry*. AAN, 230 Southern Building, Washington, D.C.

Binkley, H., and C. Hammonds. 1969. *Experience Programs for Learning Vocations in Agriculture*. The Interstate Printers and Publishers, Inc., Danville, Ill.

Cushman, H. R., C. W. Hill, and J. K. Miller. 1968. *The Teacher—Coordinators Manual for Directed Work–Experience Programs in Agriculture*. Cornell Miscellaneous Bulletin No. 91. New York State College of Agriculture, Cornell University, Ithaca, N.Y.

Drawbaugh, C. C. 1962. *A One-Year Greenhouse Laboratory Course of Study for Vocational Agriculture in Pennsylvania.* Department of Agricultural Education, Pennsylvania State University, University Park, Pa.

Hoover, N. K. 1969. *Handbook of Agricultural Occupations.* The Interstate Printers and Publishers, Inc., Danville, Ill.

Hoover, N. K. 1971. *Legal Aspects of Cooperative Occupational Experience Programs.* Department of Agricultural Education, Pennsylvania State University, University Park, Pa.

Horticultural Research Institute, Inc. 1968. *Scope of the Nursery Industry.* HRI, 833 Southern Building, Washington, D.C.

Ohio State University. 1965. *Exploring Occupational Opportunities in Ornamental Horticulture.* Center for Research and Leadership in Vocational and Technical Education, Columbus, Ohio.

Ohio State University. 1971. *Opportunities in Agricultural Occupations and the Employee Within the Business Organization.* Ohio Agricultural Education Curriculum Materials Service, Columbus, Ohio.

Pennsylvania State University. 1968. *Retail Flower Shop Operation and Management — A Teacher's Manual.* Teacher Education Series, Vol. 9, No. 1T. Department of Agricultural Education, University Park, Pa.

Pennsylvania State University. 1969. *Greenhouse Crop Production — A Teacher's Manual.* Teacher Education Series, Vol. 10, No. 3T. Department of Agricultural Education, University Park, Pa.

Pennsylvania State University. 1971. *Nursery Production — A Teacher's Manual.* Teacher Education Series, Vol. 12, No. 4T. Department of Agricultural Education, University Park, Pa.

Pinney, J. J. 1967. *Beginning in the Nursery Business.* American Nurseryman Publishing Company, Chicago.

Pinney, J. J. 1967. *Your Furture in the Nursery Industry.* Richards Rosen Press, Inc., New York.

Stone, A. A. 1965. *Careers in Agribusiness and Industry.* The Interstate Printers and Publishers, Inc., Danville, Ill.

Supervised Occupational Experience in Agriculture, Plans and Records, 4th ed. 1969. The French-Bray Printing Company, Baltimore, Md.

Teacher's Guide, Supervised Occupational Experience in Agriculture, Plans and Records, 1969. The French-Bray Printing Company, Baltimore, Md.

U.S. Department of Labor. 1965. *Dictionary of Occupational Titles,* 3rd ed., Vols. I and II. Bureau of Employment Security, Washington, D.C.

University of Alabama. 1966. *Applying for a Job.* Student workbook. Department of Agricultural Education, Auburn, Ala.

University of Alabama. 1966. *How to Make $50,000.* Student workbook. Department of Agricultural Education, Auburn, Ala.

University of Alabama. *Good Job Habits.* Student workbook. Department of Agricultural Education, Auburn, Ala.

University of the State of New York. 1968. *Ornamental Horticulture: A Guide for Planning Occupational Programs.* The State Education Department, Albany, N.Y.

Weyant, J., N. K. Hoover, and D. R. McClay. 1965. *An Introduction to Agricultural Business and Industry.* The Interstate Printers and Publishers, Inc., Danville, Ill.

TEACHING AIDS

SLIDE SETS

Exploring Turfgrass Occupations (30 slides). Department of Agricultural Education, Pennsylvania State University, University Park, Pa.

FILMS

Careers in Ornamental Horticulture (50-frame filmstrip). Vocational Education Publications. California State PolyTech College, San Luis Obispo, Calif.

Dynamic Careers Through Agriculture (16 mm; 28 min). Farm Film Foundation, Washington, D.C.

Time For Searching (16 mm; 22 min). Modern Talking Picture Service, 1212 Avenue of the Americas, New York.

TERMS TO KNOW

Agronomist	Pesticide	Retail
Arboriculture	Plant quarantine	Turf grass
Foreman	Propagation	Wholesale
Genetics		

STUDY QUESTIONS

1. List reasons why ornamental horticulture is a thriving industry with many career opportunities.

2. List the careers in ornamental horticulture where (a) a college degree is required for employment; (b) further education is suggested, but is not necessarily required; and (c) no post-high school training is required.

3. Select five horticultural careers that most interest you and write a job description for each. Tell what personal qualities an individual should possess to be successful in the careers you describe.

SUGGESTED ACTIVITIES

1. Make a horticultural career file for your department. For each career, include the following:
   (a) Job description and salary expectations.
   (b) Personal qualities required.
   (c) Training required or suggested.
   (d) Advantages of this job as a lifetime career.
   (e) Local employment opportunities.

2. Invite local horticulturists, advisory council members, school administrators, parents, and guidance counselors to a panel discussion on career opportunities in ornamental horticulture.

3. Prepare a display or make posters illustrating career opportunities in ornamental horticulture.

4. Interview persons who are employed in horticulture careers and report about your findings to the class.

CHAPTER THREE

Judging Flowers and Ornamental Plants

The ability to judge the quality of nursery and floricultural products is an important part of the training of horticulture students. Each student should be capable of recognizing the good and poor qualities of each specimen plant, landscape design, or flower arrangement. The National Future Farmers of America (FFA) Horticulture Contest is designed to test the skills of the judging teams from every state in the United States in making decisions on quality of ornamental horticulture crops and designs.

The contest is held annually for the purpose of creating a competitive atmosphere among the judging teams, who compete for recognition in their skills of judging these crops (Figure 3–1). Many state teams are selected from winners of division or area competitions held prior to the state contest. These contests are conducted to sharpen the judging skills of the contestants and to select those who are most well trained in the art of judging horticultural products. The skills learned during the training for the National Horticulture Judging Contest will be useful to all students who will later become affiliated with a horticulture business. Growers of nursery or florist crops must know what qualities are sought in their products. Retailers must also be familiar with the desirable characteristics of their plants in a finished landscape or floral design.

CONTEST ELIGIBILITY

Students who are to compete in the National FFA Horticulture Contest must meet the following requirements:

FIGURE 3–1.   High school students participating in an FFA Horticulture Contest.

1.   A student of a horticulture or vocational agriculture program must be a member of a recognized FFA chapter. Students must be in good standing with their own chapter, the state FFA association, and the national FFA organization at the time they are selected for competition on the judging team and while attending the National contest.

2.   At the time a student is selected for participation on a state judging team for the national contest, the following requirements must be met:

(a) The student must be currently attending a high school.

(b) The student must be currently enrolled in at least one certified course in a vocational agriculture or occupational horticulture program. If this is not possible in a particular school system, the student must then be following a planned course of study that includes a supervised occupational experience program having the objective of agricultural occupation training.

(c) In addition, any student who has previously participated in any official National FFA Horticulture Judging contest section (either Nursery/Landscape or Floriculture) is prohibited from future participation in that section.

(d) Contestants who enter the National FFA Nursery Contest are not eligible to enter the National FFA Floriculture Contest in the same year.

(e) Contestants who enter the National FFA Floriculture Contest are not eligible to enter the National Nursery/Landscape Contest in the same year.

Each contestant is certified for eligibility for participation in the national or state contests by the state supervisor of vocational agriculture. Whenever a student entered on a competing team is found to be ineligible, the entire team of which this student is a member will likewise be disqualified from participation in the contest.

SELECTION OF STATE JUDGING TEAMS

Individual team members who will participate in the National FFA Horticulture Judging Contest are selected by the state supervisor of vocational agriculture from competencies determined by their performance in a previously held state, area, or interstate contest. This regional contest should be held since the last national contest. Membership on the state team is determined by the state supervisor and may consist either of the highest-scoring students in the regional contest or the highest-scoring team from that contest. Any regional contests held in preparation for the national contest should be structured in a like manner, with plant material and test questions of a similar type.

CONTEST RULES

The National FFA Horticulture Contest consists of two independent sections: (1) Nursery/Landscape Contest and (2) Floriculture Contest. Contestants from an individual team are allowed to participate in only one of these contests in any given year. Three members from each school team will participate in the Nursery/Landscape Contest, while three other members will participate in the Floriculture Contest. Judging and testing in both contests will be done simultaneously, with each team represented in *both* contests. The team members who are to participate in each contest must be determined and designated on the contest certification forms.

The judging teams participating in either a state or the National FFA Judging Contest will consist of a total of six members. Three of these team members will compete in the Nursery/Landscape Contest and three others will compete in the Floriculture Contest. One alternate team member is allowed to accompany the teams in each contest section.

The contest will be conducted in such a manner as to minimize the opportunity for collaboration and to promote a competitive atmosphere among contestants. The following rules will be strictly enforced to provide the most fair judging conditions:

1.   Each team will be separated so that no two contestants from the same team will be in the same group during the contest.

2.   Contestants will not be allowed to communicate with their own or other members of judging teams. Any other assistance obtained by contestants during the contest is considered sufficient cause for elimination of the teams involved from the contest and will nullify the travel award to their team.

3.   Contestants are not allowed to handle or touch the plant material while judging during the contest. This action may be construed as an attempt to alter the plant material, making judging or identification more difficult for other contestants. The penalty for such action may result in the elimination of the team from the contest and will nullify their travel award.

4.   Persons serving as observers of the contest are permitted to attend either section. These persons will be designated by special name tags bearing the title "Observer." Any observer may not communicate with any contestant or another observer while within the contest area. Whenever an observer leaves the contest area of either section, they will not be allowed to return until the contest has been completed. Team alternatives are not to be included as observers of the contest.

5.   Final scores only will be distributed to the contestants or their coaches following the contest.

6.   Team coaches are not permitted to observe the contest while it is in session, nor may they communicate with the contest administrators prior to the contest.

INDIVIDUAL AND TEAM SCORING

Nursery/Landscape Contest

Each contestant will be scored according to the sum of the individual points accumulated from each phase of the contest. Team scores will be the accumulated points earned by all three team members. Recognition will be awarded to the high scores earned by contestants and to the top scores earned by an individual state team.

PHASE	TOTAL POSSIBLE POINTS
Individual scores	
Identification of Plant Material	350 points
Placing Classes of Plant Material and Designs	250 points
General Knowledge Examination	300 points
Practicum	100 points
	1000 points
Team scores	
Sum of individual scores from each team	3000 points

Floriculture Contest

PHASE	TOTAL POSSIBLE POINTS
Individual scores	
Identification of Plant Material	250 points
Placing Classes of Plant Material and Designs	250 points
General Knowledge Examination	250 points
Practicum	250 points
	1000 points
Team scores	
Sum of individual scores from each team	3000 points

SECTIONS OF THE HORTICULTURE CONTEST

NURSERY/LANDSCAPE SECTION

Nursery crops are judged in four separate phases: These are:

1. Identification of Plant Materials.
2. Placing Classes of Plant Material.
3. General Knowledge Examination.
4. Practicum.

Phase 1: *Identification of Plant Material*   (350 points)

Each contestant will be given a list (Form 14) containing the technical and common names of 50 nursery plants (Figure 3–2). These plants will be arranged in a random manner throughout the nursery section area and will be designated by number. Each contestant will write the number accompanying each plant specimen in the column marked "Number" next to the correct name on the official score card. Each specimen that is correctly identified will be awarded 7 points, for a total of 350 possible points for this phase of the contest. A total time of 50 minutes is given for each contestant to complete this phase. This allows 60 seconds for the identification of each specimen. An example of the Form 14 plant identification list is shown in Table 3–1. The plants listed on this form are for example only, as the plants used in various contests will not always be the same. Although the list contains 139 nursery plant names, only 50 of these would be displayed for identification during a national or state contest.

FIGURE 3–2.   Specimens for the Nursery/Landscape Contest.

TABLE 3–1: *Example of FORM 14 Plant Material Identification List*

COMMON NAME	BOTANICAL NAME	NUMBER	SCORE
	NURSERY PLANT IDENTIFICATION	Form 14	
CONTESTANT NAME		CONTESTANT NUMBER	
Glossy Abelia	Abelia × grandiflora		
White Fir	Abies concolor		
Amur Maple	Acer ginnala		
Japanese Maple	Acer palmatum		
Norway Maple	Acer platanoides		
Red Maple	Acer rubrum		
Silktree (Mimosa)	Albizia julibrissin		
Norfolk Island Pine	Araucaria excelsa (heterophylla)		
Strawberry Tree	Arbutus unedo		
Bearberry, Kinnikinick	Arctostaphylos uva-ursi		
Japanese Aucuba	Aucuba japonica		
Japanese Barberry	Berberis thunbergii		

Paper Birch	*Betula papyrifera*		
European Birch	*Betula pendula*		
Brazil Bougainvillea	*Bougainvillea spectabilis*		
Schefflera (Octopus Tree)	*Brassaia actinophylla*		
Common Boxwood	*Buxus sempervirens*		
Heather	*Calluna vulgaris*		
Camellia	*Camellia japonica*		
Blue Atlas Cedar	*Cedrus atlantica 'Glauca'*		
Deodar Cedar	*Cedrus deodara*		
Oriental Bittersweet	*Celastris orbiculatus*		
American Bittersweet	*Celastris scandens*		
Eastern Redbud	*Cercis canadensis*		
Flowering Quince	*Chaenomeles speciosa*		
Flowering Dogwood	*Cornus florida*		
Pacific Dogwood	*Cornus nuttallii*		
Red Osier Dogwood	*Cornus stolonifera*		
Spreading Cotoneaster	*Cotoneaster divaricatus*		
Rock Spray Cotoneaster	*Cotoneaster horizontalis*		
Washington Hawthorn	*Crataegus phaenopyrum*		
Italian Cypress	*Cypressus sempervirens*		
Giant Dumb Cane	*Dieffenbachia amoena*		
False Aralia	*Dizygotheca elegantissima*		
Warneckii Dracaena	*Dracaena dermensis 'Warneckii'*		
Corn Dracaena	*Dracaena fragrans massangeanna*		
Slender Deutzia	*Deutzia gracilis*		
Russian Olive	*Eleagnus angustifolia*		
Autumn Olive (Eleagnus)	*Eleagnus umbellata*		
Winged Euonymus	*Euonymus alatus*		
Wintercreeper	*Euonymus fortunei*		
Patens Euonymus (Siebold)	*Euonymus kiautschovica*		
Japanese Fatsia	*Fatsia japonica*		
Weeping Fig	*Ficus benjamina*		
"Decora" Rubber Plant	*Ficus elastica 'Decora'*		
Border Forsythia	*Forsythia × intermedia*		
Green Ash	*Fraxinus pennsylvanica*		
Gardenia (Cape Jasmine)	*Gardenia jasminoides*		
Ginkgo (Maidenhair Tree)	*Ginkgo biloba*		
Thornless Honeylocust	*Gleditsia triocanthos inermis*		
English Ivy	*Hedera helix*		
Pee Gee Hydrangea	*Hydrangea paniculata 'Grandiflora'*		
English Holly	*Ilex aquifolium*		
Chinese Holly	*Ilex cornuta*		
Japanese Holly	*Ilex crenata*		
American Holly	*Ilex opaca*		
Flame of the Woods	*Ixora coccinea*		

Dwarf Pfitzer Juniper	*Juniperus chinensis* 'Pfitzeriana'		
Keteleer Juniper	*Juniperus chinensis* 'Keteleeri'		
Blue Rug Juniper	*Juniperus horizontalis* 'Blue Rug'		
Tam Juniper	*Juniperus sabina* 'Tamariscifolia'		
Mountain Laurel	*Kalmia latifolia*		
Crape Myrtle	*Lagerstroemia indica*		
Amur Privet	*Ligustrum amurense*		
Glossy Privet	*Ligustrum lucidum*		
Sweet Gum	*Liquidamber styraciflua*		
Tuliptree	*Liriodendron tulipifera*		
Hall's Japanese Honeysuckle	*Lonicera japonica* 'Halliana'		
Tatarian Honeysuckle	*Lonicera tatarica*		
Southern Magnolia	*Magnolia grandiflora*		
Saucer Magnolia	*Magnolia × soulangeana*		
Oregon Hollygrape	*Mahonia aquifolium*		
Flowering Crabapple	*Malus species*		
Bayberry	*Myrica pennsylvanica*		
Heavenly Bamboo	*Nandina domestica*		
Oleander	*Nerium oleander*		
Black Gum (Black Tupelo)	*Nyssa sylvatica*		
Holly Osmanthus	*Osmanthus heterophyllus*		
Japanese Spurge	*Pachysandra terminalis*		
Boston Ivy	*Parthenocissus tricuspidata*		
Sweet Mockorange	*Philadelphus coronarius*		
Heartleaf Philodendron	*Philodendron oxycardium*		
Senegal Date Palm	*Phoenix reclinata*		
Japanese Photinia	*Photinia glabra*		
Norway Spruce	*Picea abies*		
Colorado Spruce	*Picea pungens*		
Japanese Andromeda	*Pieris japonica*		
Mugo Pine	*Pinus mugo mughus*		
Austrian Pine	*Pinus nigra*		
Red Pine	*Pinus resinosa*		
Eastern White Pine	*Pinus strobus*		
Scotch Pine	*Pinus sylvestris*		
Japanese Black Pine	*Pinus thunbergii*		
London Plane Tree	*Platanus × hybrida (acerifolia)*		
Mexican Frangipani	*Plumeria rubra* 'Acutifolia'		
Yew Podocarpus	*Podocarpus macrophylla*		
Bush Cinquefoil (Shrubby C.)	*Potentilla fruticosa*		
Thundercloud Flowering Plum	*Prunus cerasifera* 'Thundercloud'		
Cherry Laurel	*Prunus laurocerasus*		
Japanese Flowering Cherry	*Prunus serrulata*		
Flowering Almond	*Prunus triloba*		
Scarlet Firethorn	*Pyracantha coccinea*		

Bradford Pear	*Pyrus calleryana* 'Bradford'		
Pin Oak	*Quercus palustris*		
Red Oak	*Quercus rubra*		
Live Oak	*Quercus virginiana*		
Glenn Dale Azaleas	*Rhododendron azalea–Glen Dale*		
Catawba Rhododendron	*Rhododendron × catawbiense*		
Fragrant Sumac	*Rhus aromatica*		
Staghorn Sumac	*Rhus typhina*		
Hybrid Tea Rose	*Rosa × cv.*		
Rugosa Rose	*Rosa rugosa*		
Palmetto	*Sabal palmetto*		
California Peppertree	*Schinus molle*		
Japanese Skimmia	*Skimmia japonica*		
European Mountain Ash	*Sorbus aucuparia*		
Clevelandii Spathophyllum	*Spathophyllum floribundum* 'Clevelandii'		
Bridal-Wreath Spirea	*Spriaea prunifolia*		
Chenault Coralberry	*Symphoricarpos × chenaultii*		
Common Lilac	*Syringa vulgaris*		
Upright Japanese Yew	*Taxus cuspidata* 'Capitata'		
Dense Yew	*Taxus × media* 'Densiformis'		
Hicks Yew	*Taxus × media* 'Hicksii'		
American Arborvitae	*Thuja occidentalis*		
Oriental Arborvitae	*Thuja orientalis*		
Common Bald Cypress	*Taxodium distichum*		
Littleleaf Linden	*Tilia cordata*		
Star Jasmine (Confederate J.)	*Trachelospermum jasminoides*		
Canadian Hemlock	*Tsuga canadensis*		
Arrowwood	*Viburnum dentatum*		
Judd Viburnum	*Viburnum × judii*		
Doublefile Viburnum	*Viburnum plicatum tomentosum*		
Leatherleaf Viburnum	*Viburnum rhytidophyllum*		
Dwarf Cranberry Bush	*Viburnum trilobum* 'Compactum'		
California Fan Palm	*Washingtonia filifera*		
Mexican Fan Palm	*Washingtonia robusta*		
Chinese Wisteria	*Wisteria sinensis*		
Adam's Needle	*Yucca filamentosa*		

Phase 2:   *Placing Classes of Nursery Plant Material*   (250 points)

In this phase of the Horticulture Contest (Nursery Section), the contestants will be expected to place (judge) five classes of nursery landscape items. Four of these classes to be judged will be plant materials, while the fifth class will consist of either landscape models or landscape photographs. Each of the classes to be placed will be designated by a number and a name for identification. Each of these five classes will include four items to be placed.

The contestants will be required to place each class according to their relative merits or faults by an alphabetical designation. The four individual entries in each class will be designated as entry A, B, C, or D. Their placing is then checked on a Form 2 scorecard by each contestant (Figure 3–3). For example, if the correct placing of the entries in a particular class should be B-D-A-C, as determined by the contestant, this placing should be checked on the form. A separate form is used for each class to be judged. Each class is given a value of 50 points for a correct placement of the entries, for a total value of 250 points in this phase of the contest. Twenty-five minutes is allowed for the judging of all five classes, with 5 minutes given for each class to be placed.

STANDARD FFA PLACING CARD	Placings	Check Placing	Form 2
Contest_____	1-2-3-4		
	1-2-4-3		
	1-3-2-4		
Class Name_____	1-3-4-2		
	1-4-2-3		
	1-4-3-2		
Class No._____	2-1-3-4		
	2-1-4-3		
	2-3-1-4		
Contestant Name_____	2-3-4-1		
	2-4-1-3		
	2-4-3-1		
	3-1-2-4		
Contestant No._____	3-1-4-2		
	3-2-1-4		
	3-2-4-1		
	3-4-1-2		
Score_____	3-4-2-1		
	4-1-2-3		
	4-1-3-2		
	4-2-1-3		
	4-2-3-1		
	4-3-1-2		
	4-3-2-1		

FIGURE 3–3. Standard FFA placing cards.

Placing Nursery Plant Material

A total of four classes of nursery plant materials will be judged during the contest. These classes may consist of any of the following groups:

1. Not more than one class of groundcover plants.
2. Not more than two classes of flowering or deciduous shade trees.
3. Not more than one class of coniferous evergreens.
4. Not more than one class of broadleaved evergreen shrubs.
5. Not more than one class of deciduous shrubs.

All of these plants in any class will consist of specimens of the same species or cultivar. These plant materials may be bare-root, balled-and-burlapped, or grown in containers.

The plant material used for judging in the nursery section of the contest will be selected from the following list:

1. *Ground Cover Plants* (one class entry)
    [May be container grown]
    Winter Creeper   *Euonymus fortunei*
    English Ivy   *Hedra helix*
    Periwinkle   *Vinca minor*

2. *Flowering and Deciduous Shade Trees* (not more than two classes)
    [Balled-and-burlapped, bare-root, or container grown]
    Paper Birch   *Betula papyrifera*
    American Redbud   *Cercis canadensis*
    Washington Hawthorn   *Crataegus phaenopyrum*
    "Almey" or "Radiance"   *Malus x 'Almey' or M.*
       Crab Apple   *x 'Radiance'*
    Tuliptree   *Liriodendron tulipifera*
    Sweet Gum   *Liquidamber styraciflua*
    Pin Oak   *Quercus palustris*
    Live Oak   *Quercus virginiana*

3. *Coniferous Evergreens* (one class entry)
    [Balled-and-burlapped or container grown]
    Dwarf Pfitzer   *Juniperus chinensis 'Pfitzeriana*
       Juniper   *compacta'*
    Colorado Spruce   *Picea pungens*

Mugho Pine	*Pinus mugo mughus*
Eastern White Pine	*Pinus strobus*
Yew Podocarpus	*Podocarpus macrophylla maki*
Douglas Fir	*Pseudotsuga menziesii*
Japanese Yew	*Taxus cuspidata*
Oriental Arborvitae	*Thuja orientalis*

4. *Broadleaved Evergreen Shrubs* (one class entry)
   [Balled-and-burlapped or container grown]

Glossy Abelia	*Abelia grandiflora*
Japanese Aucuba	*Aucuba japonica*
Japanese Holly	*Ilex crenata*
Oregon Hollygrape	*Mahonia aquifolia*
Evergreen Azalea	*Rododendron azalea*
Adam's-needle Yucca	*Yucca filamentosa*

5. *Deciduous Shrubs* (one class entry)
   [Balled-and-burlapped, bare-root, or container grown]

Japanese Barberry	*Berberis thunbergii*
Crape Myrtle	*Lagerstroemia indica*
Scarlet Firethorn	*Pyracantha coccinea*
Arrowwood	*Viburnum dentatum*
Vanicek Weigela	*Weigela florida*

Point Scoring of Nursery Plant Classes

During the judging contest, the contestants are required to give final placings of the four entries in each class of nursery plants. To gain experience in placing classes, students may find it beneficial to use a point scoring system to determine the merits and faults of each entry in a class. Although an exact numerical score breakdown will not be used in the actual contest, the student may learn to rate and compare plant materials by this method. The following example of a class score breakdown for a sample class of plant material illustrates how this method is used:

JUDGING CRITERIA	MAXIMUM POINTS	SAMPLE POINT SCORING OF SPECIMENS			
		A	B	C	D
Size	20	20	15	12	15
Form	20	15	15	20	18
Density	20	19	10	20	15
Color	20	20	16	15	12
Condition	20	10	12	15	10
Total Points	100	84	68	82	70

Final placing of this class would be *A-C-D-B* (or 1-3-4-2).

Explanation of Judging Criteria

Size. Size must be proportional to container or ball and consistent with species or cultivar characteristics.

Form. Plant shape should be symmetrical and conform to the species or cultivar type.

Density. The plant should possess a full, dense branching habit having sufficient foliage to create a typical form. Deciduous plants will have a loose, open appearance, while evergreen plants will be more dense in appearance.

Color. Foliage and flowers (if present) should be healthy and true to the color of the species or cultivar.

Condition. The foliage and flowers (if present) should have good substance and be free of disease, insect, chemical, or physical blemishes.

In addition to the above judging criteria, the plant material must conform to the American Standards for nursery stock as determined by the American Association of Nurserymen. The balled-and-burlapped and container sizes must conform to the specific tables listed in the following reference:

American Association of Nurserymen. 1973. *American Standard for Nursery Stock.* AAN, 230 Southern Building, Washington, D.C.

Landscape Design Judging Standards

In addition to the placing of nursery plant material classes, an additional class to be placed in the contest includes a landscape design. This class may be represented by either landscape models or by photographs that represent the landscapes. The contestants will only give a final placing of these class entries on a Form 2 scorecard. However, for the purpose of training for the judging contest, the following judging form for evaluating a landscape class may be used:

LANDSCAPE DESIGN JUDGING STANDARD

CLASS NUMBER _____   CONTESTANT NUMBER _____

Judging Criteria	Maximum Points	LANDSCAPE NUMBER			
		A	B	C	D
Function	39				
Aesthetics	36				
Scale	15				
Presentation	10				

Total Points   100
Final Placing

1st	2nd	3rd	4th

Score _____

_____

Explanation of Judging Criteria

Function

1. The landscape plan must be functional. Each plan should be evaluated according to how well it fits (a) the climate, (b) topography and site location, and (c) the needs of the client family.

2. Function of the landscape areas is an integral part of a design. The building(s) should relate properly to (a) the public area (front yard), (b) the private area (outdoor living space), and (c) the service area (garden, storage, etc.).

3. Traffic flow around the landscape should also be functional. Consideration should be given to circulation with respect to (a) pedestrian circulation patterns and (b) automobile access and usage.

Aesthetics

1. Are accent plantings provided, both to delineate the entrance to the building and to create an attractive focal point for the outdoor living area?

2. The landscape plan should possess harmony and unity. All of the separate landscape components should be an integral part of the overall design.

3. Undesirable views should be properly screened, while attractive views from the yard or home are retained. Proper landscaping techniques are used to create interest in the yard.

4. Necessary, but unsightly areas should be properly screened from view. These might include vegetable gardens, trash cans, and air-conditioning units.

5. The landscape should provide a balance of plant varieties and materials without creating confusion. A well-balanced design offers an interesting variety of color, texture, and form without either a monotonous repetition or a confusing array of nonrelated elements.

6. The planting plan and plant material should complement the building architecture. The shapes and sizes of the plants should soften harsh building lines rather than emphasize them.

Scale

1. The planting areas and plant materials used in the design should be in scale with the building and yard.

Presentation

1. The landscape plan should be (a) simple without appearing crowded, (b) easily understood at a glance, and (c) the boundaries of the individual planting spaces and the yard are well delineated.

2. The proper marginal information is provided. This should include such items as scale, north point, location, and the name of the client.

Phase 3: *General Knowledge Examination* (300 points)

The third phase of the Nursery/Landscape Contest consists of a written examination. This examination tests the contestants' knowledge and understanding of various topics pertaining to the nursery industry. Fifty multiple-choice questions are selected from the following areas for this examination:

1. Plant Materials
2. Planting or Growing Media
3. Diagnosing Plant Disorders
4. Growing Materials (fertilizers, chemicals, growth regulators, etc.)

5. Plant Propagation
6. Plant Culture
7. Safety
8. Landscape Design

Each contestant is allowed 50 minutes to complete this phase of the contest. Each question is given a value of 6 points, for a total of 300 points in the examination.

Phase 4: *Landscape Design Practicum*   (100 points)

Each contestant will be asked to answer 10 objective questions concerning the various aspects of a furnished landscape drawing. These questions will each have a value of 10 points, and 20 minutes will be allowed for the completion of the Landscape Design Practicum. Question topics may include such items as the determination of costs of fencing, patio, or number of yards of sod required. The students should provide their own ruler (architectural scale) and a battery-operated electronic calculator (optional). Contestants are allowed to use their own copy of the following reference:

Pennsylvania State University. 1971. *Nursery Production — A Teacher's Manual.* Teacher Education Series, Vol. 12, No. 4T. Department of Agricultural Education, University Park, Pa.

FLORICULTURE CONTEST

Floriculture crops are judged in five divisions:

1. Potted plants.
2. Cut flowers.
3. Foliage plants.
4. Bedding plants.
5. Floral designs.

Each contestant will be examined for their knowledge and judging ability of these crops in four separate phases:

1. Identification of plant materials.
2. Placing classes of floral items.
3. General knowledge examination.
4. Practicum.

Phase 1:   *Identification of Plant Materials*   (250 points)

Each contestant will be given a list (Form 15) containing the technical and common names of 50 florist's plants. These plants will be arranged in a random manner throughout the judging area and will be designated by a number. (See Figure 3–4.) Each contestant will write the number accompanying each specimen in the column marked "Number" next to the correct name on the official scorecard. Each specimen that is correctly identified will be scored as 5 points, for a total of 250 possible points for this phase of the contest. A total time of 50 minutes is given for each contestant to complete this phase (60 seconds for each identification). An example of the Form 15 plant identification list is shown in Table 3–2. The plants shown on this list are for example only, as the plants used in contests will vary from one region or year. Although the list contains 124 plant names, only 50 of these will be displayed for identification during a national or state contest.

FIGURE 3–4.   Floriculture contest specimens for identification.

Phase 2:   *Placing Classes of Floriculture Crops*   (250 points)

Each contestant will judge the placing of five classes of flower crops. Each of the classes to be judged will consist of either four potted plants or four containers of cut flowers. Each class will be marked alphabetically (A-B-C-D) to allow each judge to determine the correct placing of each entry. This placing is then marked on an individual Form 2 score card. For example, if the correct placement of the class is

A-B-D-C, this alphabetical sequence is checked. A different Form 2 card is used for each judging class. Under no circumstances may any plant or flowers be touched in any way during the judging of the classes of plant materials.

The placement of floral crops is given 20 minutes for completion of the judging. Each class is placed during a 4-minute period with 50 points awarded for each class. At a state FFA flower judging contest, the placement phase will consist of a class of floral arrangements of the same design and any four classes of flowers to be judged from the following selected list:

- POTTED PLANTS
  Azalea
  Chrysanthemum

- CUT FLOWERS
  Carnation
  Chrysanthemum
  Gladiolus
  Rose

- FOLIAGE PLANTS
  Heartleaf Philodendron
  English Ivy
  Crassula
  Peperomia

- BEDDING PLANTS
  Geranium
  Petunia
  Coleus

Judging of Floriculture Crops

The judging of floriculture crops is similar to the commercial practice of grading. Each type of floral material is sorted into groups according to previously determined standards. The floral judge is required to evaluate the consistency of grading and to assess the quality of each class of plant material. The flower judge, like the commercial grader, must be able to make judgments on the quality of the crop. This requires a knowledge of the quality standards for each floriculture crop. In addition, each judge must be familiar with the cultural requirements of each crop, so valid judgments on quality may be made.

The established standards for quality of the floriculture crops is based in part on cultural perfection and commercial acceptability. Those faults that reduce the commercial acceptability of a crop, no matter what the cause may be, will in turn lower the placing of the floral group within a class. The presence of insects and disease organisms or evidence of damage caused by these pests will greatly lower the placing of any plant or flowers within a class entry. Since insects may move about freely, their presence may not be noted by all judges. Therefore, evidence of a few solitary insects on the foliage or flowers should not automatically place an entry in the lowest position for the class. However, the physical symptoms caused by either disease or insect injury should be *strongly faulted*, thus lowering the entry automatically to the last place for the class.

TABLE 3-2: *Example of FORM 15 Plant Material Identification List*

FLORAL IDENTIFICATION		Form 15	
CONTESTANT NAME ————————————— CONTESTANT NUMBER —————			
COMMON NAME	BOTANICAL NAME	NUMBER	SCORE
Aechmea	Aechmea fasciata		
Acacia	Acacia longifolia		
Agapanthus	Agapanthus africanus		
Ageratum	Ageratum houstonianum		
Anemone	Anemone coronaria		
Anthurium	Anthurium andraeanum		
Snapdragon	Antirrhinum majus		
Aphelandra, Zebra Plant	Aphelandra squarrosa		
Norfolk Island Pine	Araucaria heterophylla (exelsa)		
Florist "Huckleberry"	Arctostapholos manianta		
Asparagus "Fern"	Asparagus plumosus		
Sprengeri "Fern"	Asparagus sprengeri		
Aspidistra (Cast Iron)	Aspidistra elatior		
Wax Begonia	Begonia semperflorens		
Tuberous Begonia	Begonia tuberhybrida		
Shrimp Plant	Beloperone guttata		
Bouvardia	Bouvardia hybrida		
Schefflera (Octopus Tree)	Brassaia actinophylla		
Cabbage	Brassica oleracea capitata		
Caladium	Caladium bicolor		
Calceolaria, Pocketbook Plant	Calceolaria crenatiflora		
Calendula	Calendula officianalis		
China Aster	Callistephus chinensis		
Camellia	Camellia japonica		
Green Pepper	Capsicum annuum		
Cattleya Orchid	Cattleya cv.		
Parlor Palm	Chamaedorea elegans		
Spider Plant	Chlorophytum commosum		
Marguerite Daisy	Chrysanthemum frutescens		
Shasta Daisy	Chrysanthemum maximum		
Chrysanthemum	Chrysanthemum morifolium		
Grape Ivy	Cissus rhombifolia		
Clarkia	Clarkia elegans		
Croton	Codiaeum variegatum		

Coleus	*Coleus blumei*		
Lily-of-the-Valley	*Convallaria majalis*		
Jade Plant	*Crassula argentea*		
Cyclamen	*Cyclamen indicum*		
Cymbidium Orchid	*Cymbidium cv.*		
Cypripedium Orchid	*Cypripedium cv.*		
Dahlia	*Dahlia variabilis*		
Larkspur	*Delphinium ajacis*		
Sweet William	*Dianthus barbatus*		
Carnation	*Dianthus caryophyllus*		
Dumb Cane	*Dieffenbachia picta*		
Corn Plant Dracaena	*Dracaena fragrans massangeana*		
Red Edge Dracaena	*Dracaena marginata*		
Heath	*Erica carnea*		
Dollar Leaf Eucalyptus	*Eucalyptus pulverulenta*		
Amazon Lily	*Eucharis grandiflora*		
Poinsettia	*Euphorbia pulcherrima*		
Crown of Thorns	*Euphorbia splendens*		
Benjamin Fig, Weeping	*Ficus benjamina*		
'Decora' Rubber Plant	*Ficus elastica* 'Decora'		
Freesia	*Freesia refracta*		
Fuchsia	*Fuchsia hybrida*		
Gardenia	*Gardenia grandiflora*		
Gerbera	*Gerbera jamesoni*		
Gladiolus	*Gladiolus hybrida*		
Velvet Plant	*Gynura aurantiaca*		
Baby's Breath	*Gypsophilla elegans*		
English Ivy	*Hedera helix*		
Strawflower	*Helichrysum bracteatum*		
Baby's Tears	*Helxine soleirolii*		
Amaryllis	*Hippeastrum vittatum*		
Wax Plant	*Hoya carnosa*		
Hyacinth	*Hyacinthus orientalis*		
Hydrangea	*Hydrangea macrophylla*		
Polkadot Plant	*Hypoestes sanguinolenta*		
Rocket Candytuft	*Iberis amara*		
Impatiens	*Impatiens walleriana sultani*		
Dutch Iris	*Iris xyphium*		
Kalanchoe	*Kalanchoe blossfeldiana*		
Tritoma	*Kniphofia uvaria*		
Lantana	*Lantana camara*		
Sweet Pea	*Lathyrus odoratus*		
Easter Lily	*Lilium longiflorum*		
Statice	*Limonium sinuatum*		

Tomato	*Lycopersicum esculentum*		
Prayer Plant	*Maranta kerchoveana*		
Stock	*Matthiola incana*		
Cutleaf "Philodendron"	*Monstera deliciosa*		
Forget-Me-Not	*Myosotis sylvatica*		
Daffodil	*Narcissus pseudo-narcissus*		
Boston Fern	*Nephrolepis exaltata*		
Chincherinchee	*Ornithogalum thrysoides*		
Cyprepedium Orchid	*Paphiopedilum cv.*		
Phalaenopsis Orchid	*Phalaenopsis cv.*		
Geranium	*Pelargonium hortorum*		
Ivy Geranium	*Pelargonium peltatum*		
Emerald Ripple Peperomia	*Peperomia caperata*		
Variegated Peperomia	*Peperomia obtusifolia variegata*		
Petunia	*Petunia hybrida*		
Heartleaf Philodendron	*Philodendron oxycardium*		
Phlox	*Phlox paniculata*		
Aluminum Plant	*Pilea cadierei*		
Swedish Ivy	*Plectranthus australis*		
Buttercup	*Ranunculus asiaticus*		
Golden Pothos, Devil's Ivy	*Rhaphidophora aurea*		
Azalea	*Rhododendron indica*		
Hybrid Tea Rose	*Rosa hybrida*		
Leatherleaf Fern	*Rumohra adiantiformis*		
African Violet	*Saintpaulia ionantha*		
Snake Plant	*Sansevieria trifasciata*		
Birdsnest Snake Plant	*Sansevieria trifasciata 'Hahni'*		
Strawberry "Begonia"	*Saxifraga sarmentosa*		
Scabiosa	*Scabiosa atropurpurea*		
Christmas Cactus	*Schlumbergia bridgesii*		
Cineraria	*Senecio cruentus*		
German Ivy	*Senecio milkanioides*		
Gloxinia	*Sinningia speciosa*		
Jerusalem Cherry	*Solanum pseudocapsicum*		
Stephanotis	*Stephanotis floribunda*		
Bird-of-Paradise	*Strelitzia reginae*		
"Nephthytis" Arrow Vine	*Syngonium podophyllum*		
Marigold	*Tagetes erecta*		
Pick-a-Back-Plant	*Tolmeia menziesii*		
Tulip	*Tulipa cv.*		
Verbena	*Verbena hortensis*		
Violet	*Viola odorata*		
Pansy	*Viola tricolor*		
Calla	*Zantedeschia aethiopica*		
Wandering Jew	*Zebrina pendula*		
Zinnia	*Zinnia elegans*		

Each floral class should be evaluated on its relative merits or faults. Each entry within the class is evaluated separately and then compared with the other entrys. Cut flower entries will consist of multiple specimens (ranging from three cut flowers to one dozen in each container). With these multiple specimen entries, *uniformity* within the entry will be given the greatest emphasis. The following scale of points may be used in the evaluation of each entry within a class of cut flowers where more than one specimen is used:

Point Score Values for Cut Flowers

CONDITION OF FLORAL MATERIAL: 25 POINTS
Includes freedom from injury, bruises, or blemishes to the bloom; relative maturity of petalage; and uniformity of specimens within the entry.

PLANT FORM: 20 POINTS
Includes relative maturity of flower stem; shape and petal distribution of flower head; and uniformity of flowers within the entry.

STEM AND FOLIAGE CHARACTERISTICS: 20 POINTS
Includes stem strength and straightness; size and proportion of foliage to stem and bloom; foliage quality (color, substance, and size); and uniformity of these characteristics within each entry.

BLOOM COLOR: 20 POINTS
Includes trueness to varietal color; intensity and clarity (as a result of growing and storage conditions); and uniformity of bloom color within the entry.

SIZE: 15 POINTS
Includes size of stem, foliage, or blooms. Uniformity of size is most important within each entry. Points are deducted in relation to flowering specimens which are either undersize or oversize for the class entry.

Point Scoring Values for Potted Plants

CULTURAL PERFECTION: 40 POINTS
Includes the relative quality of the potted-plant specimen in relation to that expected for the crop as influenced by growing conditions, and freedom from damage.

FLORAL DISPLAY: 20 POINTS
Includes number and quality of blooms. Each plant should contain a large quantity of blooms for the species, which includes both mature and immature buds. This should not only account for present flower distribution around the plant, but also allow for future blooming.

SIZE OF PLANT: 20 POINTS
The size of each plant entry should be uniform among the class and of that expected for the plant species. Points are deducted for both specimens that are oversize and specimens that are undersize.

SIZE OF BLOOM: 10 POINTS
Blooms should be proportional to the size of the plant stature. Blooms should not be larger or smaller than normal for the cultivar.

COLOR OF BLOOM: 10 POINTS
Bloom color is reflected by cultural practices, growing conditions, and cultivar characteristics. Entries should be faulted for off-color blooms.

Floriculture Judging Classes

Potted Plants: Azalea Plants

Potted azalea plants may be of several types for commercial purposes. A single plant may be grown in a pot. These plants are pinched repeatedly to form a full, compact flowering specimen. Most potted azaleas are grown with several plants in each container. These produce a fuller flowering plant, which would command a greater selling price than single-stem plants. No matter which type of growth is produced, the finished azalea pot plant should have a symmetrical and vigorous branching habit. The pinching or shearing of the vegetative growth should have been done so as to produce a shape that will display the flowers to the best advantage. The two most common shapes used for potted azaleas are the flattened crown and the globe-shaped crown. The flattened crown is created by shearing the vegetative shoots in a horizontal plane at the desired flowering height. The globe-shaped crown is formed by uniformly pinching the shoots evenly around the perimeter of the plant. The shearing or pinching of the plant should be done at the proper distance from the main stem or container to allow the proper proportion of growth to the size of the container.

The flowers on the potted azaleas should be free from blemishes and pest damage. The plant should consist of from one-third to one-half fully open blooms, with the remaining flower bud potential less fully developed. These flowers and flower buds should be uniformly distributed over the surface of the plant and should be equal to the leaf surface. Any vegetative growth that extends above the flowers (*bypass growth*) is highly undesirable and should be strongly faulted.

The azalea foliage should be thick, healthy, and evenly distributed throughout the plant. Leaves should not be missing from the stem bases. These leaves should be dark green in color and free from blemishes, insect or disease pests, or nutrient-deficiency symptoms. Some spray residue may be permitted on the foliage, but excessive residue is considered to be unattractive and should be strongly faulted.

The following list of merits and faults is used when judging azalea potted plants:

MERITS

1. Flowers and buds are evenly distributed over the surface of the plant.

2. Flowers and buds are nearly equal to the leaf area of the crown.

3. Flowers are displayed to best advantage, with some bud potential both at the crown surface and within the canopy of the foliage.

4. Foliage and flowers in proportion to the size of pot.

5. No vegetative shoots (bypass growth) extended beyond the flowers.

6. Foliage is healthy, dark green, and thickly spaced along the stem axis.

FAULTS

1. Flowers are poorly distributed around the plant crown.

2. Flowers are old or otherwise in poor condition.

3. Flowers or foliage are unevenly distributed over the surface of the plant.

4. Vegetative growth (crown buds) is present beyond the flower buds.

5. More flowers are mature in proportion to bud potential.

6. Poor symmetry of plant crown; foliage is distributed unevenly over branches.

7. Plants are not at an even height or spread in a pot.

	JUDGING CRITERIA	POINTS DISCOUNTED FOR FAULTS
1.	Presence of pests or injury to plant parts.	10
2.	Poor flower bud development.	8
3.	Asymmetrical shape of plant.	8
4.	Unhealthy foliage.	7
5.	Presence of bypass growth.	6
6.	Poor balance of mature flowers to flower buds.	6
7.	Flower buds are overly mature or damaged.	5
		40

Potted Plants: Chrysanthemums

A commercial potted chrysanthemum may consist of from one to eight individual plants, grown in such a manner as to appear as a single specimen. The individual cuttings may be either pinched to create multiple stems or left unpinched (on free-branching cultivars).

Pinching should be done uniformly at the appropriate height to create a full potted plant. For judging purposes, all the entries in a class must contain the same number of plants in a pot. The potted chrysanthemum class is judged by the following criteria:

Cultural perfection. The height of the various potted chrysanthemums will vary according to cultivar, growth-retardant application, timing of pinching, and programming of the photoperiod. These cultural factors should be controlled by the grower to produce a potted plant that has a good balance between the plant growth and the size of the container. Generally, the foliage and stems should not appear to be out of balance with the container. The foliage should represent a height equal to approximately two-thirds the height of plant and pot combined. A plant that is greatly reduced in size or is too tall for the container should be strongly faulted. Foliage should extend over the entire length of the stems to the rim of the container so that the soil and pot surface are not predominant.

The shape and size of the plants in a container should be uniform and spaced adequately to produce a full specimen. A commercial potted chrysanthemum should form a circular outline when viewed from the top. When viewed from the sides, the flowers should be distributed uniformly over the upper perimeter of the crown. The outline of the plant crown will vary according to cultivar or cultural practice. This may take the form of a flat-topped or a slightly rounded crown.

The stems of each plant comprising the potted specimen should be of nearly equal length and spaced adequately to form a well-distributed floral display. The stems should be strong enough to support the foliage and flowers without the aid of staking. Weak or brittle stems indicate poor cultural conditions. The plant foliage should reflect adequate attention to pest control and careful handling, without the presence of accumulated pesticide residues. Poorly formed, dead, or diseased foliage at the base of the plant should be removed at the time of marketing.

Foliage and flowers should show no symptoms of insect, disease, or mechanical damage. Foliar symptoms of nutritional deficiencies should be strongly faulted. Flowers should be uniformly developed, with most blooms nearly at full maturity but without having an open center. Blooms having hard, green centers will not mature properly and should be faulted. Disbudding of side buds should have been done at the proper stage of development. Incomplete or recent disbudding results in an unattractive floral display.

Floral display. Blooms should be adequately spaced and uniformly distributed over the plant crown. A balance of nearly mature and immature flower buds indicates a greater flowering longevity of the potted chrysanthemum. Sparse blooming or a large number of small, crowded blooms is considered an undesirable trait for a pot plant.

Size of plant and blooms. The stems, foliage, and blooms should be in proportion to the size of container used. The potted plant should not appear to be top-heavy or stunted in growth in relation to the height of the pot. Bloom size should be consistent with the size and textural characteristics of the cultivar. Flower buds and mature blooms should be uniform and typical of the cultivar.

Bloom color. The color of the flower petals is reflected to some degree on age of the blooms, environmental conditions prevalent during flowering, cultivar characteristics, and the effects of diseases on the blooms. For judging purposes, the blooms should show no signs of fading or mechanical or pest damage, and should have the color normally expected for the cultivar.

Characteristics used in judging. The following merits and faults are considered when judging classes of potted chrysanthemums:

MERITS

1. Uniform development of stems and flowers. Plants are symmetrical, compact, and in balance with the container. A potted plant height of 38 to 46 centimeters (15–18 inches) is preferred.

2. Flowers are well distributed over the perimeter of the crown, spaced adequately for maximum development, and open blooms are well balanced by bud potential. The plant crown should be slightly rounded at the height of the flowers.

3. Plant stems are stong enough to support the foliage and flowers in an erect position without mechanical support.

4. Foliage is a healthy, dark green color. No evidence of disease or insect pest, mechanical, or nutritional damage should be present.

FAULTS

1. Stems should be uniform in height for all plants in a pot. Stems should be strong and covered uniformly with foliage to the pot rim. The plant should not be too tall or too short for the pot.

2. Flowers should not be too crowded or too loose in spacing. They should be located at the same heights on the stems and distributed uniformly across the crown of the plant. The absence of flowers, which results in poor plant symmetry or form, should be strongly faulted.

3.   Overmature, faded, or poorly formed blooms are faults that should move a plant to the bottom of the class. Hard, green centers of the flowers or irregularly shaped blooms should also be faulted.

4.   Weak, thin stems that do not adequately support the blooms are faults of the plant which reflect poor growing conditions. Stems that are too tall for the pot show inadequate growth-retardant control by the grower. Plants that have been staked or wired to support the individual stems are also faulted.

5.   Poor disbudding, failure to disbud, or evidence of recent disbudding of mature stems are faulted because they reflect improper cultural handling of the plants.

6.   Plants that show evidence of mechanical damage from mishandling, disease, or inspect pests, nutritional deficiencies, or pesticide applications are severely faulted. Any damage to blooms will move an entry to a lower position within a class of potted chrysanthemums.

	JUDGING CRITERIA	POINTS DISCOUNTED FOR FAULTS
1.	Evidence of pest damage, their presence, or pesticide burning of foliage.	10
2.	Plant stems are too tall or stems and blooms not in proportion to container.	9
3.	Plant shape is not uniform, stems are weak or broken, or flowers are not distributed uniformly over the plant crown.	8
4.	Flowers are not uniform in development, some blooms are overmature, or presence of wired stems.	8
5.	Evidence of late or poor disbudding; missed lateral flower buds that should have been removed.	5
		40

Cut Flowers: Carnations

Carnations are sold in two basic types: standards and miniatures. *Standard* carnations are produced with a single, large flower on a long stem. All side shoots and lateral buds are removed to allow maximum development of the terminal flower. *Miniature* carnations are grown with several flowers which form a spray. The terminal bloom may be removed (disbudded) to allow the lateral flower stems to elongate. Each spray will have at least three flowers and flower buds on stems that do not exceed 60 centimeters (24 inches) in length. The individual flowers on miniature carnations should not exceed 6.3 centimeters (2.5 inches) in diameter.

EXPLANATION OF TERMS USED IN DESCRIBING CARNATION QUALITIES

Sleepiness. Sleepiness is a term used to describe the physical condition of flowers when they appear to be soft, misshapen, and lack substance. This condition may be caused by ethylene gas produced by mature flowers in storage, high finishing temperatures in the greenhouse, or overmature blooms. The petal tips of these flowers turn inward toward the center, forming a cup-shaped appearance to the blooms. The flower color is often changed to a faded or bleached-out hue by this condition.

Split Calyx. The calyx of a carnation flower is united to form a cup that supports the base of the petals. When this cup-shaped calyx is split, the petals are no longer supported and will spill loosely from the calyx. The degree of splitting may vary from a minor tear along a calyx juncture to a severe splitting along the entire length of the calyx. This calyx splitting may be caused by poor temperature control, poor fertilization practices during their growth, or by cultivar characteristics. Some growers routinely place strips of cellophane tape around each calyx during flower development to reduce splitting caused by excessive pressure from the petals as the flowers mature. This practice is so common that most carnations on the commercial market will be taped. When blooms that have been taped are placed in a class with nontaped carnation heads, the taped flowers should be faulted; otherwise, the taping should be ignored when judging standard carnations.

Disbudding. The customary practice during the growth of carnations is to remove, as they develop, the lateral stems and flower buds that form below the terminal flower on standard carnations. Some lateral vegetative shoots may remain at the base of each stem after the flower is harvested. These should have been removed when the flowers were graded. Flower buds are removed from the stem below the terminal flower on standard carnations. On miniature carnations, the terminal flower bud and any excess lateral buds are removed as they develop to form a suitable spray of lateral flowers on the stem. Disbudding of flower stems or vegetative shoots should be done as early as possible after they have formed, but not so early as to result in mechanical damage to the foliage or stems. Disbudding should result in the complete removal of the shoot or flower bud and the peduncle (flower stem) without tearing or removal of adjacent leaves. The presence of peduncle stubs in the leaf axils is considered to be a fault for judging purposes.

POINTS TO CONSIDER WHEN JUDGING STANDARD AND MINIATURE CARNATIONS

Stems.   The flower stems on both types of carnation should be strong enough to support the blooms without mechanical aids. These stems should be straight and capable of holding the blooms in an upright position without excessive bending. The length of the stems should be uniform within an entry and of sufficient length to allow use in cut-flower arrangements. The side shoots that have formed along the stems should have been removed. No evidence of incomplete or missed disbudding should be evident.

Flower Form and Placement.   The flower petals should form a slight dome shape with adequate petalage to form a full center to the bloom. Most carnation cultivars form flowers having the outer petals at right angles to the calyx. Some cultivars produce other petals that are lower than a right angle, however, and should not be faulted for this reason. The optimum stage of maturity for marketing and judging carnations is when the outer petals have reached a right angle with the calyx but a sufficient number of petals remain in the center to provide substance to the bloom.

Miniature carnations should consist of from four to six flowers and buds distributed uniformly along the stem. The flowers should be borne on strong stems, but the stems should be long enough and spaced adequately to provide a loose spray for floral designing. The most terminal flowers should be well formed and nearly mature, while the lower flowers should show varying degrees of petal color. When judging miniature carnations, the form of the entire spray is to be considered.

Arrangement of Petals.   Petals should be arranged in a regular order in concentric rings, with each overlapping. The flower head should contain a continuous series of equal-sized petals which form a solid hemispherical crown across the face of the flower, without the presence of voids or skips.

Calyx.   No evidence of calyx weakness or splitting should be present. The calyx should form a perfect funnel around the petals so that the bloom face stands erect at a nearly perfect right angle to the stem. The basal bracts should be firmly attached to the calyx and should show no signs of browning. The degrees of calyx splitting that should be seriously faulted are: (a) splitting of the calyx that results in deformation of the flower or protrusion of some petals and

(b) moderate splitting, which extends the entire length of the calyx, whether the flower is deformed or only weakened.

Flower Color and Fragrance. Although the petal color is dependent upon cultivar characteristics and some carnation flower colors may be altered by food dyes, each bloom should display the optimum color and clarity. For judging purposes, the artificially dyed blooms are rarely used. Variegated petals are common among both standard and miniature carnations. These variegations in color should be uniform among the petals. Color breaking of solid-colored blooms should be faulted. The color of flower petals should be clear and brilliant, indicating that the flower has not reached full maturity. Dulled colors, dark-colored petal tips, and loss of flower substances are indications of overmaturity or sleepiness in the carnation blooms. Fragrance is not present in all cultivars so should be disregarded during judging.

MERITS

1. Flowers have a relatively flat base with the petals forming at least a right angle with the calyx. Petals form a slightly hemispherical or dome-shaped crown in the center of the flower.

2. The flower head is attached to the stem at a right angle to the calyx. The petals are firm, with fully developed petals at the perimeter and less mature petals at the center of the bloom. The petals are fully turgid, forming a flower having good substance.

3. The calyx is firm, no splits are present, and the calyx forms a funnel around the petals. The calyx supports the petals so that the shape of the flower is perfectly round when viewed from above.

4. Stems are straight, strong, and no side shoots are present. Foliage is healthy, with the leaves slightly reflexed or curled.

5. Disbudding of standards has been done properly to leave no peduncle stubs. Only one flower is located in a terminal position on the stem.

6. In addition to the preceding points, miniature carnations should have a good spray formation. The terminal flower is disbudded, along with excess side buds, to form a spray having at least two open flowers (or more) and five to seven immature flower buds. These should be uniformly spaced along the stem.

FAULTS

1. Any indication of disease or insect pests, their damage, or

injury to the plant resulting from pesticide application is strongly faulted.

2.   Signs of overmaturity which cause sleepiness (such as color fading, browned petals, flaccid petals, or shriveled blooms) should be reason to remove an entry to the last placing in a class of carnations.

3.   Splitting of the calyx may occur in degrees ranging from minor tearing to severe splits that result in deformed flowers. These splits are defined as: type 1—tearing of the calyx is less than half the length, no petal spillage; type 2—severe calyx splitting without loss of substance or deformed flower: and type 3—severe calyx splitting, resulting in a weak and deformed flower head.

4.   Poor flower form of any type. This is usually seen as an irregularity in petal arrangement, distribution, or voids in petals across the face of the flower. This may also be evidenced when the flower petals have not reached a uniform stage of maturity around the perimeter of the flower head.

5.   Stem and foliage characteristics that reduce the attractiveness of the flower. Crooked stems, presence of peduncle stubs after disbudding, or side shoots that remain will cause a carnation to be faulted. Weak stems and poor foliage are also faulted during judging.

		POINTS DEDUCTED FOR FAULTS	
	JUDGING CRITERIA	STANDARD	MINIATURE
1.	Evidence of injury from insects, diseases, or nutritional deficiencies.	8	8
2.	Blooms that are overmature, off-color, in poor condition, or show signs of sleepiness.	7	7
3.	Calyx splitting: type 3—poor flower form, skips in petalage, or severe calyx splits.	7	7
4.	Miniatures only—poor spray formation, flowers too large for use in designing.	0	6
5.	Broken stems.	6	6
6.	Weak, crooked stems or damaged foliage.	5	5
7.	Poor petal color resulting from cutting the flowers in the tight-bud stage.	4	3

8. Presence of spray residues, dirt, or other distracting material.	3	3
9. Calyx splitting: type 2—flower head not misshapen.	3	2
10. Flowers not borne at right angles to the stem.	2	0
11. Evidence of faulty or skipped disbudding.	2	2
12. Standards only—flower head is not hemispherical, with the outer petals forming a right angle with the calyx.	2	0
13. Calyx splitting: type 1—minor splits, without loss of flower form or substance.	1	1
	50	50

Cut Flowers: Chrysanthemum

Chrysanthemums grown for cut flowers may be of either of two types: standard or spray (pom-pon) chrysanthemums. *Standard* chrysanthemums are single-stem plants that produce a single flower on the terminal shoot. These plants are disbudded of lateral flower buds to allow complete development of the terminal flower. The common flower types grown as standard chrysanthemums are *incurves, fuji mums, spider mums,* and various *daisy-flowered* cultivars.

Spray-type chrysanthemums consist of several lateral flowers borne on a single main stem. The terminal flower on these stems is disbudded to allow complete development of the lateral blooms. The main objective of growers is to produce lateral flowers of these sprays on stems that are of sufficient length for use in floral arrangements. Commonly grown spray-type chrysanthemums are *decoratives, cushion, anemone, button, daisy,* and various other *pom-pon* types.

TERMINOLOGY USED WHEN JUDGING CHRYSANTHEMUM CLASSES

Standards. Chrysanthemum cultivars normally grown with a single terminal flower on a long stem.

Disbuds. A term synonymous with "standards."

Terminal Bud. A bud that will form a flower under short-day conditions. This bud is surrounded by other flower buds, which are located at nearly the same height on the stem.

Crown Buds. Buds that have begun forming a flower under short-day conditions but have ceased further development because day-length conditions are no longer optimum. These buds are generally surrounded by vegetative shoots that extend above the flower bud. If the buds that form around the terminal bud develop into mature flowers and are located above the terminal flower, they are called a *crown spray*. This condition is considered a serious fault, particularly when the lateral flowers are on shortened peduncles, forming a clubby flower spray. This type of spray is difficult to use in floral arrangements.

POINTS TO CONSIDER WHEN JUDGING STANDARD CUT CHRYSANTHEMUMS

Flower Shape.   The standard chrysanthemum may be of several types. The most typical form used for judging purposes is the incurved or "football" chrysanthemum. This type of chrysanthemum flower should have a fully globular form, with the incurved ray florets spaced uniformly over the surface to produce a firm substance to the bloom. The outer ray florets should extend slightly below the plane, forming a right angle with the stem.

Fuji or spider chrysanthemums should form a flat plane on top of a straight area. The tubular ray florets should be spaced uniformly around the outer perimeter of the flower. These should be of uniform length, free from serious mechanical damage, and without serious voids which give the flower an unbalanced appearance.

Flower Size.   The size of flower will be dependent upon cultural and cultivar characteristics. Large flowers are desirable on standard chrysanthemums, but oversized or undersized flowers for the cultivar should be faulted.

Maturity and Color.   The age of the flowers during judging is an important judging criterion. The optimum stage of flower maturity for standard chrysanthemums is when the flower petals have expanded to a degree, leaving only a few ray florets undeveloped in the center of the bloom. Blooms that show a tight green center or a loose, fully developed center should be faulted. Flower color should clearly reflect that which is typical for the cultivar. Faded or diseased florets, which indicate overmaturity of the bloom, should be severely faulted.

Foliage and Stems.   The foliage and stems should show no evidence of disease, insect, or mechanical damage. The leaves should be typical for the cultivar, showing a deep-green coloration. The flower head should be situated squarely atop the stem terminal. The flower and stem should be proportionate; with large flowers borne on long, thick, straight stems. No evidence of peduncle stubs caused by faulty disbudding should be present.

POINTS TO CONSIDER WHEN JUDGING SPRAY-TYPE
CHRYSANTHEMUMS

Flower Form. The flower form of spray chrysanthemums will vary according to the type of flower grown. Some of these most commonly found are:

Decorative. The shape of the "decorative" chrysanthemum flower should be flattened, with the center ray florets being shorter than those at the perimeter. The petals should form a circular flower, without skips or voids in the bloom. The mature flower should appear to consist of ray florets only, with few or none of the disc florets being evident at the center of the bloom.

Pom-pon. The center of the pom-pon spray chrysanthemum flower should be slightly hemispherical in shape. The flower should not appear to be flattened when viewed from the side. When viewed from above, the florets should be regularly arranged to form a round, compact flower head.

Button. The button chrysanthemum flower should form a nearly rounded shape when viewed from the side. The outer petals will extend below the junction of the stem and flower. The ray florets should be of nearly uniform development throughout the crown of the flower, with only a few of the center florets left undeveloped.

Anemone. The center of the anemone chrysanthemum consists of simple disc florets which are of a different color than the outer ray florets. These disc florets should be tightly arranged to form a compact "eye" of the flower. The outer ray florets may consist of several rows of uniformly spaced petals. These should form a rounded outline when viewed from above, without the presence of skips in the petalage.

Singles. Singles will appear as being similar to the anemone types, except the center disc florets are of the same color as the outer ray florets. But flower types should be conspicuously flattened when viewed from the side.

Flower Color.   Same as for standards.

Foliage and Stems.   Same as for standards.

MERITS

1.   Stems should be strong enough to support the flower or spray, of sufficient length for use in arrangements, and in proportion to the flower(s).

2.   Disbudding is done early, with a minimum of mechanical damage. No flower peduncle stubs remain.

3.   Flowers have reached the optimum stage of development. Flower color is bright and clear, without the evidence of fading caused by overmaturity or poor cultural conditions. Petals are uniform in size, with only a few undeveloped florets in the center of the flower.

4.   Foliage is healthy, dark green, and free from mechanical damage. Stems are straight, strong, and free from side shoots.

5.   (Standards) Flowers are borne terminally, with the blooms attached squarely. Flowers are shaped in a typical form which is consistent with the cultivar. The blooms are large but not oversized.

6.   (Spray-type) Spray is uniformly shaped on all sides of the main stem. Each individual lateral flower is borne squarely on a stem of sufficient length to form a loose, but durable spray. The flowers are located in the spray, without excessive flowering in a lower position on the stem.

FAULTS

1.   Florets falling from the flower heads. This may indicate overmature flowers or mishandling of the chrysanthemums. This is particularly a problem with large incurve standards.

2.   Evidence of disease, insects, or nutritional disorders to flowers or foliage.

3.   Weak, broken, crooked, or cracked stems, or the presence of side shoots.

4.   Flowers that are not centered on the stem, irregular petals, or promiscuous (nonuniform) ray arrangement.

5.   Présence of crown buds.

6.   Evidence of poor or late disbudding, with obvious peduncle stubs.

7.   (Standards) Flower is born on a lateral stem rather than from the terminal.

8.   (Standards) More than a single, terminal flower present on the stem.

9.   (Sprays) Any condition of the spray that would make it unfavorable. Such conditions as loose, irregular, or clubby spray formations should be faulted.

		POINTS DEDUCTED FOR FAULTS	
	JUDGING CRITERIA	STANDARD	SPRAY
1.	Evidence of injury from insects, diseases, or nutritional deficiencies.	10	10
2.	Poor flower form: Sprays—poor spray formation; standards—flower position on stem or shape of flower. Cracked neck (broken stem below the flower) on standards.	9	9
3.	Condition of blooms. Poor form of blooms. Overmature or immature inflorescence or petals. Lack of symmetry among petals. Presence of falling ray florets (standards).	8	8
4.	Lack of uniformity. Sprays—nonuniform spray formation; standards—promiscuous ray arrangement.	7	7
5.	Presence of flower buds or side shoots along the lower portion of the stem.	6	6
6.	Crooked, weak, or short stems.	5	5
7.	Evidence of poor or late disbudding.	4	0
8.	Presence of broken peduncles.	0	4
9.	Flowers or foliage off-colored or otherwise blemished. Standards—ray florets off-colored.	1	1
		50	50

Cut Flowers: Gladiolus

Gladiolus are grown from corms in fields for use in cut-flower arrangements. These flowers are a very important part of the floral designer's plant material selections. Gladiolus are used in the formation of a line in arrangements and are greatly used for large, lavish bouquets. The major advantages offered by gladiolus in design work include: (a) the spike forms a natural line, which points out the direction in a design; (b) the flowers mature from the bottom to the top of the spike, forming a natural arrangement of flowers which range from tight buds nearest the tip to fully expanded flowers at the center of the stem; and (c) the stem offers a great variety of uses, which include cutting individual flowers or even use of the foliage in a design. Gladiolus stems may be used virtually completely, with very little wastage. Gladiolus flowers may be obtained in a wide range of colors and petal patterns.

POINTS TO CONSIDER WHEN JUDGING GLADIOLUS

Stem.   The stem should be straight, stiff, and without the presence of side shoots or multiple flower stems. The stem should consist of at least one-third to one-half of its length in flowers or flower buds. The tip of the spike should not be broken and preferably should be straight.

Flowers.   The flowers should be well distributed and uniform along the stem. These flowers should all face in the same general directions, with the floret faces in a flat plane with the front of the flower. Stems having florets facing on all sides, forming a circular flower, should be considered inferior. The spacing of the individual flowers on the stem must be close enough to give the impression of a continuous flower, without skips or tightly compacted florets.

The degree of floret maturity should be optimum at the basal portion of the inflorescence, progressing to the least mature florets at the spike tip. The first bud that shows petal color should be located approximately one-third the distance from the tip, compared to the total inflorescence. Flower color should be consistent among the florets, characteristic of the variety, and free from mechanical damage or streaking caused by thrips.

Foliage.   There should be no evidence of disease, insect, mechanical, or nutritional damage. The spike tips and flower sheath (bracts) should not be browned. The length of the foliage should extend only to the level where the lowest basal floret is located.

MERITS

1.   Florets are spaced uniformly along the stem, in proportion to the length of stem. Development of flowers is progressive from the tip to the basal floret. The inflorescence contains a good balance between fully open florets, partially open florets, and immature buds.

2.   All florets are in good condition, turgid, and facing in the same general direction.

3.   The flower stem is straight, strong, free from side shoots, and the foliage is the proper length and substance. The foliage is free from damage. No leaf tip burn or thrip damage is apparent.

4.   The flower spike tip is neither bent nor broken.

FAULTS

1.   Florets are past maturity. Basal florets show signs of dis-

coloration, wilting, breakage, or dropping.

2.   Florets are not uniformly distributed along the stem, with obvious voids being present. Florets face in different directions.

3.   Florets immature (stems cut in too tight of bud) or not maturing progressively from the tip to the base.

4.   Crooked stems, inflorescence patterns, or spike tips.

5.   Evidence of disease, insect, or mechanical damage on foliage or florets. Presence of leaf tip burn, browned sheaths, or evidence of thrip damage to foliage or flowers should be severely faulted.

JUDGING CRITERIA	POINTS DEDUCTED FOR FAULTS
1. Evidence of injury from insects, diseases, nutritional deficiency, or mechanical stresses.	10
2. Poor floret or spike form. Flowers facing in different directions. Broken spike tip.	9
3. Poor distribution of florets on stem. Inflorescence is less than one-third the length of the stem.	9
4. Florets are either too tight, overmature at the base, or not uniformly mature progressively from tip to base of the inflorescence.	8
5. Stem is crooked, stem tip is bent.	7
6. Presence of side shoots or evidence of leaf tip burn.	7
	50

Cut Flowers: Roses

Commercial cut roses are one of the most popular of the florist crops for special occasion purchases and for floral designs. At the same time, cut roses are one of the more perishable flower crops grown for the retail trade. Two distinct flower types of roses are sold by florists. These are: hybrid tea (standard) and floribunda (miniature) roses. *Standard* roses are grown with long stems which bear a single, large flower. *Sweetheart* (or *miniature*) roses are grown on shorter stems and produce several flowers in a spray. The major concern of growers, shippers, and retailers is the preservation of the freshness and cut flower longevity of the cut roses. Roses are cut in the tight-bud stage and shipped to wholesalers. By the time the rose blooms are sold to consumer, the flowers have nearly opened to their optimum maturity.

TERMINOLOGY USED WHEN JUDGING CUT ROSES

Bent Neck.   A term used to describe rose stems which can no longer support the heavy flower head because of a lack of water (turgor pressure) in the vascular tissue. The flower head is bent over in the region immediately below the calyx and the petals wilt quickly because it no longer may absorb adequate water.

Bullhead.   This term is used to describe a flower bud that fails to open properly or to describe an opened flower that is malformed. The petals are shortened and wrinkled to the extent that the flower is flattened and the petals fail to expand. The flower often wilts and dies without fully expanding as it matures.

Peanut Flower.   This is simply a flower that is distinctly undersized, with fewer petals than is characteristic for the cultivar.

Peeling of Petals.   A practice of removing outer petals from a rose flower. This is done to remove damaged, dirty, or faded petals to reveal healthy, brightly colored petals. Peeling of these outer petals to cause a flower to conform to a judging class is considered a major fault.

POINTS TO CONSIDER WHEN JUDGING CUT ROSES

Stage of Flower Development.   Commercial cut roses may be judged at either of two stages of maturity which correspond to the types sold in the wholesale or retail markets. The coaches will determine which stage is to be considered before judging is begun.

Wholesale Stage of Maturity.   This stage of flower maturity coincides with the degree of petal opening found at the time of harvest and shipping to the wholesale market. The outer petals will have only slightly begun to unroll from the bud. A flower will normally open properly once cut at this stage. Some petal color should be clearly evident at this time. Large hybrid tea roses should show more petal reflexing at this stage than will the daintier sweetheart roses.

Retail Stage of Maturity.   Although this stage of flower maturity is called the bud stage, the flower petals are extended to a greater degree than in the wholesale stage. These petals and flower heads should be open to the same degree among the flowers shown in an entry. These flowers should be no more than one-fourth to one-third open. Flowers that have matured to a point where the petals have reached a half-open stage or greater are not considered for judging in these contests.

Flower Form and Size. The flower should be borne squarely upon the terminal of the stem in hybrid tea cultivars. The spray formation of miniature roses grown in sprays should be of good form and size. Most miniatures are disbudded to allow proper formation of the terminal flower bud, however. The petals should form a regular, tight spiral around the bud. The flowers should not be misshapen to form either bullheads or peanut flowers. The size and form of each bloom should be characteristic for the cultivar.

Flower Condition and Color. The condition and color of the petals should be indicative of a healthy flower that is typical for the cultivar. The color should be solid, intense, and free from blemishes. Bicolored or multicolored petals should show distinct markings which are characteristic for the cultivar. No indication of disease, insects, blemishes, spray residue, or dirt should be apparent. Fading of colors or other indications of overmaturity should not be obvious. The presence of greenish blemishes or markings on the outer petals is inherent in some cultivars, however, so should only be considered as a minor fault.

Stem and Foliage. No evidence of disease, insect damage, or nutritional deficiencies should be present. The foliage may show signs of some mechanical damage caused by the thorns without being considered more than a minor fault. However, serious tearing, blemished leaves, missing foliage, or chemical damage on the leaves must be considered a serious fault.

The stems should be strong enough to support the flowers. These stems should be as straight as possible; however, weak stems are faulted more severely than slightly crooked ones. Longer stems should be found bearing the larger flowers, with long stems considered a desirable quality of roses. Bent neck on a flower stem is a very serious fault. Side shoots and lateral flower buds should be removed in such a way as to prevent damage to the foliage, stems, or leave peduncle stubs.

MERITS

1.   Flowers are at the proper stage of maturity for judging, are of the correct size, and in the best condition. The flower color is fresh, characteristic of the cultivar, and free from blemishes.

2.   Stems are straight, strong, and the flower head is placed squarely on the terminal of the stem. Long stems are considered best, but the stem length should be proportional to the size of flower it bears.

3. Foliage and stems should be healthy, free from injury or blemishes, and of a normal color for the cultivar.

FAULTS

1. The greatest fault to be considered in cut roses is a bent neck, which indicates that the vascular tissue is plugged and water is no longer able to enter the stem.

2. Overmature, off-color, faded, blemished, or wilted blooms should be severely faulted.

3. Any obvious attempt at repairing overmature or damaged petals by "peeling" is considered a fault.

4. Misshapen blooms (bullheads or peanut flowers) or flowers of poor form should be faulted.

5. Weak, crooked, or damaged stems are faults. Also, presence of side shoots or evidence of late or poor disbudding should be considered as a minor fault.

	JUDGING CRITERIA	POINTS DEDUCTED FOR FAULTS
1.	Evidence of injury from insects, diseases, nutritional deficiencies, or mechanical stresses.	10
2.	Bent neck or weak stems. Evidence of overmaturity (flowers wilted or fully opened). Peeled petals.	9
3.	Misshapen blooms (bullheads or peanut flowers).	9
4.	Weak or crooked stems without bent neck. Poor flower placement on the stem. Evidence of poor disbudding.	8
5.	Good flower form, but evidence of petal bruising or greenish blotching on outer petals.	7
6.	Immature flowers which may not open properly.	5
7.	Obvious foliage damage caused by thorns.	2
		50

Foliage Plants

The term "foliage plant" is meant to mean any plant that is produced for its foliage attractiveness, without regard for flowers. Foliage plants may be small, delicate specimens or tall treelike plants used for room decorations. Other types may be grown on totems or allowed to trail gracefully from hanging baskets or planters. Four types of foliage plants are to be used in the Floriculture Contest for placing classes of plant material. These are:

- Heartleaf Philodendron
- English Ivy
- Crassula
- Peperomia

The foliage plant class is judged by the following criteria:

Cultural Perfection. The foliage and stems should be in proper balance with the pot or container. The plant should be attractive and well grown.

Plant Form. The plant shape will depend upon whether the variety grows upright or is a trailing type of plant. Some differences in shape will occur among plant types and methods of training, such as trailing types trained to a totem.

Upright types of foliage plants should be compact and symmetrical in appearance. The stems should be strong enough to support the foliage. Special supporting materials may be used if added support is beneficial to the appearance of the plant.

Trailing foliage plant types will not form as nearly compact, symmetrical potted specimens. These plant types are most often displayed in hanging basket planters. The trailing plants should uniformly cascade over the pot rim and reach a length that is in proportion with the container.

Foliage. Foliage plants are grown for their leaf characteristics. Since these leaves are the most prominent features, they must display the characteristic color(s) and be free of spray residues or pest damage. Sufficient foliage should be present to provide a continuous mantle around the plant.

Characteristics used in judging. The following merits and faults are considered when judging classes of foliage plants:

MERITS

1. The foliage is properly colored for the variety. The foliage color is bright, clear, and free from nutritional or cultural blemishes.

2. The form is characteristic for the plant variety. Foliage is symmetrical and uniformly distributed over the plant. The stems and foliage are in the proper proportion to the container.

3.   Foliage is a healthy color. No evidence of disease or insect pest, mechanical, or nutritional damage should be present.

FAULTS

1.   Any deviation from the merits listed above may be considered faults. Such features as poor plant form, lack of symmetry in the plant, poor proportion of plant to container, improper foliage color, and blemished foliage will cause a specimen to be faulted.

JUDGING CRITERIA	POINTS DISCOUNTED FOR FAULTS
1. Evidence of pest damage, presence of pests, or pesticide burning of foliage.	10
2. Lack of symmetry of foliage and poor proportion of foliage to container.	10
3. Foliage is not properly colored, smaller than normal for variety, or sparse.	9
4. Weak stems, unsupported stems (when required by variety), or improper use of supports.	8
5. Foliage is soiled or presence of pesticide residue.	7
6. Mechanical damage to foliage caused during shipping.	6
	50

Bedding Plants

The term "bedding plant" is meant to mean any annual, biennial, or perennial plant which is normally seed propagated and grown outdoors during the warmer months. The vegetatively propagated geraniums and vegetable plants are also included in this group of plants. Only ornamental plants which are grown normally as annual bedding plants are considered in this class for judging purposes. The bedding plants to be considered are:

- Geranium
- Petunia
- Coleus

The bedding plant class is judged by the following criteria:

Cultural Perfection. The appearance of the plant should indicate the skill of the grower. A proper balance should exist between healthy foliage and flower buds or blooms (except for coleus). Geranium plants should have a good proportion among the plant foliage, flowers, and the container. The correct size of a geranium plant is one that has a plant twice the size of the container.

Floral Display. Blooms on petunias and geraniums should be adequately spaced and uniformly distributed over the plant crown. A good balance of nearly mature bloom and immature flower buds should be present to indicate flowering potential. Although the flowers should be borne above the foliage, the pedicels should not be elongated to leave the foliage appearing to be separated from the flowers.

Characteristics used in judging. The following merits and faults are considered when judging classes of bedding plants:

MERITS

1.   The flowers and flower bud potential of petunias and geraniums should be equal to nearly one-third of the leaf area. The plant should be proportional to the size of container (if present) or should be of suitable size for bedding.

2.   Foliage color should be characteristic of the variety. Geraniums should display the dark zonation on the foliage. Coleus plants should show a uniform and characteristic foliage color pattern.

3.   Flowers of petunia and geranium should be borne above the foliage without excessive pedicel stretching.

4.   Flower color is uniform on flowering types and faded or overmature flowers are not present.

FAULTS

1.   Poorly formed flowers, plant shape, or flowers located beneath the foliage would be considered as faults.

2.   Improper foliage coloration (coleus) or dull, dingy foliage indicates poor growing conditions.

3.   Overmature blooms, poor flower bud potential, or elongated flower pedicels make the plant unattractive.

JUDGING CRITERIA	POINTS DISCOUNTED FOR FAULTS
1. Evidence of pest damage, presence of pests, or pesticide burning of foliage.	10
2. Lack of foliage symmetry, loose or open appearance of foliage, poor plant shape.	10
3. Flowers overmature, nonuniform, or poor bud potential.	9
4. Plants too tall, leggy, or weak stems.	8
5. Poor foliage color, leaves are undersize, irregular foliage markings.	7
6. Dirty foliage or presence of pesticide residue.	6
	50

Floral Design Judging Standards

In addition to the placing of florist plant material classes, an additional class to be placed in the Floriculture Judging Contest includes floral designs. Four floral arrangements of similar design will be placed. The floral arrangements will be judged on the following criteria:

Design.   The design elements of color, texture, line, and form are used together to create a planned relationship in the arrangement.

Balance.   Proper gradation of flower sizes and a visual stability exists within the arrangement.

Harmony.   Scale is used to create an attractive arrangement. The arrangement is constructed with a blending of color, texture, and properly sized blooms and foliage.

Unity.   All parts of the arrangement blend together to create a successful design.

Originality.   The design possesses a special quality that makes it stand apart. The visual line of the arrangement leads the eyes to the focal point to create rhythm. Colors, textures, and shapes of flowers are blended properly to create a design that stands out from the others.

The contestants will give a final placing of these class entries only on a Form 2 scorecard. However, for the purpose of training for the judging contest, the following judging form for evaluating a floral design may be used:

FLORAL DESIGN JUDGING STANDARD

CLASS NUMBER _____ CONTESTANT NUMBER _____

Judging Criteria	Maximum points	A	B	C	D
		DESIGN NUMBER			
Design	30				
Balance	15				
Harmony	18				
Unity	19				
Originality	18				

Total Points   100
Final Placing

1st	2nd	3rd	4th

Score _____

Phase 3: *General Knowledge Examination* (250 points)

In this section each contestant will answer questions concerning the field of floriculture. This phase of the contest is used to test a contestant's ability to show general knowledge and understanding of a wide array of the various areas of floriculture. Fifty multiple-choice questions are selected from the following areas:

1. Diagnosing plant disorders.
2. Growing materials (fertilizers, chemicals, growth regulators, etc.).
3. Plant propagation.
4. Plant culture.
5. Floral design.
6. Planting and planting media.
7. Plant materials (anatomy, physiology, taxonomy, etc.)
8. Safety.
9. Salesmanship.

Each contestant is allowed 50 minutes to complete this pase of the contest. Each question is given a value of 5 points, for a total of 250 points possible in this phase of the contest.

Phase 4: *Practicum*   (250 points)

The floriculture practicum is designed to test a contestant's ability to use the general knowledge required for employment at a greenhouse or retail shop. Each student must complete each area of this practicum. The practicum consists of three parts: (a) flower arranging, (b) planting cuttings, and (c) salesmanship.

Flower Arranging   (84 points)   Each contestant will design a flower arrangement having a retail value of $10.00. The arrangement is intended for display on a 4-foot-high bookshelf in a living room. The materials that will be provided for this design include: gladiolus, spray chrysanthemums, foliage greens, container, floral foam block, and tape. The wholesale prices for materials will be provided. From these prices the contestant must submit an itemized bill for the arrangement, which includes: materials, overhead, wages at $3.50 per hour, and a delivery charge of $1.00. Each contestant *must* provide his/her own knife, florist shears, and a pencil.

Planting Cuttings   (83 points)   Contestants will be tested on their ability to plant rooted chrysanthemums in pots. Each contestant will be provided 20 rooted cuttings, from which five will be selected. Other materials provided include: a 5½- or 6-inch azalea pot, a drainage piece, a correctly moistened 1–1–1 soil mixture, label, and pencil.

Salesmanship   (83 points)   Contestants will be graded on their ability to receive and record a telephone order. A telephone and an order form will be provided. A contest official will grade each contestant's telephone presentation, while serving as the customer, and will also evaluate the written order.

SELECTED REFERENCES

Janick, J. 1972. *Horticultural Science,* 2nd ed. W. H. Freeman and Company, Publishers, San Francisco, Calif.

Kalin, E. W. 1975. *A Manual for Flower Judging.* Pi Alpha Xi Collegiate Honorary and the Department of Horticulture, Washington State University, Pullman, Wash.

Ohio State University. 1973. *The Nursery Worker—Part I, Basic Plant and Soil Science.* Department of Agricultural Education, Columbus, Ohio.

Pennsylvania State University. 1970. *Greenhouse Crop Production—A Teacher's Manual.* Teacher Education Series, Vol. 10, No. 3T. Department of Agricultural Education, University Park, Pa.

Pennsylvania State University. 1968. *Landscape Design.* Teacher Education Series, Vol. 9, No. 3. Department of Agricultural Education, University Park, Pa.

Pennsylvania State University. 1971. *Nursery Production—A Teacher's Manual.* Teacher Education Series, Vol. 12, No. 4T. Department of Agricultural Education, University Park, Pa.

Pennsylvania State University. 1968. *Retail Flower Shop Operation and Management—A Student Manual.* Department of Agricultural Education, University Park, Pa.

Urbanic, C. E. 1973. *The Nursery Worker—Part II, Nursery Practices.* Department of Agricultural Education, Ohio State University, Columbus, Ohio.

Wotowiec. P. J. 1974. *The Greenhouse Worker,* 2nd ed. Department of Agricultural Education, Ohio State University, Columbus, Ohio.

Wyman, D. 1972. *Wyman's Gardening Encyclopedia.* Macmillan Publishing Co. In., New York.

TERMS TO KNOW

Asymmetrical	Density	Observer
Balance	Disbudding	Scale
Calyx	Floret	Size
Color	Floriculture	Spike (Flower)
Condition	Form	Terminal Bud
Coniferous	Function	Uniformity
Criteria	Harmony	Unity
Deciduous		

STUDY QUESTIONS

1. List reasons why horticultural judging contests are an important part of training for students.

2. What are the requirements of eligibility for students who wish to participate in a state or national FFA horticultural contest?

SUGGESTED ACTIVITIES

1. Study, review, and practice the knowledge and skills necessary for competition in the horticultural contest.
2. Make a plant identification file.
3. Make sample test question cards (with the answers on the back) that can be used to review for competition.

PART TWO

Greenhouse Production and Marketing

PART TWO

CHAPTER FOUR

The Greenhouse

Greenhouse structures are man-made devices constructed for creating conditions favorable to the growth of plants. They are designed to allow the growth of plants during seasons or in climates that would ordinarily be unfavorable. The greenhouse must, therefore, provide adequate insulation from cold weather while allowing for an abundance of direct sunlight and artificial heating.

Florists' crops require for growth a proper temperature during all seasons, adequate light, and ample moisture and humidity. The rays from the sun provide both heat and light to greenhouse plants. As these rays pass through the greenhouse covering, some are in the form of heat. This heat energy is absorbed by the plants and the soil. The heat from the plants and soil does not radiate back through the covering as readily as it entered; therefore, it becomes trapped within the structure. Ventilation is then required for removal of this excess heat and to provide fresh air to the growing crops.

The earliest greenhouses were constructed in an attempt to extend the natural growing seasons of fruit and vegetable crops. It is believed that the ancient Roman civilization used trenches in the ground covered by translucent mica sheets for growing herbs and food crops during colder weather. These early greenhouses were heated directly from the sunlight or rotting manure was laid on the floor to create heat. These structures were similar in design to the more modern coldframes used for starting seedlings and protecting tender plants from frost.

During the fifteenth century English gardeners constructed wooden shelters around their orange trees to protect the tender blossoms in cold weather. No glass was used as a covering; rather, wooden shutters were opened or closed as the temperature fluctuated. These were heated by large stone fireplaces, which heated the air that surrounded the trees. This style of greenhouse gave rise to the name "hot house," which greenhouses are sometimes called. One of these "orangeries" during the nineteenth century in England once covered as many as 400 small orange trees.

The first greenhouses to be constructed in the United States were simple "lean-to" structures set against houses during the late eighteenth century. These were constructed from glass sashes propped up against the south wall of the houses. Heat was provided to these greenhouses both from the sun and through flues that connected the greenhouse with the fireplace in the home. These primitive greenhouses eventually gave rise to the large, free-standing structures which were used for the commercial production of florists' crops in the United States.

Commercial greenhouse ranges did not become an important part of the country's economy until the industrial development after the Civil War. By 1880, there were nearly 5000 commercial ranges, which covered a total of 38,823,247 square feet. Most of this area was under glass greenhouses, although some production was in coldframes, hotbeds, and under lath houses. The first commercial greenhouse provided cut flowers and vegetables for the local area only. Large greenhouse ranges were established in the more populated areas of the country as growth and expansion moved westward. Since flowers and vegetables were highly perishable, the marketing area for them was limited to a small region surrounding the greenhouses.

Modern commercial greenhouses bear little resemblance to those earlier growing structures. The advent of jet transportation, refrigerated railroad and truck cargo carriers, and a sophisticated interstate highway system have enabled growers to market their flowers in areas several thousand miles distant from their greenhouses. The rising costs of labor have forced growers to use more advanced methods of technology to provide better plants at lower prices. Glass greenhouses are now more efficient in producing high-quality crops. In more recent years the use of fiberglass and polyethylene coverings has revolutionized the production of commercial florists' crops. These changes have made possible the production of superior cut flowers and potted plants at reasonable prices to the consumer.

GREENHOUSE STRUCTURES

Greenhouse styles vary according to the types of crops to be grown, material used as a covering, size of the greenhouse range, and the cost of construction materials used. Regardless of the materials used in their construction, various greenhouse arrangements may be found including detached houses, gutter-connected houses, and Dutch houses.

Detached or *separated greenhouses* are free-standing structures that are not connected to other greenhouses. These greenhouses may be constructed in a truss frame or in a hemispherical (quonset) shape (Figure 4–1). They are covered with glass or plastic; however, glass glazing is used only on A-frame truss houses. The detached greenhouse is designed to offer the best light, ventilation, and temperature control for the crops. This style of greenhouse separates individual crops for better environmental control. When the greenhouses are built in a detached fashion, more area is required and some valuable land is wasted between the houses.

FIGURE 4–1. Quonset-style polyethylene-covered greenhouses arranged as detached houses.

Gutter-connected greenhouses are constructed with their side walls joined. These are less expensive to construct when two or more houses may be attached together, since no waste space is lost between them. The interior walls may be of cheaper construction materials or they may be eliminated entirely (Figure 4–2). The houses in the center of the range are protected from wind and heat-radiation losses by those on the sides. Therefore, heating these greenhouse ranges requires less fuel. Temperature control of crops requiring the same climate is easiest when the gutter-connected design is utilized. However, when markedly differing temperature regimes are required between houses, the side walls must be enclosed to prevent heat losses to cooler houses. The gutter area of the houses creates a physical barrier to light along the benches nearest this area. The shading effect from the gutters on plants nearest the sides of the greenhouse may have a noticeable influence on plant growth. Growers usually attempt to keep shade-tolerant plants in these locations to minimize crop delay in production schedules. Ventilation of large gutter-connected greenhouse ranges is difficult because side ventilation is not feasible in the center houses. They must be cooled by fans placed at the ends of the greenhouses.

FIGURE 4–2. Gutter-connected glass greenhouses constructed on a hillside.

FIGURE 4–3. Dutch-style glass greenhouses have a lower roof profile than conventional greenhouses.

Dutch greenhouses are actually clear-span ridge-and-furrow (gut-ter-connected) greenhouses having twice as many roof peaks as con-ventional glass greenhouses (Figure 4–3). Researchers have found that larger houses produce better growing conditions. These green-houses are wider and often longer than standard houses. The width (up to 40 feet) of this style of greenhouse is free of all support posts, allowing better utilization of space within the structure. The roof supports are constructed of aluminum or steel beams to allow elimi-nation of the support posts in the middle of the structure. The ridges on the roof members have been reduced in height, allowing twice as many ridges to be placed on the ranges. The result of this modification permits a greater capturing of incident sunlight by the glass, without the requirement for added glazing material. A large Dutch house may be constructed with high side walls and large clear-span areas to facilitate the use of labor-saving equipment and machinery within the greenhouse (Figure 4–4).

The framework used to support the greenhouse structure is determined to some degree by the glazing material used. Glass-covered greenhouses must be durable enough to support the immense weight of the glazing. When polyethylene sheeting is used, however, very lightweight framing may be used. The parts of a greenhouse may be learned easiest from a diagram of a conventional glass greenhouse. The naming of these parts and an understanding of their functions is important for every greenhouse employee.

FIGURE 4–4. A clear-span, gutter-connected greenhouse range provides a large area for crop production under the roof.

The framework of the glass greenhouse consists of support posts encased in the foundation. These posts support the roof trusses at about 10-foot intervals along the length of the greenhouse. The eave or gutter (on gutter-connected houses) is attached to the top of each post in a line along the length of each side of the house. The ridge of the structure is located at the peak or highest point of the length of the greenhouse. Glazing bars are placed at frequent intervals along the length running from the ridge to the eaves. These glazing bars support the glass and provide a method for attaching the glass to the roof. The weight of the glass on a greenhouse roof necessitates additional support along the glazing surface, particularly in a wooden structure. This added support is provided by purlins which run the length of the greenhouse midway between the eave and ridge. A support truss of steel is added at frequent intervals to provide the strength required to support the entire superstructure of the greenhouse.

Quonset-style polyethylene or fiberglass covered greenhouses are of much simpler design, since little weight is added by the plastic. These houses are supported by bow-shaped ridges (ribs) which form an arc over the top of the structure. These bows may either run from side to side at the ground level or they may be placed on top of raised sidewalls to provide additional head room. Pipe-frame ridge caps and purlins are either bolted or welded onto the bows to add support to the structure. The sophisticated truss system required for supporting glass-covered structures is not required when polyethylene or fiberglass is used.

CONSTRUCTION MATERIALS FOR GREENHOUSES

The primary requirement of any greenhouse superstructure is that it have adequate structural strength to support the roof under maximum load. The structural strength (load) of a greenhouse includes:

1. The weight of the superstructure and glazing material.
2. The weight of snow and ice.
3. The lift caused by wind.
4. The temporary weight of workers who must maintain the glazing surface.

Construction materials that fit these capabilities are wood, black iron or galvanized tubular steel pipe, and aluminum.

Wood is the least expensive supporting material but is bulky and requires high maintenance to preserve its structural integrity. The wood most often used for greenhouse construction is generally redwood, but cypress and pressure-treated softwoods may also be used for framing a greenhouse. Wooden greenhouses were once used exclusively for constructing glass-covered structures. These are now used more for plastic covered houses, where weight is less of a factor in support. The major drawback to wooden-frame glass greenhouses is that short spans are required to support the superstructure. This means that more support posts, purlins, and trusses are required, since wood is quite heavy in relation to its structural strength.

When glass is used as glazing on a wooden-frame greenhouse, the glass is placed on painted, puttied bars, and then covered with bar caps for protection against rotting of the wooden bars. Rotting generally occurs at the gutters and along the bars where the glass brads (nails) are set. Routine maintenance of the glazing surface is required when wooden bars are used. The underside of the roofing bar must be kept clean to prevent clogging of the drip grooves. These grooves are placed in the bars to carry off condensation water from the glass to the gutter. The bars require a thorough cleaning with a steel bristle brush frequently (every 1–2 years) to remove moss, dirt, and old paint. The bars are then repainted with a good-quality white paint. White paint is used to reflect light within the greenhouse and to minimize shadows.

The glass must be removed from the greenhouse roof about every 8–10 years to clean the upper side of the bars and to reseal the glass. When the glass is removed, the bars are thoroughly cleaned, repainted, and new calking applied. The glass is reset on the bars and secured with glazing bar caps. Although this is a strenuous and time-consuming maintenance operation, it is necessary for the prevention of leaks during rain and for heat conservation.

Aluminum glazing bars have nearly replaced wooden bars on glass greenhouses. These are lighter in weight, yet stronger than wood. The aluminum bars require practically no maintenance, since painting is not required. Larger panes of glass are used on the aluminum structure, therefore reducing the shading factor from the roof. Although aluminum glazing bars require little routine maintenance, the glass must be reset periodically to prevent serious leakage of water. The aluminum expands during warm weather and causes the glass to slip out of place. This condition must be corrected before serious leaks occur.

COMPARISON OF CONSTRUCTION STYLES

It is necessary to compare the various greenhouse construction styles and materials to determine the fuel efficiency for heating and the growth of plants in each structure. From an economic standpoint, quonset-style, polyethylene-covered houses are by far the least expensive to construct and outfit for heating and cooling. These generally consist of 42-foot lengths of galvanized steel pipe which are bent into bows having a width of 25 to 30 feet across the base. These ribs are covered with two layers of ultraviolet-light-resistant plastic sheets in 6-mil thicknesses (Figure 4–5).

Glass greenhouses are by far the most permanent structures. If they are constructed with triple-strength glass panes and aluminum glazing bars, they will require very little maintenance for up to 20 years. This is in sharp contrast to polyethylene coverings, which require replacement every 2 to 3 years as the result of weathering and deterioration of the plastic caused by ultraviolet light. Glass houses have very low thermal transmittance ratios (about 4 percent for a new, tight house). These houses can be heated more efficiently than single-layer plastic-covered houses because they retain solar heat better and will seal themselves in freezing weather.

A double-layered plastic house is initially much tighter than a glass house, but the thermal transmittance factor is about 16 percent (percent heat loss compared to applied heat). Plastic houses do, however, form a condensate layer (water droplets) on the lower surface of the plastic film, which will reduce heat losses. The double-layered covering on plastic houses will substantially reduce the thermal transmittance associated with a polyethylene cover. Further heat losses can be prevented by inflating the two layers by use of a small centrifugal fan. This fan keeps the two layers separated by a layer of air. Using this air-inflation technique, a greenhouse may be heated at a savings of up to 50 percent over a house having only a single layer of covering.

Many fiberglass products are available for greenhouse coverings. The fiberglass most used is polyvinyl resin materials which are reinforced with glass fibers (known as fiberglass-reinforced plastics). These are available in either corrugated or flat sheets. The strength of these plastics is derived both from the corrugated surface and from the weight of the resin. The fiberglass products combine the light weight and unbreakable qualities of the sheet films with the durability and easier maintenance of glass houses (Figure 4–6).

Ultraviolet light will react with the base resin of the fiberglass panels and cause a "blooming" or fraying of the surface. The exposed fibers will eventually collect dirt, which reduces the light transmission qualities of the covering. The better-quality fiberglass-reinforced plastic products are coated with up to 15 percent acrylic-modified polyester or comparable product to reduce this ultraviolet light decomposition.

When it becomes necessary to refinish a fiberglass surface that has become seriously bloomed, the panels must be thoroughly cleaned. The remaining loose fibers are removed with coarse steel wool and a trisodium phosphate solution. Once the panels are adequately cleaned, the surface is allowed to dry for several hours in the bright sun. The panels are then recoated with an acrylic refinishing solution applied with a brush. These acrylic refinishers extend the life of the fiberglass panels but do not bond well to old surfaces.

FIGURE 4–5.   The interior of a quonset-style, polyethylene-covered greenhouse provides ample space for crop production at an economical price.

FIGURE 4–6.   A greenhouse covered with reinforced fiberglass plastic panels. (Courtesy National Greenhouse Company, Pana, Illinois)

GREENHOUSE BENCHES

Greenhouse benches are simply growing platforms used primarily to raise the plants to a level where they may receive maximum light and escape the cold-air drafts at floor level. Air movement around the growing plants is improved with raised benches, since air is allowed to flow below and above the growing surface. When steam or hot-water pipes are placed beneath the benches, the heat helps to warm the soil and stimulate root growth of the crop plants. During the summer, air is free to move under the benches. This helps to keep the benches dry and reduces disease problems.

Raised benches also help to increase the drainage out of soil and pots. The wooden slats used for constructing most benches are spaced to permit fast drainage. Benches make greenhouse sanitation easier, since pasteurization of soils is more easily accomplished when the soils are in benches. Benches also increase work efficiency for the workers because the crops are placed at a convenient working height and the crops are confined to areas with walk space between them.

Bench Arrangements

The arrangement of benches in the greenhouses determines the efficiency of workers who must maintain the crops. The grower must maximize the space devoted to growing area while allowing easiest access by the workers. The most common bench arrangements found in commercial greenhouse ranges are lengthwise benching, crossbenching, and peninsular benching designs (Figure 4-7).

Benches that run the entire length of the greenhouse are called *lengthwise* benches. These are the least efficient for workers, who must walk the entire distance of the greenhouse to reach the opposite side of the bench. This arrangement is used most often for cut flower crops grown in ground benches. The entire bench is planted to the same crop (such as roses) to make management more efficient. Often a system of mechanical conveyors or carts is used for carrying flowers to save employee time when these benches are used.

Benches constructed across the width of the house instead of lengthwise are of the *crossbenching* design. These are arranged with walks between benches and along the side walls of the greenhouse. Although this bench arrangement allows easy access to the plants by workers, much space is lost in production by the added walkways.

A modification of this crossbenching system is the *peninsular* design. A wide central walkway is provided for use of mechanized carts or wagons. Smaller aisles provide access to each bench along the greenhouse. This system of benching provides a maximum amount of production bench space, while allowing convenience for the employees who must maintain the crops.

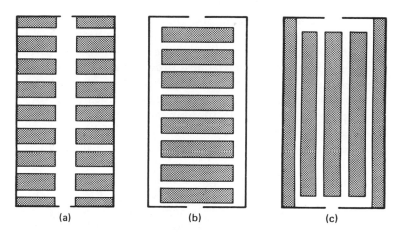

FIGURE 4–7. Some greenhouse benching arrangements: (a) peninsular, (b) cross-benching, and (c) lengthwise benching.

Types of Benches Used in Greenhouses

Various styles of greenhouse benches are utilized in the commercial production of florists' crops. The bench style used is determined to some extent by the nature of the crops being grown, the degree of mechanization employed by the greenhouse, and the quantity of plants that must be maintained in the ranges. Some bench styles that may be found in commercial greenhouse ranges are raised benches, ground benches, pallets, and mat-watering benches.

Raised benches. *Raised benches* are used for both cut flower and potted plant production (Figure 4–8). The height of the raised benches is usually 24–30 inches (61–76 centimeters) for potted plants. The width of the bench varies with its height. A 5-foot width is considered maximum for handling potted plants from both sides of the bench. Narrower benches are used for cut flower crops (such as snapdragons), so workers can reach to the top of the plants for harvesting through support wires. When cut flowers are grown in raised benches, the soil depth should be at least 6 inches, with adequate drainage provided from the bench.

Walkways should be at least 18 inches wide for cut flower benches. For potted plants, the side aisles should be at least 24 inches in width. The central aisle should be as wide as 48 inches to allow passage of carts and wheelbarrows.

Most raised benches are constructed to provide rapid drainage. Various materials may be used to provide a growing surface. Wood is most commonly used but will rot rapidly unless treated with a preservative. Copper naphthanate is used most often as a wood preservative for greenhouse benches. This material is either soaked into the wood or brushed onto the surface to aid in rot resistance. The copper in this material also provides a fungicidal action, so treatment with this material will disinfect the wood before the crops are planted. The common wood preservatives used for construction purposes, such as pentachlorophenol and creosote, should not be used in greenhouses. These materials cause foliage burning and possible death to plants that come in contact with them.

The bottom boards for benches made from wood are spaced ¼ inch apart for drainage. If this space is not allowed, the wood will swell upon wetting and completely seal the bench. Corrugated transite (an asbestos and cement product) is much more durable and costly than wood. When the transite is used, it is best to pitch the bottom slightly from one side and allow adequate drainage space at the lower side. The most rapid drainage is provided by benches constructed from 1-inch mesh welded-wire fabric. This support medium can be used only for potted plant crops. The wire mesh allows rapid drainage and provides excellent air movement around the plant foliage. Since water does not stand in the bench, slugs and snails are rarely present and the air movement aids in disease prevention.

FIGURE 4–8.   An inexpensive raised bench for bedding plants for potted crops.

Ground benches.   *Ground beds* are used primarily for cut flower and greenhouse tomato production. The benches are either placed on the ground or excavated into the soil to allow proper drainage. The purpose of raising crops in ground benches is to allow easy access by workers who must work on plant stems and flowers at eye level.

If the ground beds are constructed of concrete, there should be a V-shaped bottom which slopes to the center and provides drainage. These benches are constructed so that drain tiles can be laid down the center, with pea gravel poured to the level of the tile for rapid drainage of overflow water. If the ground beds do not have concrete bottoms, they should have 4–6 inches of crushed rock and drain tiles placed for subsoil drainage. At least 6 inches of soil is required on top of the gravel for plant growth. These ground beds drain better and plants are damaged less easily when the sides of the beds are raised above the greenhouse walkways. The maximum width of cut flower ground benches is generally 3 feet.

Pallets as benches.   Raising potted plants on *pallets* is similar to growing the plants directly on the ground. The entire floor surface is covered with crushed rock to provide a clean working surface. The plants may be potted in a central work area. These are then transported to the growing structures on pallets (wooden platforms) designed to be carried by fork-lift trucks. The pallets with the plants may be shoved together as tightly as possible to achieve maximum use of the greenhouse space. These plants are then watered and fertilized using soluble liquid fertilizers delivered by overhead sprinklers (Figure 4–9).

FIGURE 4–9. Bedding plant crops are often grown directly on the ground, when black polyethylene and gravel are used as a mulch to prevent weed growth.

Mat-watering systems.   Benches designed for use with *mat-watering* systems are similar in design to other raised benches (Figure 4–10). The bottom of the bench must be flat and absolutely level for the proper function of the watering system. Flat transite sheets are used most often to provide a watertight surface. A sheet of polyethylene is placed over the floor of the bench to prevent the loss of water. The fibrous mat designed for use in this system is placed over the plastic and smoothed to eliminate folds in the surface. A drip irrigation tube delivery system is placed over the mat. When a slow flow of water is delivered through the irrigation tubes, the mat becomes saturated. Potted plants on the mat will receive a constant supply of water from the base of the pots. This watering system is especially adapted for plants that require careful watering to prevent foliage injury (such as African violets and gloxinias).

FIGURE 4–10. A mat-watering system being used for African violet production.

Painting and Preserving Greenhouse Equipment

Greenhouses are painted not only for the purpose of making them more attractive, but to prolong the life of the wood and metal construction materials. The greenhouse worker will be involved with the routine maintenance of the greenhouse surfaces and benches when not occupied with the plant production duties. A basic knowledge of painting and preserving methods will be valuable when later decisions are required.

When selecting a paint for use on wood in a greenhouse, it is necessary to select a high-quality white exterior-grade house paint. The high temperatures and high moisture contents of the air make the selection of paints an important part of maintaining a greenhouse. The exterior house paints that are best for greenhouse use are called "breather" paints, since they allow air movement through the painted surface. This free air movement prevents the blistering and scaling often seen on old painted surfaces.

Paints containing a titanium–lead mixture are tolerant of moisture and more durable than those containing only lead. The latex (water-base) paints rated for exterior use also provide air circulation through the painted surfaces. These are applied easily by brush and are easy to clean up after use. The new paints containing a fungicide are mildew-resistant. These paints are especially beneficial for use in greenhouses, where the high moisture levels favor mildew growth.

Some flat-grained or porous wood surfaces require a "primer" coat to seal the grain before the finishing surface is applied. Oil-base paints are often used as self-sealers, since the oil penetrates into the grain of wood. These primer coats should always be free from zinc. Generally, new wood will require three coats of paint, a single application of primer and two thinly applied coats of finish paint. When the surfaces have weathered to the point where the wood surface is barely visible, a single coat of finisher paint is all that is necessary to restore the surface to its original luster and protection.

Metal support pipes used in greenhouses will eventually have to be painted to prevent rusting. Black iron pipe will rust sooner than galvanized metal. These iron pipes or beams must be painted with a metallic zinc primer. The zinc from the paint will, however, cause toxicity to plants with which it comes in contact, so care must be taken to protect them. Galvanized surfaces will resist rusting for a period of time. However, in a greenhouse environment, this galvanized coating may eventually be lost and rusting will appear. When this happens, the surfaces should be thoroughly scraped with a wire brush and a zinc primer paint applied. The metallic surfaces may then be painted with either a white oil-base or latex-base exterior paint.

GREENHOUSE ENVIRONMENTAL CONTROLS

Greenhouses are constructed for the purpose of admitting the light required for the growth of plants. Since growing plants require high light during the day, the greenhouse roof cannot be insulated against heat losses during cold weather. A greenhouse must maintain the optimum temperatures for growth of the crops while being exposed to wide outdoor temperature extremes.

Greenhouses are heated by solar energy admitted through the roof and by an auxiliary heating system. Heat is lost from a greenhouse by conduction, radiation through the roof, or by the use of a greenhouse ventilation system.

The largest heat losses will occur at night in cold winter months. Conversely, during the hot days of summer, the excess heat which is trapped by the greenhouse must be removed by high capacity exhaust fans or by the use of a cooling system.

The heating and cooling systems of a greenhouse must maintain uniform temperatures throughout the structure at all seasons of the year. Some types of greenhouse heat distribution units are steam boilers, hot water heaters (boilers), and unit heaters.

Steam boilers. *Steam boilers* are used to heat large greenhouse ranges because steam can be forced over greater distances without losing its heat. The steam rises from the top of the boiler to the greenhouses. The steam releases its heat in the pipes in the greenhouses, changes to water, and flows back to the boiler through return pipes to be reheated. Steam is hotter than the water in hot-water systems, so fewer pipes are required to heat a greenhouse. One of the major disadvantages of steam heating systems is that the steam is under high pressure in the pipes. This pressure corrodes the pipes, necessitating frequent repairs, installation of expensive heavy-walled pipes, and creates a danger to employees. When a steam pipe ruptures, hot steam may be spewed for great distances. The major advantage of steam heating systems is that steam is available for soil pasteurization in the greenhouse.

Hot-water systems. *Hot-water systems* provide more even heat than steam units. The pipes are not as hot, so there is less danger of plant damage near them. The hot water is constantly circulating in a hot-water system; therefore, there is heat distributed even though the boiler is not operating. With steam it is necessary to keep the water boiling to keep steam in the pipes. Many greenhouses are equipped with steam boilers that operate a hot-water system. The steam boilers are more durable and may be converted to produce steam whenever it is required for soil pasteurization.

Unit heaters. *Unit heaters* provide greenhouse heating to individual structures where pipe heating is not used. The units may operate independantly from gas or oil fuel sources or may be attached to steam or hot-water systems. The unit heaters are then used to heat the air and distribute the heat by fans. Unit heaters are designed to heat small areas (such as single greenhouses) rather than entire ranges (Figure 4–11).

Heating Greenhouses Through Pipes

It is important to supply heat within a greenhouse in a manner that best offsets the heat losses. In most localities this requires a perimeter heat system to account for wall losses and another system for heat loss through the roof. Perimeter heat pipes are particularly important to prevent cold floors and to ensure that snow is melted in the gutters of adjoining houses (Figure 4–12).

Temperature uniformity and efficient use of added heat are the two important goals in greenhouse heating. Because warm air rises, it is only natural that a hot spot exists in the gables of the greenhouse

FIGURE 4–11.   Unit heaters are used to warm greenhouses and create air circulation for proper crop growth. (Courtesy National Greenhouse Company, Pana, Illinois)

FIGURE 4–12.   Hot water is circulated through the greenhouses in pipes located along the walls, overhead in the gables, or under the benches.

and a corresponding cold spot exists on the floor. The temperature of the cold spot can be controlled by additional heating so that temperatures will be adequate for plants growing in ground beds or benches. To accomplish this, however, means that some of the additional heat will rise to the roof and similarly raise the temperature of the hot spot near the roof. Air circulation patterns occur in greenhouses even when conditions appear to be ideal. These air patterns vary in intensity and direction according to the placement of the radiant heating pipes.

When heating pipes are located along the side walls (perimeter) only, heated air rises rapidly to the eaves of the greenhouse. This hot air warms the side walls and gutters before broadening as it slowly proceeds horizontally to the greenhouse center. Here, it meets a similar air mass from the opposite side wall. Both air masses turn downward to the plant and bench level and return to the area of the side walls and coils.

Air circulation patterns created by heat from pipes beneath benches or overhead differ considerably from those at the side walls. Coils placed overhead or beneath the benches cause a verticle circulation pattern. Air rises rapidly upward from the pipes, broadens, and cools as it moves slowly downward in an irregular pattern. When side wall-mounted coils are combined in a heating system with those beneath the benches, the air circulation pattern is dominated by the side wall pipes. However, the cold air that returns to the side coils is mixed with the warm air from under the benches. This reduces the amount of cold air located at the plant height on the raised benches.

Heating with Unit Heaters and Convection Tubes

In recent years considerable interest has developed in the use of horizontal unit heaters with attached perforated polyethylene convection tubing for heating greenhouses. The theory behind the convection-tube heating system is to provide a means of circulating the air within the greenhouse volume. Therefore, the warm air in the gables is mixed with the colder air near the floor, producing a moderate temperature suitable for plant growth. When a convection-tube circulation system is coupled with horizontally mounted unit heaters and underbench pipe heating, maximum heating efficiency is achieved.

Ventilation During Summer and Winter

Winter ventilation of a greenhouse is quite different from summer ventilation and cooling. In winter the air flow must be turbulent, with small amounts of air required; whereas in summer the air flow should be smooth, with large quantities of air being moved. The winter ventilation system is an important part of the climate control system. It must have ample air flow capacity to maintain a proper heat balance by removing the excess solar heat on mild sunlit days. It must also be able to introduce cold winter air into the greenhouse without producing cold drafts on the plants. This requires very thorough mixing of the cold outside air with the warm inside air before it reaches the plant level. It is important that all parts of the greenhouse are at the same temperature. To achieve this, the ventilating system must distribute the air uniformly throughout the greenhouse.

The use of exhaust fans and side-wall ventilators may be used for summer ventilation of the greenhouse. However, during the winter months this system would quickly pull the greenhouse temperature to a level equal to that of the outdoor air. This would result in plant damage and higher fuel bills. The convection-tube ventilation system provides a means for introducing cold air into the greenhouse while heating it to the desired temperature by the unit heaters. The damp air is then removed from the greenhouse by use of the exhaust fans located in the end or side walls.

The convection-tube ventilation system consists of a specially constructed pressurized fan with a perforated plastic tube attached. The tube runs the length of the greenhouse above the plants and is closed at the far end. When the fan is in operation, the tube is inflated and air is forced through the perforations. The air pattern from the tube is directed along the roof surface to the side walls and back to the center of the house. Here, the air meets a similar current from the opposite wall. The air is then pulled back to the end of the greenhouse, where it passes through the fan for recirculation. The convection-tube system may be used for heating, cooling, or air circulation of a greenhouse when coupled with an inlet shutter, heaters, exhaust fans, and a wet-pad cooling system.

The convection-tube system is designed to control the environment of the greenhouse by automatically operating the various components required for raising or lowering temperatures. When the temperature is ideal for the growth of the crops, the side ventilators are closed, heaters are off, and no cooling units are in operation. The convection tube simply recirculates the air in the greenhouse.

When the temperature falls below the desired operating temperature of the greenhouse, either one or both unit heaters are switched on automatically to provide heat. The heated air is distributed through the convection tube to evenly warm all parts of the greenhouse. If the greenhouse temperature reaches a level above the desired heat balance, especially on warm days, the side ventilators are opened and at least one exhaust fan is operated to pull outside air into the house. When small amounts of cooling are required, the cooling thermostat simultaneously opens the motorized shutter situated behind the convection fan and energizes one of the exhaust

fans. The vacuum produced by the exhaust fan draws fresh, cool outside air in through the shutter. The air is propelled down the tube
and is rapidly mixed with the warm greenhouse air.

When the relative humidity (moisture content) of the air in the
greenhouse reaches a high level, plant diseases may become a problem for the grower. The humidity may be reduced by introducing
cooler outside air having a lower water vapor content. During cold
weather this outside air must be heated before it is allowed to reach
the plant level. This requires the use of both the fresh air supply
through the motorized shutter and the heating system working together. The motorized shutter is opened to admit the cool, dry air.
This air is heated and then distributed through the convection tube.
At the same time the warm, humid air is being evacuated through
the exhaust fan located at the opposite end of the greenhouse.

Summer Cooling of the Greenhouse

There are several methods by which the temperature of a greenhouse can be lowered. Older greenhouses are equipped only with
roof and side-wall ventilators. These greenhouses are cooled because
the hot interior air naturally rises, so cool air is drawn in through the
side ventilators. Air movement across the roof ventilators induces a
negative pressure on the air in the house, causing a constant air circulation through the greenhouse. Exhaust fans can also be added at
the sides or ends of the greenhouse to provide greater air exchange
with the outside environment (Figure 4–13).

The shortcoming of each of these methods for cooling a greenhouse is that the cooling effect is limited by the temperature of the
outside air. Without fans, there are hot spots in the house where
there is only limited air movement. Temperatures on hot summer
days may surpass 90°F (32°C) even when exhaust fans are in operation. Most plants cease growth, or are permanently damaged, by air
temperatures of 90 ° F or above.

Heat is removed from the surrounding air as water changes from
the liquid state to the vapor state. The principle of heat exchange
through the evaporation of water allows adequate greenhouse cooling
through the introduction of cool air to the greenhouse atmosphere.
Modern greenhouse cooling is accomplished by utilizing the ability
of water vapor to accept heat.

Greenhouse temperatures often become as high as 20° F (11° C)
hotter than outside air temperatures. This is caused by heat being
trapped in the gables from sunlight. Wet-pad cooling can maintain a
temperature nearly 10°F (5.6° C) cooler than the outside air tempera-

ture (Figure 4–14). The result is greenhouse temperatures that are 20 to 30°F cooler than conventional uncooled greenhouses.

The advantages to the use of wet-pad cooling systems are many. Growers may have profitable production of high-quality crops year-round. Conventional noncooled houses must be emptied in the summer. Cooling of the greenhouses aids in keeping the crops on schedule. Many flowering plants will not form flowers properly at high temperatures. Higher light intensities may be maintained for proper plant growth without the added high greenhouse heat. Better crops can be grown faster when ample light is available. However, most greenhouses will require the addition of summer shading to protect the plants from "sunburning" of the foliage and to provide an added cooling effect.

FIGURE 4–13.   Exhaust fans are placed along the sides or ends of greenhouses to pull the hot air from the interior during the summer months.

FIGURE 4–14.   Greenhouse structures are cooled during the hotter periods of the year by placing shading material over the roof and by the use of wet pad cooling systems.

Older houses that are not equipped with pad cooling systems are cooled using a side ventilator system and fans. (Figure 4–15.)

The wet-pad cooling system consists of large-volume exhaust fans and a correctly sized continuous wet-pad system. The pads must be uniformly porous so that air will flow evenly through them and absorbent, so that they retain some water. Pads are commonly constructed of shredded aspen wood (called excelsior). The pad system is composed of a number of wire-frame panels filled with the excelsior pads. These are placed in a continuous line along the length or width of the greenhouse. A recirculating water system keeps the pads sufficiently saturated to achieve good evaporative cooling efficiency. A product sometimes used in place of excelsior pads is a self-supporting material made from cellulose that has been impregnated with antirot salts and wetting agents.

FIGURE 4–15.   Side ventilator system and fans. (Courtesy National Greenhouse Company, Pana, Illinois)

The wet-pad system requires periodic attention to maintain adequate cooling of the greenhouse. The pads may become clogged with algae unless an algicide is added to the tank water. This application should be made once a month to prevent excessive algae growth. Under the weight of the water applied to the top of the pad, the excelsior will slip and leave gaps in the pad system. Since the air flow is directed through the path of least resistance, little cooling will result when gaps and holes are present. When holes are discovered, they should be filled with excelsior. Gaps between the ventilator and the pad system should also be corrected by releveling the pads or by covering the holes with plastic sheeting.

HOTBEDS AND COLDFRAMES

Various forms of hotbed and cold-frame structures are used by growers for starting seedlings of nursery stock, rooting woody ornamental plant cuttings, and precooling florist plants under natural conditions. The purpose of these structures is to grow plants during cool weather when outdoor conditions are unfavorable. Use of the less expensive hotbeds frees valuable greenhouse space during the high production seasons. These frames are used primarily for bedding plants during the spring season to harden off the plants when they are placed outdoors to toughen them before they are marketed. The plants are tender and so must be protected from frosts.

Cold frames use only sunlight as the heat source. As light energy enters the frame through the glass or plastic covering, it is accumulated as heat and stored in the soil. This heat is reradiated from the soil during the night. The stored heat is generally sufficient to prevent freezing of plants in the cold frame during mild frost periods. During colder weather, or when very tender plants are to be grown in the frames, a heating source is used in addition to the solar energy. Electric heating cables similar to those used in propagation beds are the most commonly used heating devices. Sometimes hot-water or steam radiant pipes are buried in the soil beneath the beds for emergency heating on cold nights. Growth frames that are heated by artificial sources are called hotbeds.

Cold frames or hotbeds are constructed of wood, brick, transite, or concrete blocks. The depth of the front wall is generally 2 feet to facilitate larger plants, such as lilies or azaleas, which are to be precooled. The back wall of the frame is placed about 1 foot higher than the front so that water will drain easily from them. The sloped face of the frame is generally directed toward the south or west directions to catch the maximum amount of solar energy. Although the beds may be of any convenient size, the standard bed width is 6 feet. A standard-sized glass sash (3 × 6 feet) may then be used to cover it.

The hotbeds or cold frames may be constructed easily and inexpensively to save valuable greenhouse space during the high production periods of the spring or autumn months. These beds require rather high labor, since the sashes must be constantly raised or lowered according to environmental fluctuations. Watering of the plants in the frames is also rather time consuming. Many commercial greenhouses and nurseries equip these frames with mechanical devices that automatically water the plants, apply mist to cuttings, and raise or lower the sashes when temperatures fluctuate.

SHADEHOUSES

Shadehouses are used for the production of plants in warm climates or during summer months. Nurserymen use these structures for the growth of hydrangeas and azaleas during the summer months. Foliage plant growers in the extreme southern regions of the country may grow their crops entirely under open-shade structures. These shade structures make excellent holding areas for field-grown stock while it is being prepared for shipping to retail outlets.

The shadehouses are most often constructed as a pole-supported structure and covered with either lath (lathhouses) or polypropylene shade fabric. A typical structure may be constructed with 12-foot lengths of 4- × 4-inch posts set into concrete to a depth of 3 feet. These support posts are set 10 feet apart and tied together by 2- × 6-inch boards. Snow fencing is then rolled across the top of the structure to form an open roof shading. The south exposure of the structure is also covered with snow fencing to provide shade from the sun and provide protection to the plants.

GREENHOUSE EMPLOYEES

The greenhouse employee should become familiar with all the parts and functions of the greenhouse structures where he/she works. The employee must be able to perform routine maintenance on the structures and should be able to recognize when this maintenance is required. The greenhouse is a highly refined plant factory. The production of high-quality plants cannot be accomplished unless all the equipment used is in good working order.

The employees in a commercial greenhouse will have to be capable of making repairs on the glazing surfaces, such as replacing broken glass panes or recovering plastic houses. The heating pipes and heating equipment must be constantly kept in good operating condition. When large steam boilers or hot water units are used for heating, a knowledge of their operation and maintenance is necessary in case the manager is not available when a malfunction occurs. The operation of a large commercial greenhouse requires skills in carpentry, plumbing, heating, and electrical wiring. This occupational field may be highly motivating and rewarding to an ambitious employee who enjoys working with plants and machinery.

SELECTED REFERENCES

Ball, V. (ed.). 1975. *The Ball Red Book,* 13th ed. George J. Ball, Inc., West Chicago, Ill.

Laurie, A., D. C. Kiplinger, and K. S. Nelson. 1968. *Commercial Flower Forcing,* 7th ed. McGraw-Hill Book Company, New York.

Mastalerz, J. W. 1977. *The Greenhouse Environment.* John Wiley & Sons, Inc., New York.

Nelson, K. S. 1967. *Flower and Plant Production in the Greenhouse,* 2nd ed. The Interstate Printers and Publishers, Inc., Danville, Ill.

Nelson, K. S. 1973. *Greenhouse Management for Flower and Plant Production.* The Interstate Printers and Publishers, Inc., Danville, Ill.

Ohio State University. 1974. *The Greenhouse Worker.* Ohio Agricultural Education Curriculum Materials Service, Columbus, Ohio.

TERMS TO KNOW

Algicide	Lengthwise Benching
"Blooming"	Mat Watering
Coldframe	Pallets
Convection Tubes	Peninsular Benching
Convection-Tube Ventilation	Quonset-style Greenhouses
Corrugated Plastic	Relative Humidity
Crossbenching	Shadehouse
Exhaust Fan	Thermal Transmittance
Glazing Bars	Transite
Greenhouse Range	Ultraviolet Light
Hotbed	Wet-Pad Cooling System

STUDY QUESTIONS

1. Describe the purpose of the earliest greenhouses. Tell how they were constructed and heated.

2. Compare the types of construction materials used for greenhouses by listing the advantages and disadvantages of each.

3. Contrast the various styles of greenhouse construction by listing the advantages and disadvantages of each type.

4. Select one type of benching which you believe to be good and tell why you would prefer this method to others.

5. List and give a brief description of the equipment that is available for environmental control in greenhouses.

SUGGESTED ACTIVITIES

1. Visit local greenhouses to observe the construction and environmental control equipment.

2. Invite a specialist in greenhouse construction (or a commercial greenhouse owner) to speak to the class.

3. Construct models of the various types of greenhouses using balsa wood, plastic, and wire.

CHAPTER FIVE

Plant Production in the Greenhouse

Flower and plant production in the greenhouse is becoming increasingly more technical as new information concerning the culture of florists' crops is obtained. Greenhouse production of crops is a fascinating occupation and there is a constant need for well-informed greenhouse employees. A person who seeks employment at a commercial greenhouse should have a genuine interest in the production of greenhouse crops and a sound concept of plant growth requirements.

The commercial production and selling of cut flowers and potted plants is the function of the *floriculture industry*. The individuals involved with these aspects of floriculture are called *flower growers*. Commercial flower growers produce their crops year-round, since flowers cannot be stored or preserved for long periods of time. This practice requires a great amount of skill on the part of the grower and specialized growing structures (greenhouses) in most parts of the United States.

PLANT PROPAGATION

Seed Propagation

Many of the crops grown by the flower grower are started from seeds. Seeds contain a tiny plant (embryo), a food supply (endosperm), and a seed coat. The size of the seed determines the amount

119

of food supply that is available to a germinating plant. Seeds may be rather large or extremely small. Some seeds used by flower growers are so small that they appear to be dust. These extremely small seeds require considerable care during their germination period. Small seeds are generally sown on top of the germinating medium and are not covered. This practice allows the young plants to emerge quickly and manufacture their own food by photosynthesis. Larger seeds, such as tomato seeds, are large enough to supply food for a germinating plant while it is growing out of the soil medium. Therefore, these larger-seeded plants are germinated by sowing the seeds more deeply in the seed flats and by covering them with a small amount of soil.

Seeds are sown in specially prepared germinating flats that provide a soil depth of about 3 inches (7.6 centimeters). The soils used for seed germination may either contain some topsoil, or they may contain no soil at all (artificial soils). The mixture that is most commonly used for commercial seed germination contains equal portions of peat and vermiculite. This soil mixture is screened to provide small particles that will ensure good contact between the seed and soil for uniform moisture. Small seeds require the finest soil mixtures. All flats and soil media should be sterilized before they are used for seed germination, since the environment provided the germinating seed is also ideal for disease organisms. Sterilization of soil is accomplished by steam, electric heaters, or gas fumigants. Plastic flats and tools are cleaned in a diluted bleach solution. The commercially prepared artificial germinating mixes are already sterilized, so they may be used without further handling.

Seeds are sown in rows in the seed flat so that air will circulate around the seedlings. This practice also reduces the spread of disease organisms in the flat, if any are present. Seeds should be sown very carefully to be certain that they are not sown too thickly. The germination flat is watered very carefully the first time, to avoid washing the seeds and soil from the flat. A fine mist spray nozzle should be used on the end of a hose, or the seed flat may be *subirrigated* by placing it in a shallow tub containing 2 inches (5 centimeters) of water. The first watering should uniformly wet the medium. As soon as water appears at the surface of the soil, it is removed from the tub to prevent floating the seeds from the flat. A fungicide recommended for the prevention of "damping-off" diseases is often included with this first watering. This treatment reduces the opportunity for soil-borne diseases to kill the germinated seedlings.

The watered seed flats are then placed in a location that will favor proper seed germination. It is customary for the seed flats to be covered in some fashion, since no other water should be required until germination has taken place. When flat panes of glass are available, these may be placed over the flats in the greenhouse. The seed flats may also be wrapped in clear polyethylene bags to conserve moisture. If the flats are to be kept in a greenhouse during this period, it is necessary to place newspaper over the tops of the glass or plastic (Figure 5–1). If this is not done, the temperature of the air inside the flat will become excessively high, resulting in the death of the seeds.

FIGURE 5–1.   Plant seeds are often germinated in flats covered with a sheet of plastic film to keep the seedlings moist, and with newspapers to prevent overheating of the soil from sunlight.

FIGURE 5–2.   Newly germinated seedlings ready for transplanting to bedding containers.

The seed flats should be inspected periodically to determine whether they are adequately moist, if any diseases may be starting, and whether germination is beginning. The first signs of germination are indicated by a slight heaving of the soil, a light-colored crook-shaped stem extending from the seed, followed by the evidence of seed leaves being pushed out of the soil. Disease organisms may be seen as a mold growth on the surface of the soil. Whenever the seeds begin to germinate or evidence of disease is found, the covering of the flat should be raised slightly to allow air circulation to the seeds (Figure 5–2). From this time on it is important that the seed flats be inspected often and water added to the soil mix by a fine mist spray whenever it is required. The seed flats should never be allowed to dry out after germination has begun. Once the seeds have germinated uniformly, the cover may be removed from the flat.

Most seeds are germinated at a temperature of 70°F (21°C), although others have different requirements. Soil heating cables are generally used to ensure proper heating in a greenhouse. Some seeds germinate best under lights, while others germinate best in the dark. Those that require dark are either covered by soil (if large enough for this practice), or the flats are placed in a darkened room during germination. All germinated seeds require light as soon as the seedlings emerge. These germinated flats are placed in an area that is shaded from the sun for a few days before they are moved to benches in full sun.

Transplanting

Once the seedlings are well established in their seed flats, they will require high light and good air circulation. The seedlings will also require a light fertilization as soon as they are placed on the greenhouse bench. During this phase of growth, the temperature in the greenhouse needs to be reduced from 70°F (21°C) to the growing temperature required for each plant type. Generally, the temperature required is 55–60°F (13–16°C) at night with a 10°F rise in temperature during the day. The soil should never be allowed to dry completely during this period of growth. When watering is done, the soil medium should be thoroughly soaked. Gradually space the watering frequencies further apart as the plants near transplanting time. This is done to make the seedlings less brittle and less likely to be broken when handled.

The seedlings are transplanted (replanted) to individual pots or benches as soon as they are large enough to be handled by their first set of true leaves. The seedlings are gently lifted from their individual rows by slipping a wooden pot label under their roots. The seedling roots often grow together rather thickly, and it is necessary to carefully separate the individual plants. Only a small area of the seed flat should be removed at one time to prevent the seedlings from drying rapidly. Once each of the individual plants has been separated, it should be quickly placed in a pot or flat for further growth (Figure 5–3).

The transplanting container may hold a single plant, or it may hold several plants, depending on its purpose. A tomato may be transplanted to a peat pot and grown singly. Other containers may be large enough to hold 6 or 12 plants each (Figure 5–4). These are called *community packs*. Some florists' flowering plants will be transplanted

FIGURE 5-3. An individual
seedling is placed in each hole
of a flat of peat pots.

FIGURE 5-4. Many different
types of containers are used
when growing bedding plants.

(shifted) into larger pots several times before finally being sold as a finished pot plant.

The soil used for the transplanting medium is generally coarser than was used for germination. This soil mix should also be sterilized, as should all pots and equipment used during transplanting to reduce diseases. The seedling is placed into a prepunched hole in the soil with a *dibble*. The dibble is inserted into the center of the soil of the pot, making a hole having tapered walls (Figure 5-5). The roots of the seedling easily slip into this hole and the soil may be tamped slightly around them. The planted pots are then carefully watered to settle the soil and ensure good soil-to-root contact. These plants are then placed in a well-lighted area of a greenhouse for further growth (Figure 5-6).

Asexual Propagation

Many florists' crops may be propagated from their stems or other vegetative parts, instead of from seeds. When given the proper conditions, a mature stem or other plant part may form roots and shoots. It is possible to produce many new plants from a single parent plant by this method. This is important to the flower grower because each

FIGURE 5-5.   Bedding plant methods.

(a)   A dibble machine places individual holes in each pot for the transplanting of seedlings.

(b)   These holes make the insertion of the seedling root system into the soil much easier.

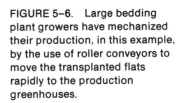

FIGURE 5-6.   Large bedding plant growers have mechanized their production, in this example, by the use of roller conveyors to move the transplanted flats rapidly to the production greenhouses.

of the newly formed plants will be exactly the same as its parent. This is called *clonal propagation*.

The parent plants used for obtaining the cuttings are called *stock plants*. These plants are kept in an isolated area of the greenhouse and are given special care to keep them healthy and growing vigorously. Cuttings are taken periodically from these stock plants by removing the stem tips to a length of 3–4 inches (7.6–10 centimeters) on each stem. Stem-tip cuttings generally produce a new plant fastest, since a well-developed shoot is already present. For some plants the stem may be cut into sections containing one or two leaves.

These *leaf-bud* cuttings then must form roots as well as shoots in the propagating bed (Figure 5–7).

FIGURE 5–7. Vegetative propagation methods: leaf bud cutting (left) and tip cutting (right).

The Propagation Bed

During the period of time while the cutting is forming roots, it must be provided with moisture through its leaves. Since the cutting was severed from its parent plant, it no longer has roots to absorb water. The purpose of the propagation bed is to provide a rooting medium and a high-humidity environment around the leaves. The high humidity is provided from an automatically controlled mist system. Nozzles that provide a fine spray of water are suspended above the propagation bed (Figure 5–8). These nozzles are controlled either by time clocks or other electronic devices that allow only a small amount of moisture to be directed over the cuttings periodically during the day. Only enough water is applied to keep the leaves moist while the plants are forming roots.

The rooting medium for propagation beds may consist of any growing medium, but the most widely used is an equal mixture of horticultural peat and perlite. The depth of the rooting medium is at least 4 inches. Under this rooting medium is placed a thermostatically controlled heating cable to provide a warm rooting zone in the medium. Heating the bed from below the medium stimulates faster rooting of cuttings.

Cuttings are made by removing the upper portion of the stem tip from just below a *leaf node*. The stem that remains above the leaf node on the stock plant is also removed to prevent a site for disease entry on the parent plant. A diagonal cut is generally made at the base of the cutting to maximize the area for roots to form. No leaves

FIGURE 5–8. A mist propagation bench in operation.

are removed from the cutting because they must provide all the food to the plant until roots have formed. Many cuttings will root satisfactorily without the use of rooting *hormones*; however, the rooting powders stimulate somewhat faster rooting.

These powders also contain a small amount of fungicide, so diseases will be kept to a minimum in the propagation bed when these are applied to the cuttings. It is possible to use too much rooting hormone on the cutting, resulting in the burning of the cut stem and a site for disease entry through the killed tissue. These powders should be dusted onto the cutting bases, rather than dipping the cuttings, to prevent the spread of disease from one plant to the others.

The cuttings are *stuck* in the propagation bed as closely as possible without allowing the leaves to overlap. Before sticking the cuttings, the bed is well watered and firmed with a weighted board. Lines are cut into the rooting medium with a knife drawn along a straight board to locate the rows for the cuttings. These lines are placed 5 inches apart along the bed. Cuttings are stuck to a depth of 2–6 inches (5–15 centimeters), depending on the type of cutting used.

When all the cuttings are stuck, the bed is again well watered to settle the medium around the bases of the cuttings. The newly stuck cuttings require a misting for 4–6 seconds each minute until the roots begin to form. Once rooting has begun, the misting interval may be reduced until no more mist is required. When the cuttings have formed adequate roots to sustain the plants without requiring mist, they are ready to be potted in their own containers. The rooted cuttings are lifted from the propagation bed by slipping a knife or spatula under the roots, while supporting the cuttings with the other hand. There is no advantage to be gained by leaving the cuttings in the propagation bed until the cuttings are heavily rooted. In most cases the plants grow better in their own pots when removed from the propagation bed with only a few newly formed feeder roots. It is best to remove the cuttings when a majority of the cuttings are rooted and either restick or discard those cuttings which have failed to root.

GREENHOUSE GROWING MEDIA

Soil Mixes for Florists' Crops

Soil mixes used for greenhouse production of potted plants and cut flowers are highly modified mixtures of soil, inorganic, and organic materials. When topsoil is included as a portion of the mixture, it is generally combined with other materials to improve the water-holding capacity and aeration of the potting soil. Many greenhouses do not use topsoil as an additive to the soil mixes, but rather use a combination of these organic and inorganic components as an *artificial* soil mix. When managed properly as to watering and fertilization practices, these artificial mixes grow crops that are equal to those grown in topsoil.

The most commonly used *organic* additive to a soil mix is sphagnum *peat moss*. Peat moss is a desirable component of an artificial medium because (1) it is light in weight, (2) it stores water well when it is wet, (3) it provides some internal drainage and good aeration for the roots, and (4) it can both provide some nutrients to the plant and store some nutrients. Peat moss is usually free of most disease organisms and weed seeds when it is used directly from the bale. The peat moss is often very difficult to mix with other soil additives when it is dry, so it is generally put through a soil shredder and then wetted before mixing with other materials. Various wetting solutions may be used for easier wetting of peat moss in soil mixes.

There are many *inorganic* additives that are used in commercial greenhouse soils. These include (1) perlite, (2) vermiculite, (3) haydite, (4) sand, and (5) gravel. All of these inorganic materials are used to provide proper drainage and air spaces to the root zone of the pot or bench. Today's greenhouse soils generally consist of either peat and perlite or peat and vermiculite as the primary components. In some cases, haydite, sand, or gravel are added to provide weight to a pot.

Perlite is an expanded volcanic rock material that acts very much like sand or gravel except that it is very light in weight. It allows very little nutrient storage capacity for the soil mix. *Vermiculite* does store appreciable amounts of fertilizer and also contains *magnesium* and *potassium* for use by the growing plant. Vermiculite is also very light in weight when used in a soil mix.

For some crops, hardwood barks and pine barks may be used in addition to peat. Pine bark is used particularly in a medium for growing azaleas. The pine bark provides the proper aeration to the plant roots, while maintaining the correct soil acidity for the growth of these plants.

Whenever top soil is excluded from a soil mix it is necessary to closely watch the moisture content and fertilizer status of the mix. Soil mixes containing 50 percent (by volume) of peat moss cannot be allowed to dry completely. If this should occur, it is difficult to rewet the soil medium. Since no soil is included in the artificial mixes, fertilizer must be added constantly during the growth of the plant.

POTS AND CONTAINERS FOR FLORISTS' CROPS

Many florists' crops are grown and sold in plant containers. The types of pots and containers used in a greenhouse vary in size and shape, depending on their intended purposes. Most flowering pot plants are grown in round pots, while bedding plants may be sold in community packs (Figure 5–9). The standard florists' pots used are:

Standard pots.   equal in width and height

Azalea pots.   three-fourths as high as the width

Bulb pot.   half as high as the width

Rose pot.   one-and-one-half times as tall as it is wide

FIGURE 5–9.   Pots are constructed in various shapes and sizes for production of different types of crops.

When a newly rooted cutting is first planted, it is placed in a smaller-sized pot. The smaller pot allows the plant to become established in a smaller area, without a large soil volume which collects water. Once the plant roots have filled the small pot, the plant may be *shifted*, or repotted, into its finishing pot for sale. In some cases, a plant may be shifted into increasingly larger pots several times before it is finally sold.

The type of pot used depends on the nature of the crop to be grown. A *standard* pot has a rather narrow base, so a tall plant may tip over if planted in it. Potted chrysanthemums and many other flowering plants are most often grown in *azalea* pots, because the base on this pot is wide enough to give the plant stability when it is fully grown. *Bulb pans* are very shallow containers that were developed for such bulb crops as tulips, daffodils, and hyacinths (Figure 5–10). These plants are very shallow rooted and thus require very little soil in their pots. The deep *rose pots* are adapted best to deeply rooted plants requiring a large rooting area.

FIGURE 5–10.   A bulb pot has a low profile for growth of the shallow-rooted bulb crops.

The pots used for most flowering plants are constructed of either clay or plastic. The use of clay pots in a greenhouse has diminished greatly in the past years, since they are heavier than plastic pots. Clay pots may be sterilized by steam to reduce diseases. However, plastic pots must be cleaned with a liquid disinfectant, as steam will melt plastic pots. Clay pots contain many tiny holes in the walls that allow the free passage of air into and out of the soil medium. Since plastic pots do not "breathe" as do clay pots, the greenhouse worker must be careful to avoid overwatering plants in such pots. All containers used for potted plant production should have drainage holes at their bottoms. When planting and during later stages of growth, these drainage holes can sometimes become clogged. It is necessary to clear these holes when this occurs to be certain that water will flow freely from the pots.

Spring bedding and vegetable plants are grown in a wide variety of containers. The most common types of containers used for these plants are the *community packs* and *peat pots.* A community pack may be made of plastic or of a fiber material. These containers are intended to hold several plants together in the same soil area (usually 6 to 12 plants). The seedling plants are transplanted directly to these community packs, where the plants remain until they are sold.

Peat pots are used to grow the plants in the greenhouse and later planted "pot and all" into larger pots or in a garden. These pots will disintegrate in the soil, allowing the plant roots to penetrate the peat fiber. When a seedling is placed in the standard 2¼-inch peat pot, it will develop to a larger size than those placed in a community pack. These plants do not have to compete for moisture and nutrients. Peat pots will dry rapidly under greenhouse conditions, because the peat fiber acts as a wick to pull water from the soil medium. Careful attention needs to be given to the proper water requirements of plants grown in these containers (Figure 5–11).

FIGURE 5–11.   Root formation of a seedling in a peat pellet, a type of peat pot.

FERTILIZING AND WATERING GREENHOUSE PLANTS

Greenhouse plants require an added source of fertilizer to maintain their proper growth. Researchers have found that plants require certain elements in forms that are usable and in optimum concentrations for growth. Sixteen elements have been found to be essential for plant growth. These are:

MAJOR ELEMENTS REQUIRED*		
FROM AIR	FROM SOIL	
Carbon	Nitrogen	Calcium
Hydrogen	Phosphorus	Magnesium
Oxygen	Potassium	Sulfur

*These elements are required in large amounts by plants.

MINOR ELEMENTS (TRACE)—ALSO FROM SOIL	
Iron	Copper
Zinc	Manganese
Boron	Chlorine
Molybdenum	

Nitrogen, phosphorus, and potassium are generally supplied by commercial fertilizers and are referred to as the *fertilizer elements*. Calcium and magnesium are the main components of agricultural lime. These are called the *lime* elements. *Dolomitic lime* contains both calcium and magnesium, so this form of lime is used most in greenhouses. Sulfur is found in fertilizers along with other major elements in forms such as *superphosphate*. This fertilizer provides both needed fertilizer salts, phosphorus and sulfur. The *minor elements* are also known as *trace* elements, because they are required in very small amounts in the rooting medium. When field topsoil is used as a soil medium, ample trace elements are usually available to the growing plant. However, when artificial media are used, trace elements need to be added to the mix to provide for the proper nutrition of the plants.

Dolomitic lime is generally added to a soil mix before the crop is planted, for two reasons: (1) lime furnishes needed calcium and magnesium to the plant, and (2) lime adjusts the acidity of the soil to a level for good plant growth. The acidity of the soil is referred to as the *soil pH*, which is determined from a soil test. Most plants grow best when the soil mix is kept at a pH level of 6.0–6.5. Dolomitic lime is added to the soil media while it is being mixed. When peat is used as a major component of the soil mix, *1 pound of lime* is added for each *cubic foot of peat* used to counteract the acid nature of the peat. As an alternative to this method, 1 pound of lime per cubic yard of soil is added for each 0.1-pH unit rise desired in the soil mixture. For example, if the soil test of the mix shows that the pH is 4.0 and you desire a pH of 6.5, then 25 pounds of lime are to be added to each cubic yard of the mixture (1 cubic yard equals 27 cubic feet).

Phosphorus is added to soils in the form of superphosphate. This fertilizer is included during the mixing, along with the dolomitic lime. The standard rate for mixing superphosphate is to add *five pounds* of the fertilizer to each *cubic yard* of soil. When properly mixed, this fertilizer addition will generally provide adequate phosphorus and sulfur for the entire growth of most crops in a greenhouse. If the superphosphate is not added as a premix in the soil medium, then phosphorus must be included as a regular fertilizer application from some other source.

Fertilizer Injector Systems

Most greenhouse operators fertilize their crops with injector systems using soluble liquid fertilizers. This is done to reduce the amount of labor required to feed and water the crops grown in the greenhouse. These automatic watering and fertilizing systems consist of (1) water distribution lines that deliver water to each pot or to a bench and (2) a device (*fertilizer proportioner*) that mixes the correct amount of fertilizer solution with the water (Figure 5–12).

The principle of the soluble liquid feeding system is such that whenever the plants require water they are also given low amounts of fertilizer at the same time. When a plant is growing actively, it requires more water and food. These plants may be watered and fed several times during a day. The fertilizers used in these systems are mixed in a *stock solution tank* at a higher concentration than will be delivered to the plants. The fertilizer proportioner then combines the concentrated fertilizer with the water in the desired concentration to be used by the growing plants.

A fertilizer proportioner is a mechanical device that introduces the concentrated stock solutions of fertilizer into the pipe lines or hoses used for watering the crops. These proportioners work on the principle of maintaining a specified dilution ratio as water passes through the device. There are two basic types of proportioning devices available. They may operate by changes in water pressure as it passes through the proportioner, or they may contain different-sized cylinder pumps that accurately measure the water and fertilizer solutions as they pass through the pump. Both types of proportioners have fixed dilution ratios which determine the concentration of stock fertilizer solution that may be used with them. The *lower* the dilution ratio of the proportioner, the *larger* the stock solution reservoir required to provide enough fertilizer to water the plants.

For most greenhouse operations, the proportioner dilution ratios used range from 1:12 to 1:500. If the proportioner has a dilution ratio of 1:100, this means that for each 99 gallons of water applied, *1* gallon of stock fertilizer will be mixed with it to deliver *100* gallons of fertilizer and water mixture to the plants. The fertilizer solution in a stock container prepared for a 1:200 proportioner would be *10 times* more concentrated than one prepared for a 1:20 proportioner. In other words, 10 times as much fertilizer would be added to the stock container for the 1:200 proportioner.

FIGURE 5–12.   A large capacity fertilizer proportioner in a commercial greenhouse.

Preparing the Fertilizer Stock Solution

The amount of fertilizer applied to a crop is measured in terms of *parts per million.* This means that when 100 parts per million (ppm) of any fertilizer is desired, the stock solution is prepared in such a way that the fertilizer proportioner will mix 100 parts of the fertilizer for each million parts of fertilizer and water that comes from the end of the hose.

A fertilizer bag is marked with the fertilizer analysis in the bag. Bold printed numbers on the front label identify the type of fertilizer analysis included. These numbers appear in three parts, separated by dashes. The *first* number identifies the percentage of *nitrogen* added, the *second* number tells the percentage of *phosphoric acid* (P_2O_5 provides phosphorus), and the third number tells the percentage of potash (K_2O provides potassium). For example, a bag that is labeled as 15–10–30 will contain *15 percent* nitrogen, *10 percent* phosphoric acid, and *30 percent* potash (Figure 5–13).

Most greenhouse crops are fertilized with every watering at a rate of 100 to 300 ppm of *nitrogen.* When calculating the amount of fertilizer to add to a stock solution, the nitrogen content is the figure

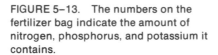

FIGURE 5–13. The numbers on the fertilizer bag indicate the amount of nitrogen, phosphorus, and potassium it contains.

used most often. If you were to prepare a stock solution of 15–10–30 at a rate of 100 ppm, you would add enough of this fertilizer to the stock container to provide 100 ppm of nitrogen to the plants. For more detailed information about mixing fertilizer solutions, see Table A–1 in Appendix A.

Applying Water to Crops

There are many different methods for watering crops, ranging from simple hose watering to sophisticated automatic delivery systems. No matter how the water is applied, this function is one of the most important operations in the greenhouse. The plants should never be allowed to wilt severely or they may not recover fully. On the other hand, the plants should not be overwatered either, for this practice causes the plants to become weak and tall, or the roots may be killed from lack of air. The person responsible for watering in a greenhouse must always be aware of the soil-moisture conditions of the crops.

When the plants are to be watered by a hose only, the job requires a great deal of time to perform properly. Each pot in a bench must be carefully watered to be certain that adequate water is applied (Figure 5–14). When a plant is properly potted, a space of ¾–1 inch remains between the soil and the top of the pot. This space is filled with water and is allowed to flow through the soil medium. Enough water should be applied to allow at least a 10 percent leaching at each watering. Leaching means that the water flushes the pot free of unused fertilizer salts which remain. At least 10 percent of the water applied must run out of the bottom of the pot. Bench crops are watered in the same manner, except that it is more difficult to determine whether proper leaching is occurring. Generally, it is necessary for the applicator to dig down into the soil periodically to determine whether the water has reached the bottom of the bench.

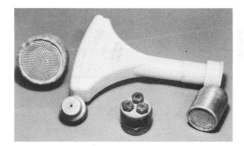

FIGURE 5–14.   When water is applied to crops by hand, various nozzles are placed on the end of the hose to control the flow of water.

Most soluble liquid fertilizers used with proportioners contain a dye (blue or green) to indicate whether the fertilizer is being mixed with the water. The water applicator should always be watching to be certain the water coming from the end of the hose contains this dye. If it does not appear to be colored, several items should be checked:

1.   The hose may be twisted so that the pressure of the water is reduced. This may cause some proportioners to malfunction.

2.   The strainer in the uptake hose from the fertilizer stock container may be clogged. The strainer may be washed off and replaced to regain fertilizer uptake to the proportioner.

3.   If the first two situations do not appear to be the problem, it is possible that the fertilizer stock solution may have been mixed at the wrong rate, therefore causing the fertilizer to be applied at too weak a concentration. Discard this fertilizer mixture and make up a new stock solution to be certain that it was prepared properly.

Watering should be done early enough in the day to allow the foliage on the plants to dry before the sun sets. Wet foliage during the night causes an increased humidity in the greenhouse atmosphere. This condition promotes the outbreak of diseases on the plants. Generally, it is best to finish all watering of the plants by midafternoon to prevent high nighttime humidity. Certain florists' plants do not tolerate the wetting of their foliage at all. African violets and gloxinias, among others, produce brown spots on their leaves when cold water is allowed to stand on their leaves (Figure 5–15). These crops should be watered in a manner that will reduce the chances for wetting their foliage.

FIGURE 5–15.   Spots on the foliage of an African violet caused by the splashing of cold water.

Automatic Watering Systems

The purpose of the automatic watering devices is to reduce the amount of handwatering required for the production of the crops. These watering systems are designed so that each potted plant or those plants growing in ground benches will receive the same amount of water (and therefore fertilizer). Most of these are connected to master control centers that electronically control the frequency and duration of the watering on each bench. Such control centers make it possible to water several benches in one greenhouse simultaneously (Figure 5–16).

The most common watering system used for potted plants is called the *tube watering system* (Figure 5-17). Individual, small-diameter leader tubes are connected to a larger header hose. Each pot is watered by a single tube. When this method is used, each pot receives the same amount of fertilizer and water, while the foliage on the plant remains dry. These tubes occasionally become plugged, therefore causing them to fail to deliver water to the pots. This will cause a few plants on the benches to become wilted while the rest of the plants are not. A careful check of the plants should be made daily to be certain that all tubes are working correctly.

Cut flower crops are grown in benches without the use of pots. The most common method for watering these benched crops is by *soaker hoses*. These hoses allow water to drip from along their length, therefore saturating the soil. It is necessary to occasionally hand water these benched crops to be certain that adequate fertilizer and water is in contact with the plant roots. When these hoses become plugged with dirt, areas of the bench may not receive adequate water. These plugged hoses will need to be replaced.

FIGURE 5–16.   An electronic control center that regulates the frequency and duration of water application to the crops in a greenhouse.

FIGURE 5–17.   Individual potted plants may be watered by use of the tube watering system.

Spring bedding plants grown in flats are watered best by an overhead sprinkler system (Figure 5–18). These water the plants by spraying fine water droplets onto the flats from above the plants. This method of watering greatly reduces the amount of hand watering required in a greenhouse. Other watering systems are also commonly used in greenhouses. It is important that you become familiar with the operation of any type of system being used and learn to recognize when the system is not watering the plants correctly.

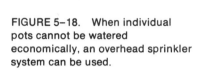
FIGURE 5–18.   When individual pots cannot be watered economically, an overhead sprinkler system can be used.

GREENHOUSE PEST CONTROL

The greenhouse is an ideal location for the establishment and rapid increase of weeds, diseases, and insect pests. Much of the control of these pests in the greenhouse is through prevention. Strict sanitation within the greenhouse should be practiced at all times. The temperature and humidity of the greenhouse is also controlled to reduce certain pest problems. When these problems do occur, the pests may then be controlled through the use of pesticides.

An important defense against the establishment of weeds, diseases, and insects in greenhouses is by the proper sterilization of greenhouse soils to be used. This may be accomplished by cooking the soil mixes with either live steam, electric heating devices, or by chemical sterilants. The most common method used is *steam sterilization*, where steam is available for heating. A steam line is connected to a bench and enclosed with a heat-resistant cover (Figure 5–19). Steam is allowed to penetrate the soil until a temperature of 160–180°F (71–82°C) is reached. After 30 minutes at this temperature, the steam is turned off and the soil is allowed to cool. It is important that only clean tools be used in the newly sterilized soils when planting to prevent the reintroduction of pests. Steam sterilization of soils will eliminate most of the serious disease, insect, and weed problems in that soil for some time. However, these pests are easily introduced if strict sanitation is not practiced during the growth of the crop.

FIGURE 5–19.   A bench is prepared for steam sterilization of the soil.

Weeds

Weeds should not be allowed to grow in any location inside or around the outside of a greenhouse. Weeds not only produce seeds that are easily moved to the crops being grown, but also provide a haven for the breeding of insects and diseases. All weeds should be controlled by removing them whenever it is practical. Weed seeds are also easily moved into a greenhouse on the shoes worn by workers. It is a good practice to keep all shoes clean and *never* place your feet on any bench surface that contains a growing crop.

Diseases

Diseases are a serious problem in a greenhouse, but may be reduced when proper care is taken. In order for a disease pathogen to become established, the proper conditions must exist in the greenhouse. Generally the high humidity existing in the greenhouse can easily start an outbreak of disease. If the greenhouse is allowed to cool rapidly, without adequate air exchange, the disease organisms grow rapidly. Temperatures should be carefully checked to prevent this cause of diseases. Watering plants late in the day will contribute to the addition of moisture to the air, so late watering should be avoided. Once a disease has become established, it is very difficult to control it, since most of the plants could be ruined in just one evening. Foliar diseases, such as botrytis, leaf spots, rusts, and mildews, are generally found to occur from poor control of the greenhouse humidity, temperature, and air circulation (Figure 5–20).

Correct sanitation methods may also aid in preventing diseases in growing crops. All dead leaves and plants should be removed from the pots or benches as soon as they are found. Employees working around the crops should be careful to use only clean or sterile tools and pots. Water-hose ends and other watering devices should not be allowed to touch the ground when not in use. Dirty hoses will rapidly spread diseases in the greenhouse. Hoses should be carefully rolled up and stored where they will not be in contact with sidewalk soil.

FIGURE 5–20.   The disease called "Botrytis" may severely damage a crop in one night when the greenhouse humidity becomes too high.

Insects

The greenhouse is a perfect location for insects to live and multiply. It is virtually impossible to grow a commercially acceptable greenhouse crop without carrying out an effective insect control program. Insect pests gain entrance into a greenhouse in various ways. During the summer, open doors and ventilators provide an easy access. Some pests are brought in on plants or with soil, tools, or even on the workers themselves. If a pest population can be detected before it can spread, there is a much better chance that the pest can be controlled. A routine check should be made daily to locate any problem before insects can cause serious damage. When looking over the plants, pick up the pots and look closely for insect damage, eggs, insect droppings, or the insects themselves under the leaves or on flowers.

An important method for controlling insects is the correct identification of these pests. First, you should learn to tell the differences in damage caused by chewing, sucking, and mining insects. Some insects only damage flower buds or flowers. Others mainly affect roots or stems. Slugs and snails are not insects, but they can also cause serious damage to plants. They are hard to find on plants, but you can follow their trails of mucous slime to their resting places. They are found most often in cool, dark locations, such as under pots or benches, when they are not feeding.

To correctly control insect pests in a greenhouse, it is necessary to learn to identify them and to recognize the damage caused by each.

Aphids. Each aphid is capable of producing at least 50 daughters in one life span. Each daughter will begin reproducing in about 1 week from the time it is born. Aphids are *sucking* insects and cause damage by producing severe wilting when large numbers are found on a plant. Aphids also spread certain plant virus diseases when they feed, and can cause much more serious crop injury in this manner. Aphids are easily spotted when large numbers exist, since they are not usually capable of flight.

Whitefly. Whiteflies are one of the most common and difficult insect pests to control in a greenhouse. The adults are tiny white-bodied insects which gather on the underside of the foliage, where they feed by *sucking* plant juices. When the foliage is disturbed, they fly in a cloud around the plant. The adults lay eggs on the undersides of the leaves. When the whitefly feed on the undersides of the leaves, they excrete a sticky honeydew waste. Black sooty mold grows on this honeydew and can be readily detected. If the mold growth becomes excessive, it may completely cover the leaves and in extreme cases will kill the plant.

Spider mites. The "two-spotted spider mite" attacks a wide range of greenhouse crops. These are not true insects, for they have eight legs. They are easily identified, when in large populations, by their masses of webbing over the foliage and flowers of the plants. Individual mites may not be seen without the aid of a hand lens or microscope. Mites also *suck* plant juices, mainly from the undersides of the leaves. The infested leaves appear to have a rust-brown or mottled coloration from the upper surface. When the leaves are viewed from the undersides, it is possible to see the mites moving along the leaf. Often the presence of the mites is accompanied by the growth of black sooty mold as well (Figure 5–21).

FIGURE 5–21.   The two-spotted spider mite can become a serious pest problem for greenhouse managers. They are often best seen by observation of the webbing on the tips of the plants.

Mealy bugs.   These tiny insects are often observed feeding at the terminals of the plant shoots or in the axils of the leaves (point where the leaf attaches to the stem). They appear to be covered with a cottony mass. This covering sheds insecticides and water rather easily, so control must be directed at the young mealybugs that have not developed this cottony mass. Adult female mealybugs lay their eggs under this covering. After the eggs hatch, the young insects leave the protection of the covering to seek a new feeding location. At this stage in their development, they are most susceptible to control by insecticides.

Scale.   Scale insects are similar to the mealybugs, since they also develop a covering that protects their bodies when feeding by sucking plant juices. Control of scale insects is directed at timing insecticide treatments to coincide with the migration of the young, unprotected *crawlers*.

There are many other insects that may infest greenhouse crops. Those mentioned previously are considered to be the most common and most serious of the insects found in greenhouses. Other insects that may be found are caterpillars (young moths or butterflies), leaf miners, borers, cutworms, and thrips, to name a few. If you find an insect that you cannot identify, or spot damage occurring to plants which does not seem to be caused by the more familiar insects, be sure to call this to the attention of the crop manager. It is important that the insect populations be controlled before they spread throughout the greenhouse.

PROPER HANDLING OF PESTICIDES

The pesticides used for controlling insects and disease organisms are among the most lethal chemicals known. This situation is amplified by the fact that the chemicals are applied to plants in an enclosed space, the greenhouse, where little exchange with the outside air is possible. These conditions, the enclosed space and dangerous chemicals, subject the pesticide applicator to a multitude of dangers. These chemicals may be applied safely to the crops, without risking the illness or death of the applicator, only when the proper safeguards against contamination are practiced.

Read the Pesticide Label

Before mixing any pesticide, take the time to read the instructions printed on the label. This label instructs the user about the *special safety measures* needed, the *rates* to be used, the *crops* that can be treated, and the *pests* that are controlled, among other instructions. An important part of the label is the *signal word*. These tell how toxic a chemical is to the user and indicates the type of precaution required for their safe handling. The signal word statement on a label may be one of the following:

SIGNAL WORD	TOXICITY	SYMBOL
Danger—poison	Highly toxic	Skull and crossbones
Warning	Moderately toxic	None
Caution	Low toxicity or nearly free from danger	None

When a label states that the pesticide is either highly toxic or moderately toxic, the applicator should be fully protected before even opening the container. The user should wear rubber *gloves*, rubber *boots*, a long rubber *raincoat*, a *hat* that covers the back of the head and neck, long-legged *pants* that are made of closely woven fabric, *goggles* or face shield, and a *gas mask* with an air tank (Figure 5–22). Although not all pesticides used on greenhouse crops are dangerous to greenhouse personnel when used safely, all of these chemicals should be treated as if they were lethal. Learn to respect these pesticides for their potential dangers.

FIGURE 5–22.   Whenever pesticides are applied in a greenhouse, the worker should be properly clothed and protected.

Pesticide Application Equipment

Pesticides are applied to florist's crops in several ways. The most common methods for pesticide application in the greenhouse are by *hand sprayers* or pressure sprayers, *aerosols,* and *fogging devices.* The hand sprayers generally are carried by the applicator and have a volume capacity of not more than 4 gallons of pesticide liquid. The hand sprayers and pressure sprayers deliver sprays that are mixed in water. The pesticide label tells the type of formulation that is contained in the bag or bottle. *Liquids* and *wettable powders* are those formulated for use in hand sprayers.

When mixing pesticides for use in hand sprayers or high-pressure sprayers, the applicator should wear protective clothing. The sprayer tank is filled to half its volume with water, then the required amount of pesticide is added. The water and chemical are thoroughly mixed by stirring before filling the sprayer with water to the required level (usually indicated by a mark on the tank). Most pesticide labels give instructions for the mixing of 1 gallon of the pesticide. However, in some cases the instructions are for larger volumes.

For example, if the directions state that *1 pint of pesticide is to be mixed in 100 gallons of water*, this must be calculated for the smaller sprayer. If the capacity of the sprayer is only 3½ gallons, the calculations are:

$$1 \text{ pint}/100 \text{ gallons} = 1 \text{ teaspoon/gallon}$$
$$(1 \text{ teaspoon/gallon}) \ (3\frac{1}{2} \text{ gallons}) = 3\frac{1}{2} \text{ teaspoons/tank}$$

If the pesticide instructions say that *1 pound of chemical is added to 100 gallons of water*, the calculations are as follows:

$$1 \text{ pound}/100 \text{ gallons} = 1 \text{ tablespoon/gallon}$$
$$(1 \text{ tablespoon/gallon}) \ (3\frac{1}{2} \text{ gallons}) = 3\frac{1}{2} \text{ tablespoons/tank}$$

Once the pesticide is fully mixed in the sprayer, the top is secured and sealed. The hand sprayer must be pumped full of air to create the pressure that delivers the spray. The sprayer should be pumped to a *pressure of 30 pounds per square inch*, if a gauge is attached. If no gauge is present, the plunger should be pumped 30–35 times, or until it is very difficult to push the handle to the locking position. The sprayer is now ready to use, but the contents should be shaken occasionally during the spraying to make certain the chemical remains thoroughly mixed. A high-pressure sprayer develops the pressure required for spraying from an electric or gas-powered engine, so pumping is not required.

The spray is directed at the foliage in a manner that completely covers the plants with liquid. An application is first made from underneath the foliage with the spray nozzle pointed upward. It is important that a good coverage of the pesticide is applied to the underside of the leaves, since most insects are found there. The next pass of the spray is directed over the top of the plants from above. Enough chemical is applied to cause a slight runoff of the spray from the leaves. It is important that the plant leaves be covered uniformly, without applying too much or too little spray.

When the spraying operation is completed, the spray equipment and all measuring spoons or bottles should be thoroughly cleaned. The sprayer should be flushed with water several times, with the last flushing placed under pressure by pumping the closed sprayer and evacuating the water through the spray nozzle. Be sure to change your clothes and wash your hands and face thoroughly with hand soap and water before resuming other work in the greenhouse. The newly sprayed plants should not be handled for several hours after they have been sprayed. In some cases, as indicated on the pesticide label, no worker should enter a greenhouse that has been sprayed without wearing protective clothing for 24 hours. Whenever a greenhouse is fumigated or treated with a dangerous chemical, warning signs should be posted to prevent the accidental poisoning of workers.

Aerosols and *fogging devices* are used in greenhouses to apply a gaseous vapor of pesticides to the plants. These are applied from special equipment designed for this purpose and require specifically formulated pesticides. An *aerosol* is premixed in a pressurized bottle and is used in this manner to create a fine spray vapor in the greenhouse. A *fogging device* generates a fog from an internal combustion engine that heats the pesticide as it is sprayed into the exhaust pipe of the machine. The heated vapors create a fog in the greenhouse and penetrate the foliage of the plants (Figure 5–23). Both aerosols and fogs are applied to completely closed greenhouses which remain sealed overnight. The greenhouses must be ventilated early the next morning following pesticide application before normal work may resume in these houses.

FIGURE 5–23.   The fogging machine produces a very dense fog of insecticide in the greenhouse.

Whenever pesticides are to be used in a greenhouse, certain procedures should be followed for their correct and safe application:

1. Know what the causitive agent is before attempting to control it. Determine whether damage is caused by an insect, disease, nutrient deficiency, or some other cause.

2. Be certain that the chemical selected for control is registered for that particular use. The pesticide label will name the pests controlled and on which crops it may be used.

3. Be sure that the chemical is mixed at the labeled rate when preparing the spray mixture. Double-check your calculations.

4. Wear protective clothing and avoid breathing any pesticide fumes or dusts.

5. Clean up thoroughly after spraying. Clean all utensils, the sprayer, and especially, clean yourself afterward.

6. Return all pesticides to a specially designed storage area. Keep this area locked when not in use.

SUMMARY

Commercial greenhouses are used for the production of potted plants, tropical foliage plants, and cut flowers by flower growers. The operation of these growing structures requires a considerable amount of labor and knowledge to produce high-quality crops. A high degree of skill is needed by the greenhouse workers to carry out the duties of maintaining the daily requirements of the growing plants.

Greenhouse crops are propagated from either seeds or vegetative cuttings. Seed propagation is a major part of the spring bedding plant production, as well as with other potted plant and cut flower crops. A greenhouse worker becomes actively involved with all phases of seed propagation, including sowing, germination, transplanting, and growing the finished plants. A basic knowledge of the requirements for germinating seed, such as preventing diseases and obtaining fast growth, are necessary to work in a commercial greenhouse. Some of the techniques needed are learned best by actually doing these jobs. This training allows you to acquire the skills firsthand.

Vegetative propagation is used when a plant can be reproduced most economically by cutting a plant into several pieces, each of which will grow to look exactly like the parent plant. This method of propagation is used extensively for the production of such crops as poinsettia, chrysanthemums, many tropical foliage plants, and other florist's crops. Often this method is the only practical procedure available for reproducing plants, because seeds are rarely formed on some plants or the seeds do not produce a plant that is like the parent.

Florists' crops are grown in highly modified soil media. Recently, soil as we often think of it has been eliminated altogether from these growing media. The artificial soils, as they are called, consist of equal parts of peat and perlite, peat and vermiculite, or some combination of other organic and inorganic ingredients. Since soil has been eliminated from these growing media, a high degree of skill is required to provide all of the needed growth conditions to the crops. The growing plants must be constantly attended to provide adequate water, fertilizer, and proper environmental conditions.

The watering and fertilization of the greenhouse crops may seem to be the most mundane of all possible tasks, yet these are the two most important duties for a greenhouse worker. The growing plants must never be allowed to dry excessively and require constant additions of fertilizer for rapid growth and maintenance of healthy foliage. When the plants are watered by hand-held hoses, this job is doubly important. The worker doing the watering must see to it that all plants are watered sufficiently and uniformly. The water line must be constantly watched to see that fertilizer is being applied, as indicated by the presence of the dye in the water. When the fertilizer stock container is emptied, it is the duty of the water applicator to prepare a correct fertilizer formulation to replace it.

The second important aspect of hand watering is that this employee is able to observe the plants on a daily basis. His knowledge of the growth conditions of the plants may prevent serious losses to the greenhouse business. When watering, the employee is able to spot diseases or insect pest populations before they have spread widely through the greenhouse. Even when automatic watering systems are used extensively in a greenhouse, the employees should be constantly observant of potential problem areas of the crops.

Pest problems cannot always be avoided, even though early detection prevents serious outbreaks. When the problems do exist, the best control is through the use of pesticides. These chemicals are applied by pressure sprayers, fogging devices, or aerosol applicators. In some cases other methods are used to apply pesticides to the growing crops. Whenever these lethal chemicals are handled, it is imperative that the user exercise extreme caution to avoid endangering himself/herself or coworkers. In all cases, the pesticide applicator should see to it that all labeled instructions are followed.

SELECTED REFERENCES

Ball, V. (ed.). 1975. *The Ball Red Book*, 13th ed. George J. Ball, Inc., West Chicago, Ill.

Drawbaugh, C. C. 1962. *A One-Year Greenhouse Laboratory Course of Study of Vocational Agriculture in Pennsylvania.* Department of Agricultural Education, Pennsylvania State University, University Park, Pa.

Edmond, J. B., T. L. Senn, F. S. Andrews, and R. G. Halfacre. 1975. *Fundamentals of Horticulture*, 4th ed. McGraw-Hill Book Company, New York.

Hoover, N. K. 1969. *Handbook of Agricultural Occupations*, 2nd ed. The Interstate Printers and Publishers, Inc., Danville, Ill.

Laurie, A., D. C. Kiplinger, and K. S. Nelson. 1968. *Commercial Flower Forcing*, 7th ed. McGraw-Hill Book Company, New York.

Mastalerz, J. W. (ed.). 1966. *Bedding Plants*. Agricultural Extension Service, Pennsylvania State University, University Park, Pa.

Mastalerz, J. W. 1977. *The Greenhouse Environment*. John Wiley & Sons, Inc., New York.

Nelson, K. S. 1967. *Flower and Plant Production in the Greenhouse*, 2nd ed. The Interstate Printers and Publishers, Inc., Danville, Ill.

Nelson, K. S. 1973. *Greenhouse Management for Flower and Plant Production*. The Interstate Printers and Publishers, Inc., Danville, Ill.

Ohio State University. 1974. *The Greenhouse Worker*. Ohio Agricultural Education Curriculum Materials Service, Columbus, Ohio.

Oldale, A., and P. Oldale. 1975. *Plant Propagation in Pictures*. David & Charles, Devon, England.

Peters, R. B. 1968. *Peters Fertilizers*. Robert B. Peters Company, Inc., Allentown, Pa.

Pirone, P. P., B. O. Dodge, B. D. Rickett, and H. W. Rickett. 1970. *Diseases and Pests of Ornamental Plants*, 4th ed. The Ronald Press Company, New York.

Wright, R. C. M. 1975. *The Complete Handbook of Plant Propagation*. Macmillan Publishing Co., Inc., New York.

TERMS TO KNOW

Clonal Propagation	Heating Cable	Stock Plant
Damping-off	Leach	Subirrigate
Dibble	Leaf Node	Topsoil
Dolomitic Lime	Peat Pot	Toxic
Embryo	Perlite	Trace Elements
Endosperm	Soluble	Transplant
Fertilizer Injector	Standard Pot	Vermiculite
Foliar Disease	Sterilize	Wettable Powder
Germination		

STUDY QUESTIONS

1. Describe the steps required when sowing and watering seeds in a seed flat.
2. Describe the disease-preventive measures to be taken when germinating seeds.
3. Explain the term "asexual propagation."
4. Explain the advantages for each of the following when used in a soil mix: peat moss, perlite, vermiculite, haydite, sand, and pine bark.
5. List five containers that are used for plant production and explain what each is designed to do.
6. List the nine major elements required by plants for growth. Where are these elements derived?
7. How much dolomitic lime should be added to a cubic yard of soil to raise the pH from 5.6 to 6.5?
8. Define the term "soluble liquid feeding."
9. Explain the advantages for using fertilizer injectors in greenhouse production systems.
10. What is the purpose of leaching soil?
11. Give three reasons why liquid fertilizer may not be getting to the plants during the watering process.
12. List three types of automatic watering systems.
13. Work the following fertilizer problem:
    You are to prepare a 100-ppm solution of a 30–10–10 fertilizer in a 5-gallon container. The fertilizer injector to be used has a dilution ratio of 1:16. How much fertilizer should be added to the container to make up the 5 gallons of fertilizer solution?

14. What general precautions should be taken to prevent diseases and insect problems in a greenhouse?

15. What important information will you find on a pesticide label?

16. What precautions should be taken whenever pesticides are used?

17. You have been instructed to mix an insecticide for use in the greenhouse. The pesticide label instructs the user to mix 1 pint of the liquid pesticide in 100 gallons of water. How much of this pesticide will you mix in a 3½-gallon sprayer?

SUGGESTED ACTIVITIES

1. Cut open different types of seeds and locate the *embryo, endosperm,* and *seed coat* for each.

2. Plant several types of seeds and compare their germination time and rate of growth.

3. Sow seeds in both sterilized and unsterilized soil mixes. Compare their growth and development.

4. Compare the growth of seedlings grown in the greenhouse with those grown in a darker classroom.

5. Propagate plants by taking cuttings and rooting them in a mist bed.

6. Make a display showing different types of soil mixes. List the advantages and disadvantages for each type of mix.

7. Visit a greenhouse to observe plant propagation, fertilization, insect and disease control, and watering practices.

8. Compare the growth of seedlings in peat pots, peat pellets, clay pots, and plastic pots.

CHAPTER SIX

Specialized
Crop Production

The greatest demand for greenhouse flower crops is during the major holidays and during the spring season. Many of the flowering plants are grown for sales on a year-round schedule, such as chrysanthemums, roses, carnations, and African violets. Other crops are grown for marketing at a specific holiday or season of the year, such as poinsettia, hydrangea, Easter lily, and bedding plants. Each of the crops grown for the florist trade requires specific cultural conditions, and all plants are not grown in a similar manner. This chapter will describe how some of the more important greenhouse potted plants and cut flower crops are grown commercially.

CHRYSANTHEMUMS

Chrysanthemums are grown as either potted plants or for cut flower use. The growth requirements and timing of the flowering are similar for both, even though the cut flower types are grown in soil-filled benches rather than in pots. Whether a chrysanthemum is to be grown as a cut flower or as a potted plant is determined by the growth and flower characteristics of the cultivar. Some chrysanthemums produce their flowers on taller stems and may have large flowers. These types are selected for cut flower production because the long stems are more suitable for use in flower arrangements by florists. The naturally short-stemmed cultivars are most often grown as potted chrysanthemums. (See Figures 6–1 and 6–2.)

FIGURE 6-1. White pot mum 'St. Moritz'. (Courtesy Pan American Plant Company)

FIGURE 6-2. Standard semi-incurve chrysanthemum 'May Shoesmith'. (Courtesy Pan American Plant Company)

Planting Chrysanthemums

Chrysanthemums are propagated from rooted cuttings. These are usually obtained from a commercial propagator who specializes in the production of chrysanthemum cuttings. Some greenhouses obtain nonrooted cuttings from the propagator and root these themselves (Figure 6-3). Regardless of the type of cutting used, it is necessary to plant the cuttings as soon as they arrive.

FIGURE 6–3. Chrysanthemum cuttings are rooted in pots under mist.

FIGURE 6–4. Chrysanthemum cuttings are placed in pots with a light layer of soil over the roots. The cuttings are placed at an angle to the outside lip of the container so that growth is directed away from the pot center.

The standard method for planting potted chrysanthemums is by placing five cuttings in a 6-inch azalea pot. These cuttings are arranged in equal spacing around the outside lip of the pot, so the growth of the plants will be directed away from the center (Figure 6–4). Chrysanthemums produce most of their roots near the upper surface of the soil (shallow-rooted). For this reason the cuttings should be planted only to a depth that will cover the roots. Deeper planting will prevent the necessary drainage required for proper growth.

Chrysanthemums grown for cut flowers are planted in a similar manner, with the exception that they are placed in soil-filled benches (Figure 6–5). Cuttings are placed 8 inches (20 centimeters) apart in rows and planted with a minimum amount of soil covering the roots. When these are watered for the first time, the plants should fall over

FIGURE 6–5.   Chrysanthemums grown for cut flowers are planted in soil-filled benches.

slightly, requiring resetting of the cuttings. If the cuttings do not do this, they have been planted too deeply. This first water application should contain a dilute fertilizer. This is necessary for the proper growth of new roots into the pot or bench soil medium. During the first week of growth, the newly planted cuttings should be protected by a shade covering, and frequent applications of mist on the foliage may be necessary to prevent severe wilting. At the end of the first week in the greenhouse, the chrysanthemums will be well established in the soil and may be given daily fertilizer applications.

Flowering of Chrysanthemums

Chrysanthemums produce flower buds during periods of short daylight hours (during long night periods.) The plant growth is controlled by the number of hours of darkness, called a *photoperiodic* response. When chrysanthemums are grown under conditions of daylight longer than 14½ hours per day, the resulting growth will be vegetative. The plants will continue to grow new stems and leaves, but no flowers will be formed. When the length of the daylight hours are less than the critical day length of 14½ hours, flower buds are produced. Once the short-day treatment is begun, it is important that it be continued nightly until the flowers are mature; otherwise, the plants may again resume the vegetative growth and cease to form flowers.

Both potted and cut flower chrysanthemums are grown in two phases: *vegetative* and *flowering*. During the vegetative phase of growth, the plants are given a long-day treatment. The long days are provided by growing the plants in a greenhouse which is lighted for 4 hours each night. By turning the lights on from 10:00 P.M. until 2:00 A.M. each night, the plants will produce new shoots, roots, and foliage but will not form flowers. This long-day treatment is necessary for the proper development of stems before the flower buds are allowed to form. The long-day treatment (lighting at night) is generally given for a period of 1 or 2 weeks to potted chrysanthemums. Those grown for cut flowers may require up to 4 weeks of long-day treatment to produce the long stems required by florists.

At the end of the vegetative period of growth, the plants are placed under conditions favoring flower bud formation. The night lighting is stopped and the plants are covered nightly with a light-proof cloth, so no light may enter the bench. By covering the plants from about 5:00 P.M. until 9:00 A.M., the plants will begin forming flower buds. This dark-cloth treatment must continue nightly until the chrysanthemum flowers begin to open and petal coloring is evident (Figure 6–6).

FIGURE 6–6. When the short-day photoperiod is interrupted regularly during the chrysanthemum flower bud development, the buds may not form properly. This condition leads to the formation of "crown buds."

Pinching Chrysanthemums

Potted chrysanthemums are grown in a manner to produce several flowering stems on each cutting that was planted. This is accomplished through *pinching* each plant by removing the top 1/2–3/4 inch of the terminal shoot of each cutting with the fingers. This treatment causes the side buds (lateral buds) to form new flowering stems. Most potted chrysanthemums are pinched about the time that the short-day treatment is begun. Chrysanthemums grown for cut flowers are generally not pinched, since pinching will reduce the length of the stems.

Height Control of Potted Chrysanthemums

Chemical growth regulators are applied to potted chrysanthemums to control plant height. Several of these chemicals are recommended for height control of chrysanthemums, but the most commonly used product is called *B-Nine*® (also *Alar*).® This chemical is applied as a foliar spray at the rate of 0.25 percent (see instructions on package label) 2 weeks following the start of the short-day treatment. A second application may be necessary for taller-growing cultivars several weeks later. Naturally short growing chrysanthemums will not require a growth-retarding chemical application.

The growth retardant is applied as a fine spray to the upper surface of the leaves on each plant. Equal amounts of chemical are applied to each plant to be certain that uniform plant growth will result. This chemical requires several hours to be absorbed by the foliage, so the plants are watered thoroughly before spraying them with the chemical and then not watered again for 10–12 hours.

Disbudding Chrysanthemums

Both potted chrysanthemums and standard cut flower types are disbudded during the flowering phase of growth. This treatment is done so that only the terminal flower will be allowed to mature. The lateral buds are removed from the stem for a distance of 2–3 inches (5–7.6 centimeters) down from the terminal bud (Figure 6–7). The timing of the disbudding is somewhat critical, as the side buds must not be allowed to become too long after they have formed. Generally, the disbudding is done 3–4 weeks before the plants are scheduled to flower. At this time the lateral buds are large enough to remove without the risk of damaging the terminal bud. When the taller potted mums are grown, the second growth-retarding chemical application is made following the disbudding treatment.

Some cultivars of potted and cut flower chrysanthemums are

(a)   The lateral buds closest to the terminal flower are removed.

(b)   The terminal flower develops faster and becomes larger when the lateral flower buds are removed.

FIGURE 6–7.   Disbudding chrysanthemums.

grown to produce several flowers on each stem. These types are called "sprays" or pompons. Rather than removing all the side shoots from these plants, the greenhouse worker removes only the *terminal* flower and allows the side shoots to develop into flowers (Figure 6–8). The method used for disbudding a chrysanthemum is determined by the nature of the flower and by its intended use.

FIGURE 6–8.   Spray chrysanthemums are disbudded only by the removal of the terminal flower bud to allow side shoot flower formation.

Flowering Response and Scheduling

Chrysanthemum cultivars are not all alike in their development from vegetative growth until flowering. Some may produce mature flowers in as little time as 6 weeks after the short-day treatment is begun, while others may require up to 15 weeks for flowering. The taller-growing cut flower cultivars generally require the longer periods for flower development. Chrysanthemum cultivars are described by their response period required for flower development. For example: an *8-week* chrysanthemum cultivar will require 8 weeks to produce a mature flower from the start of short days. This information is helpful in scheduling chrysanthemum crops at a commercial greenhouse.

When scheduling a crop of pot chrysanthemums, for example, the grower will determine when he desires a crop to be in flower. He will then determine which cultivars he will be growing, depending on which colors and types of flowers are desired. By counting backward from the flowering date the same number of weeks listed as the *week response* for that cultivar (found in the chrysanthemum catalog), the grower is able to determine when the dark treatment is to be begun. Additional time is added for vegetative growth (1–2 weeks of long days) to determine the planting date for the crop. If a 10-week response cultivar had been selected, it would be planted 12 weeks prior to the expected flowering date. This would allow for 2 weeks of vegetative growth and 10 weeks for flower development (Figure 6–9).

FIGURE 6–9. When the flowers have reached maturity, the chrysanthemum plant is wrapped in a plastic sleeve for shipment.

Fertilizing Chrysanthemums

Chrysanthemums are considered to have a high requirement for fertilizer during the entire period of their growth. The plants are generally given a light application of fertilizer each day while they are in the greenhouse. The most common method for fertilizing chrysanthemums is by use of a soluble fertilizer delivered through an injector system. Potted chrysanthemums are generally grown on a fertilizer analysis having a 30–10–10 ratio (30 percent nitrogen, 10 percent phosphoric acid, and 10 percent potash). Other fertilizer analyses may be used in different greenhouses. The rate of application of the fertilizer depends on the rate of growth of the plants. Lower rates are applied during the early stages of growth. The fertilizer rates are increased as the plant develops and begins the flowering processes. For more information concerning the scheduling of chrysanthemums, see Table A-2 of Appendix B.

POINSETTIA

Poinsettias are grown as potted plants for sales during the Christmas season (Figure 6–10). Each year the demand for poinsettia increases and plants are now available from Thanksgiving until Christmas. Some poinsettia are also being sold in limited quantities at Easter and Mother's Day. Although the growth requirements for the poinsettia are rather exacting, this plant may be grown with relative ease.

FIGURE 6–10.   Poinsettia bracts and flowers.

The colored portion of the poinsettia which gives the plant its distinct character is actually comprised of colored leaves called *bracts*, rather than flowers. The true flowers of the poinsettia are quite small and are found at the tip of each flowering stem.

Planting Poinsettia

Poinsettias are obtained from cuttings produced by commercial propagators who maintain healthy stock plants (Figure 6–11). These may be obtained as either rooted or unrooted cuttings, but most growers prefer to root these themselves (Figure 6–12). The cuttings are rooted under mist and then planted in 6-inch azalea pots. Most poinsettias are planted with three cuttings in each pot and the plants are pinched to make a well-branched potted plant. Other growers prefer to eliminate the pinching of the cuttings so that the colored bracts will be larger.

FIGURE 6–11.   Stock poinsettia plants are grown under ideal conditions to provide cuttings for potted poinsettia plants.

FIGURE 6–12.   These poinsettia cuttings are being rooted in pots under an automatic mist system.

The soil medium used for growing poinsettia should be well drained to allow the water to pass through the medium quickly. A soil mixture containing peat and perlite is satisfactory. Poinsettia plants are very susceptible to root-rotting diseases, so they must not remain in wet soil for long periods of time. For this reason it is important that the watering practices used on poinsettia be watched very carefully, especially during the early stages of growth of the plants.

Flowering of Poinsettia

Flowers are formed on poinsettia during short-day conditions, in a manner similar to chrysanthemums. The plants are first grown under conditions of long days (short nights) to establish growth of the roots, stems, and leaves. This is done by turning on lights in the middle of the night. The plants will not form flowers during these long days. At the end of this vegetative growth phase, the lights are turned off and a short-day treatment is begun to form flower buds.

Because the poinsettias are being grown for the Christmas season, the daylight hours are normally short enough that the plants receive the proper dark period for flower bud development in the late autumn months. Many growers do not use the dark cloth shading to cover the poinsettias for this reason. However, stray light from car headlights, parking lots, or night lights will cause a delay in flower bud development. Whenever light may enter the greenhouse where poinsettias are being flowered, it becomes necessary to shade the poinsettias with a lightproof cloth material.

Controlling Height of Poinsettia

Even under nearly ideal growth conditions, poinsettias often become too tall for a suitable pot plant. Many growers also propagate many of their own cuttings, so they may have some poinsettias that were first planted in August and others that were planted in September. The earlier planted cuttings will naturally be taller than those planted later. When these poinsettias are sold, they must be all the same height in the pot. Other factors causing the plants to become too tall are cloudy weather, high temperatures, and overwatering.

Growth-retarding chemicals may be applied to the poinsettias to slow the growth and make the finished plants more uniform in height. Several chemicals are registered for use on poinsettia, but the

most commonly used chemicals for this purpose are B-Nine (Alar)®
and Cycocel.® B-Nine is used as a foliar spray as is done with chrysanthemums. Cycocel is most often applied as a soil drench to the
pots of poinsettias. Regardless of which chemical growth retardant is
used, they must be applied before the short-day treatment is begun.
If these chemicals are applied to the plants after this period, the
bracts will be smaller than is desired and may affect the flower color.
Both of these chemicals produce a darker green leaf color on the
treated poinsettias.

Scheduling Poinsettia Flowering

Because the length of days become shorter earlier in northern
states than those in the south during the fall, scheduling of poinsettias for all areas of the United States is not the same. Poinsettias
are generally planted in pots by middle or late August. The cuttings
are *pinched* 2 weeks later by removing ½–1 inch (1.3–2.5 centimeters)
of the terminal growth (Figure 6–13). During this vegetative growth
period, the plants are grown with night lighting from 11:00 P.M. until
2:00 A.M. each night. The lighting at night is stopped and dark-cloth
treatment (short days) is begun by the first week of October for
flowering in mid-December. Growth regulators should be applied before October 15, no matter when the shading is begun. The bracts
should be turning color by Thanksgiving, if they are flowering on
schedule for mid-December. At this time, the dark-cloth shading may
be stopped, since stray light will no longer affect the flowering of the
poinsettias. If earlier flowering is desired, appropriate adjustments
must be made to the schedule to allow for proper timing of the
blooms (Figure 6–14).

In recent years, poinsettias have been grown for off-season sales,
such as for Valentine's Day, Easter, and Mother's Day. This may be
done by adding about 2 weeks of long-day treatment (night lighting)
to the schedule used for Christmas flowering. Poinsettias also differ
in response to flowering time, as was shown for chrysanthemums.
Most require from 8 to 10 weeks to develop flowers from the start of
the short-day treatment. A poinsettia having a *10-week response*
which is to be flowered for Valentine's Day (*e.g.*, February 7) will require *10 weeks* plus *2 weeks* to determine the planting date for a
single stemmed plant. This would mean that these poinsettia should
be planted during the week of November 16. If these same plants
were to be grown as branched plants (pinched), they will require an
additional 2 weeks of growth. These pinched plants which will flower

for Valentine's Day must be planted during the week of November 2, or 14 weeks before they are scheduled to flower. For sample schedules for poinsettia production, see *Table A-3* of *Appendix C.*

(a)   The terminal before pinching.

(b)   The terminal of the shoot is removed by carefully breaking it off.

(c)   New lateral shoots form below the pinched terminal and will produce a fuller, bushier plant.

FIGURE 6–13.   Pinching poinsettias.

FIGURE 6–14. Employees place decorated poinsettias in plastic and paper sleeves to protect the fragile bracts during delivery from the greenhouse. (Courtesy DuPage Horticulture School)

FLORISTS' FORCING AZALEAS

Azaleas that are forced into flower in greenhouses for the florist trade are handled in a similar manner as those grown for landscaping purposes. Forcing azaleas are generally flowered in 6-inch pots from December through May, although some may be flowered at any time of the year. Landscape azaleas are generally more winter hardy than forcing azaleas and are flowered only for spring sales. Since florists' azaleas are forced into bloom in greenhouses, they are sold throughout the United States, especially during the spring holiday season. Landscape azaleas are limited in sales to regions where the plants are adapted to climatic conditions.

Florists' azaleas are a very popular potted plant and the demand for them is especially good each spring. In past years, greenhouse operations produced their own azalea plants, from propagation to salable plants. The trend in recent years is toward the forcing of vegetative liners which have been produced by commercial propagators of azaleas. Now a greenhouse operator in the northern or midwestern states may purchase started plants that will be ready for forcing in his greenhouse range. Azaleas are easily forced into bloom in greenhouses after the plants have been subjected to a period of warm temperatures to initiate flower buds, followed by a suitable period of cold temperatures to overcome the rest period in the buds. Since azaleas require a cold temperature treatment to cause flowering, they are referred to as a *thermoperiodic* plant. By selectively regulating the timing for bud initiation and cold treatment, the grower can force azaleas for any season of the year.

Propagation of Azaleas

Most forcing azaleas are propagated and grown into liners (vegetative plants) by commercial azalea propagators located in the warmer regions of the United States. The largest azalea-growing nurseries are located in western coastal states, along the gulf of the southeastern states, and upward along the eastern coastal states. These regions are ideal for azalea production, because the mild winters allow a longer season for plant growth. These nurseries produce the bulk of the azaleas that are forced into flower for sales in the other parts of the country.

The majority of the azaleas grown in the United States are propagated from terminal shoot cuttings. Cuttings are taken only from healthy stock plants that have been grown under controlled conditions. The cuttings are 3–5 inches (7.6–13 centimeters) in length and should have 5–8 leaves on the terminal shoot. Cuttings are generally taken and rooted when the current season's growth is mature, from July to October. Many growers are now finding that azalea cuttings root best when the cuttings are rooted from September through March. After the cuttings have been removed from the stock plants, they are treated with a fungicide solution, a rooting hormone, and then rooted under mist.

Rooting hormones have been found to stimulate the rooting of azalea cuttings. The most common rooting hormones used are IBA (indole butyric acid) and NAA (naphthalene acetic acid). These chemicals are often combined with fungicidal powders (such as Captan) in commercial rooting powder formulations. After treatment of the bases of the azalea stem cuttings, they are stuck in a mist propagation bed. The medium used in the propagation bed must be well drained, sterile, and porous, so that roots may penetrate the medium readily and be removed easily once the plants are well rooted. Some growers use combinations of peat and sand, vermiculite and sand, or peat and perlite. Other nurserymen are now using individual propagation blocks made of compressed peat to root the azalea cuttings. Use of the individual blocks eliminates the need for disturbing the root systems of the cuttings when they are removed from the mist bed.

Liner Production of Azaleas

Rooted azalea cuttings are most often planted to plastic containers for the purpose of increasing size or grade. These are first placed in 2¼ or 3 inch pots and kept in either a greenhouse or an outdoor nurse bed for the first growing season. These liners may be sold at any stage of growth or size, ranging from the small rooted liners, vegetative plants in their finishing pots, as budded plants that must be precooled for flowering, or as finished plants that have already been precooled. Azalea plants that will be sold as budded or finished plants will be transplanted to their finishing-size pots (usually 4-, 5-, or 6-inch azalea pots) and may remain in the nursery for up to 3 years to reach the desired size.

The growing medium used for potting azalea liners is most often sphagnum peat moss. It is most popular because it is acid in reaction (azaleas require an acid soil for growth) and retains moisture well. Other media used by some nurseries include bark, peanut hulls, sawdust, or other organic materials.

During the growing period of the liners at the nursery, periodic pruning or shearing is required to develop salable plants. By this method the terminal growing points are removed, causing the stems to branch and form a more compact, uniform appearing plant. Since it is desirable to keep these plants in a vegetative condition, the terminals are removed every 6 weeks to avoid flower bud formation. Flower buds will be initiated on new shoots within 6–7 weeks after the shoots are pinched under the proper temperature conditions. The presence of the flower buds prevents further vegetative growth and restricts the size of the plant. Pruning of azalea liners is done with shears, knives, or by hand pinching. Some nurseries are now using chemical pruning agents which are sprayed onto the foliage. These chemicals selectively kill the terminal shoot but allows growth of side shoots.

When the azalea plants have reached the desirable size for their grade, they may be allowed to form flower buds. Since flower buds require approximately 6–7 weeks to form on new growth following pinching, timing of the last pinching is very important. Many azalea cultivars form flowers best during autumn when the daylight hours are short (9–11 hours/day of sunlight). When buds are to be initiated at other periods of the year, artificially shortened daylengths may be obtained by darkening the greenhouse or nursery beds.

Flowering of Azaleas

Azaleas may be made to set flower buds at any stage of growth under the proper conditions. However, most azaleas are flowered for the Christmas through May season. Many greenhouses will receive vegetative azalea liners in the spring previous to the date they are to be flowered. During the summer months, the azaleas will be pruned twice (at 6-week intervals) to develop full, compact plants. During late summer and early fall, flower buds will naturally form. Further growth of the flower buds, following initiation, requires a period of cool temperatures to break flower bud dormancy.

Azaleas may be cooled naturally in cold frames outdoors to break bud dormancy. This is possible in colder climates, but temperatures must be maintained below 50°F (10°C) and above freezing for a 6-week period for best results. Many greenhouse operators use lighted walk-in refrigerators for precooling azaleas, since the temperatures may be maintained more easily. The use of refrigerators also makes it possible to have flowering azaleas at any season of the year.

Forcing Azalea Flowering

Azaleas are forced into flowering, following the 6-week cooling period, in the greenhouse at 60–65°F (15.6–18.3°C.) For most azalea cultivars this requires about 5–7 weeks. Many greenhouse growers will purchase the budded plants, which may be cooled at the greenhouse, to reduce the amount of space occupied by the plants during the summer months. Azaleas may also be purchased from wholesale nurseries as precooled, budded plants. Small greenhouse operators prefer this type of plant, since no cold frames or refrigerators are required. These plants may be placed in heated greenhouses and forced into flower as soon as they are received.

POTTED EASTER LILY

The Easter lily's popularity is greatest during the Easter season, since it is the recognized flower for this holiday (Figure 6–15). Some lilies are grown for use as cut flowers in arrangements, but the major production of Easter lilies is as potted plants. Commercial lily production is somewhat complicated by the fact that the dates for Easter vary from March 21 to April 22 during different years. These ranges of dates makes timing of this crop difficult from one year to the next.

FIGURE 6–15.   The Easter lily is the recognized flower for Easter sales.

Lily Cultivars

The lily cultivars presently grown for potted plant production were derived from the Japanese lily known as 'Giganteum.' Several of these cultivars are presently grown in the United States. The forcing lilies grown as potted plants for Easter are Ace, Croft, Georgia, and Nellie White. *Ace* lilies are the most popular for forcing by growers. This cultivar is medium tall in height and is more floriferous (produces more blooms) than the other lily cultivars. *Croft* is still a popular forcing cultivar, although it may develop a disorder called "leaf tip burn" when grown with improper fertilization. Croft lilies are somewhat similar to Ace and Nellie White cultivars, so are grown less often. *Georgia* lilies are grown more often for cut flower production, since they are a taller stemmed cultivar. They may be used for potted plant production when height control chemicals are used. *Nellie White* lilies are excellent for potted plant production. This cultivar has excellent plant characteristics and forces easily for flowering at Easter.

Propagation of Easter Lilies

Most of the propagation of Easter lily bulbs is done on the west coast of the United States. Lily bulbs are propagated from the bulb scales, which are removed and lined out in fields in early summer. The following fall the scales are lifted from the field beds. The small bulblets that have formed at the base of each bulb scale are removed. These bulblets are then replanted and allowed to grow for 1 or 2 seasons to produce the forcing-size bulbs used for flower production.

When the bulbs are dug from the field to be shipped to greenhouses for forcing, they are first graded into various size categories. The larger bulbs produce the most flowers, so are considered to be the best for forcing. These bulbs are also the most expensive, so many growers purchase medium-sized bulbs for their greenhouses. Generally, bulbs that are 7–8 inches (18–20 centimeters) in circumference are most often used for forcing lilies.

Precooling Lily Bulbs

Easter lily bulbs require cold-temperature treatment for timing the flowering on schedule. The precooling treatment may be given to cased bulbs in refrigerated chambers or after the bulbs have been potted. Most small greenhouse operators purchase bulbs that have been precooled by the propagator and are ready for forcing on arrival. These are called *commercial case-cooled bulbs*. The bulbs are planted as soon as they arrive, or they may be stored in the case for a few days in refrigerated coolers at 35–40°F (1.7–4.4°C).

Some growers prefer to precool their own Easter lily bulbs at the greenhouse. The most common methods used for precooling bulbs at the greenhouse include natural cooling, home case-cooling, and use of the controlled-temperature forcing schedule. *Home case-cooled bulbs* have not been precooled before they are shipped to the greenhouse. When the bulbs arrive, the cases of bulbs are immediately placed in refrigerators at a constant 40°F (4.4°C) temperature. These bulbs must remain in the refrigerator at this temperature for 6 weeks. Bulbs that are to be precooled by the *natural cooling method* are placed in pots as soon as they arrive. The pots are watered and then placed in cold frames or cold greenhouses to receive natural temperatures during late fall and early winter. The bulbs must receive 1000 hours (6 weeks) of temperatures that average 40°F (4.4°C). The soil must be kept damp at all times or the cooling will not be effective. The soil should not ever be allowed to freeze, or bulb damage will result.

Bulbs that are to be scheduled by the *controlled-temperature forcing method* are potted as soon as the non-precooled bulbs arrive. Instead of being placed in cold greenhouses, the potted lilies are placed in greenhouses held at 62–65°F (16.7–18.3°C). The lilies are kept in the greenhouse for 3 weeks to form roots. The greenhouse temperature is lowered to 40°F (4.4°C) at the end of the rooting period or the pots are moved to refrigerators. The potted lilies must be kept moist during the entire precooling treatment for it to be effective. The lilies are returned to a 62–65°F greenhouse about 110 days prior to the scheduled date of blooming.

Planting Lilies

Regardless of the method used for precooling lilies, the bulbs should be planted in the same manner. Lilies are normally planted in 6- or 7-inch (15- or 18-centimeter) standard-size pots. Lilies require excellent drainage to prevent root-rotting diseases, so a handful of sterilized gravel is placed in the bottom of the pot. About 1 inch of soil mix is placed on top of the gravel before the bulbs are placed in the pot. A single bulb is placed with the nose up in the center of the pot and soil mix added up to the rim of the pot. The soil mix is then pressed firmly to a level of 1 inch below the pot rim. Deep planting of the bulbs provides adequate drainage and allows the lily stem to root to provide anchorage for the plant. It is very important that all the bulbs be planted at exactly the same level, since potting depth determines to some extent the days required from planting until blooming.

The soil medium used for planting Easter lilies should provide very good drainage, yet be heavy enough to support the plant when it is in bloom. Sterilized field soil is used to provide weight. It is amended with sand or other coarse material and with peat moss, in equal proportions. Lime and dry fertilizers may be added to the soil mix prior to planting. Superphosphate is *not* normally added to lily potting soils, since the fluoride found in this fertilizer may cause the leaf-tip-burn problems found on some cultivars. When lilies are watered, the pots should receive enough water to allow adequate leaching through the bottom of the pots. They should not be watered again until the soil is dry to the touch, but before the foilage begins to wilt.

Forcing Flowering of Lilies

Easter lilies are removed from cold temperature precooling refrigerators or cold frames approximately 110 days before the scheduled date of flowering. The date that most lilies are scheduled to be in bloom is the week before Easter, so that they will be in full flower for holiday sales. Therefore, forcing must begin in late November for early Easter dates or late December for later Easter dates. Lilies that are ready for forcing are moved to greenhouses held at 60°F (15.6°C) at night and 70°F (21°C) during the day.

Flower bud initiation occurs when the new plant shoot is 4–6 inches (10–15 centimeters) in height. During this time the soil must never be allowed to dry out, the temperature must be maintained at the desired setting, and fertilization should be carefully controlled. The flowering schedule for the crop is maintained by adjusting the daytime temperature either up (to speed up growth) or down (to slow growth) to control the rate of growth after mid-February. Generally, experience and close observation by a skilled grower is required to make adjustments in the lily flowering schedule. Normally, the lilies will flower at the scheduled time when the temperature is held at the prescribed settings for 110 days.

Height Control of Lilies

Some lily cultivars normally produce taller flowering plants than is desired for sales. Poor growing conditions, particularly cloudy days during winter, may also cause the stems to become tall. Plants that have become taller than is desirable may be treated with chemical growth retardants which are registered for this purpose. The amount of chemical applied and the method used (soil drench or foliar spray) will depend on the amount of height reduction desired. The taller growing cultivars, such as Croft and Georgia, require greater height control than do Ace and Nellie White lilies. The plants are treated when the stems are 5–9 inches (12.7–23 centimeters) in length. These chemicals will reduce the length of the subsequent stem growth without reducing the number or size of the blooms.

Handling Flowering Lilies

Easter lilies are ready for sale from a wholesale greenhouse when the first bloom is in the white, puffy stage. These blooms have not yet opened but the bud is pointed downward at a 45-degree angle (Figure 6–16). This bloom will open easily in a retail store but will ship with less possible damage than an opened flower. Plants that are coming into flower earlier than the scheduled date may be held in lighted refrigerated coolers for up to 2 weeks at 32°F (0°C) without serious damage to the plants or blooms. Those plants which have opened flowers in the coolers should be removed and sold for early sales, since the open blooms will not store well at this low temperature. The foliage should be sprayed with a fungicide as a protection from diseases before they are placed in the coolers. The plants may be removed one day before full bloom is desired. They are then placed in a shaded greenhouse to protect the blooms from full sunlight.

FIGURE 6–16.   Easter lily plants are ready for sale from a commercial greenhouse when the buds have pointed to a 45-degree angle and are in the white, puffy stage.

GROWING CARNATIONS

The commercial production of carnations is very important to the florist industry of the United States. Cut carnations are one of the most used flowers for florists' cut flower arrangements (Figure 6–17). Most of the carnations grown for the florist market are grown in the western states and in Colorado, although many urban area greenhouses also produce locally grown cut carnations. Greenhouses located in the northern states and those in the upper elevations of mountains also contribute a large proportion of the carnations sold in the United States. Some of the carnations sold in the United States are imported from South America.

Many different carnation cultivars are grown for the cut flower market. *Standard* carnations are grown with a single flower on a long, sturdy stem. Most of the carnations grown are of this type. Many of the major cultivars of carnations grown today were derived from the cultivar 'William Sim.' This was a red cultivar that gave rise to other colors of flowers through "sports." A *sport* is a variation in growth habit found on a single stem or flower of a plant. The various Sim cultivars are widely grown today.

Miniature carnations are also popular with the florist trade, although not so popular as are standards. These are smaller plants, having 5–8 small blooms in a cluster. The miniature carnations offer short-stemmed blooms for corsages and small table arrangements.

Carnation blooms are grown and marketed the year around, but heaviest demand for this cut flower occurs during the major florist holidays of Christmas, Valentine's Day, Easter, Mother's Day, and for weddings in June and August. Once the carnation plants begin to set flowers, the production of blooms is almost continuous throughout the year. The experienced crop manager must be capable of keeping a steady supply of blooms ready for the market at all times and also be capable of providing the increased numbers of blooms demanded for holiday sales.

FIGURE 6–17.   An arrangement of standard white carnations.

Carnation Propagation

The propagation of carnations for forcing is done almost entirely by commercial propagation firms located in various areas of the country. Stock plants are maintained by these firms to provide a steady supply of vegetative cuttings which are disease- and insect-free, true to type for color and form, and always available to the greenhouse trade. Propagation of carnations is by terminal shoot cuttings. These may be sold as unrooted cuttings (which will be rooted by the carnation grower) or as rooted cuttings after a suitable period in a mist propagation bed. These cuttings normally root in 2 weeks under mist. Most carnation growers prefer to purchase rooted cuttings, so that they may be planted in the production beds as soon as they arrive.

Planting Carnations

Carnation production benches are normally replanted every 1 or 2 years. After the second year in a bench, the plants become too old to produce high-quality blooms and often contract one of several serious vascular diseases. For this reason, new cuttings are used to replace the old blooming plants on a regular schedule. Carnations may be planted at any time of the year, but most growers find it more convenient to plant during the month of June. Plants started in June will produce a heavy crop for the following Valentine's Day season, which brings a high price for the blooms.

During the first few months of growth, the new carnations are trained to build strong and vigorous plants. They are planted in a loose, well-drained soil mix containing sterile soil, peat, and coarse material for drainage. The cuttings are planted shallow to avoid root- and stem-rotting problems. During the first week after planting, the new cuttings may require frequent misting to prevent excessive wilting. The cuttings are fertilized as soon as they are planted and started on a regular fertilization schedule 1 week later. During this time a light shading is applied over the plants to aid in their establishment in the bench.

Training and Timing for Carnation Production

Carnation plants are pinched by removing the terminal 1–2 inches (2.5–5.0 centimeters) of the main stem on each cutting after at least 6 inches (15 centimeters) of new growth has occurred in the bench. This generally requires 6–8 weeks from the time the cuttings were planted. Growers may use various techniques for training the newly established carnation plants. These include (1) not pinching, (2) single pinch, (3) using the pinch-and-one-half method, and (4) double pinching.

When carnations are to be grown at close spacing (2 × 4 inches in the bench) for a fast first crop, they are not pinched. These carnations will bloom much sooner than those that are pinched after planting. Carnations that are given a single pinch after planting will produce a heavy crop, followed by a steady decline in numbers of blooms cut. This decline in production is followed by a similar peak in production. The single-pinching method of training carnations is used when a high yield is desired for a specific holiday, with a steady production in blooms during the low demand periods. Generally, two complete crops of flowers can be cut from single-pinched plants in 38–47 weeks following the planting of rooted cuttings. Flowers scheduled for blooming during the winter months require the longest time to finish because of the lower light intensity available.

The term "pinch-and-one-half" refers to pinching the terminal of the cutting first (as described for the single-pinching method), then later pinching only half of the breaks (lateral shoots) which have grown from the first pinch. This pinching method provides a high production peak of blooms, followed by a long period of declining but steady yield and another peak in production. More time is required to produce the first crop from this method of training, but higher yields between the peaks in production are realized. Double-pinched cuttings produce flowers in two definite cycles during the season, with little flower production between these peaks in yield.

Continuous Cropping of Carnations

Carnation plants that are to remain in production for 2 or more years require pruning during the late spring or early summer to rejuvinate the plants. The height of the plants is reduced by selectively removing all weak stems and by cutting the stems to a height of 12–15 inches (31–38 centimeters) above the soil line. This may be done gradually to allow some plants to remain in production for flowers during the summer. Plants pruned in this manner will be back in production 5–6 months later. Those plants which are cut back in mid-May are usually in good production for November and Christmas sales. By gradually cutting back carnation plants during the summer, the production of flowers during the following winter is spread out over a longer period.

Temperature Control for Carnations

Carnations require a cool growing climate and high light intensity for the best flower production. Carnations grown during the summer months should be kept as cool as possible by the use of wet pads and exhaust fans. Where temperatures may not be kept cool during summer, the carnations are generally cut back and not allowed to flower until the outside temperatures are naturally cool. The best growth and flowering occurs when the night temperature is kept at 50–55°F (10–13°C) at night with daytime temperatures no more than 10–15°F higher.

When the greenhouse temperature is allowed to rise above the recommended levels, the flowers formed are often misshapen and of inferior quality. A common problem that results from high night temperatures is "calyx splitting." The calyx is the portion of the flower that surrounds the petal while it is a bud. As the flower opens, the petals emerge rapidly at high temperatures, causing the calyx to split and no longer support the petals (Figure 6–18). Growers will often place a strip of adhesive tape around the buds before they open to keep this problem from occurring.

FIGURE 6–18. Improper growing conditions can cause the incidence of split calyx on carnation flowers. The flower on the left has the calyx taped to prevent this problem.

Disbudding Carnations

Standard carnations are grown with a single flower on a long stem. This is achieved by the removal of all lateral (side) flower buds as they form on the stem (Figure 6–19). Miniature carnations are grown as a "spray" with 5–8 smaller flowers on a shorter stem. These carnations are disbudded by the removal of the terminal flower only (Figure 6–20). This procedure allows the lateral flower stems to elongate and provides a more attractive stem of flowers. Disbudding is done at regular intervals in the carnation greenhouse. The unwanted buds are removed as soon as this can be done without causing damage to the bud that is to form a flower. This is usually done about 4 weeks before the bloom will be harvested.

FIGURE 6–19. A carnation
shoot ready to be disbudded by
removal of the lateral flower
buds.

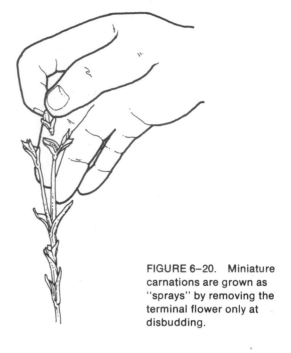

FIGURE 6–20. Miniature
carnations are grown as
"sprays" by removing the
terminal flower only at
disbudding.

Harvesting and Grading of Carnations

Carnations are cut at the proper stage of development so that the
blooms will open properly and remain fresh as long as is possible for
the consumer. Carnations that are to be shipped to distant markets
are harvested when the center petals have extended to form a hemi-
sphere. These flowers are still rather tight, but will open satisfactorily
once they are placed in refrigeration by the wholesale florist. Miniature carnations are harvested when the top two blooms are fully open
and petal color may be seen on the third bloom. Carnations that are
to be sold to a local market are cut when the petals are fully opened
and extended. If the flower buds are cut too early, they may not open
properly for the florist.

Carnation blooms are generally harvested three times a week
during the cooler months of the year. During warm weather, the
blooms may be harvested daily. Carnation blooms are removed from
the plants at a point that will provide the proper length of stem to
meet grading standards (22 inches for top-grade standard flowers)
and still retain new shoots for later flower production.

The stems are usually cut at a point so that 3 or 4 lateral shoots remain on the plant. These will produce the next blooms starting 5 months later. The stems are removed by breaking them at a node rather than by cutting. This prevents the transfer of vascular diseases from one plant to another in the bench by cutting tools. When the stem is broken at a node having a lateral shoot, the stem is broken in the opposite direction to avoid damaging the young shoot.

As soon as the cut carnations are removed from the production benches they are taken to a headhouse for grading, packaging, and refrigerated storage. The flower stems are graded according to length of stem, quality of flower, and size of blooms. The accepted grades set up by the Society of American Florists is placed on a label that accompanies each bunch of carnations to be shipped.

GRADE TAG COLOR	MINIMUM FLOWER SIZE (DIAMETER)	MINIMUM STEM LENGTH (INCHES)
Blue	2¾	22
Red	2¼	17
Green	Any size	10
White	All with split calyx	Any length

Carnations are sorted into groups of 25 flowers having the same grade, tied, and placed in refrigerated storage at 40°F (4.4°C). Any blooms that are blemished or do not meet the minimum grading standards are discarded. Flowers that are refrigerated are first placed in clean containers that are partially filled with water and a floral preservative. Carnations are shipped from the greenhouse as soon as possible. They are placed in heavy cardboard boxes lined with newspapers. Rolled paper pillows are placed under the heads of each bunch to protect the stems from breakage. The blooms are laid in opposite directions as they are layered in the box. During warm weather, crushed ice may be added to the box for added cooling.

Some carnations are colored (dyed) before they are shipped. There is some demand for carnations colored in blues, oranges, lavenders, and golds, which do not occur naturally. These may be obtained by placing white carnation flower stems in warm water mixed with the desired food dye. These containers are then placed in a warm room so that the stems will absorb the colored water. After 12–15 hours the white petals will have become the same color as the dye. These flowers are then shipped in the same manner as are other cut carnations.

CUT ROSE PRODUCTION

Commercial cut rose production is one of the most important of the floriculture industry (Figure 6–21). The major areas of the United States for rose production are in California, Colorado, and many areas of the north near urban markets. Roses are generally produced by specialized growers in large greenhouse ranges. Roses are not as popular as a cut flower as they once were. This is due in part to their poor keeping quality in the home as compared to carnations and chrysanthemums.

FIGURE 6–21. An arrangement of 'Forever Yours' roses.

The major types of roses grown for cut flowers are the hybird tea and the floribunda roses. *Hybrid tea* roses are grown with a large terminal flower on a long stem. *Floribunda* roses (also called *sweetheart* roses) are grown on shorter stems and may be grown as a spray, with several smaller blooms on each stem. The floribunda roses have a longer cut flower life than the hybrid tea roses, so they are becoming increasingly more popular by florists. Floribunda roses will generally produce twice as many cut stems per plant as a hybrid tea rose.

Roses are grown the year around, but heaviest demand (and the highest prices for the blooms) is during the major florist holidays of Christmas, Valentine's Day, Easter, and Mother's Day. The greatest demand is for red roses, so at least 40 percent of the roses grown are of this color. Other rose flowers grown are of pink, yellow, and also white in color. White roses are in greatest demand for weddings, particularly during the months of June and August. A few lavender roses are sold, as are blooms of various shades of peach and tangerine. Bicolored and multicolored blooms, as seen in garden roses, are not popular in the florist trade.

Rose Propagation

Rose plants are propagated by commercial nurseries in the southern and western United States. The rose plants are produced from grafted rootstocks grown in fields. During the first season in the field, rootstock cuttings are stuck in rows for rooting and to produce new top growth. The following spring the desired hybrid tea or floribunda rose cultivar is bud-grafted onto the rootstock. A single bud is placed approximately 4 inches (10 centimeters) above the soil line and wrapped with a budding rubber strip. When the bud graft is well healed, the top of the rootstock is broken above the graft to force the growth of the scion (grafted bud). The following season the scion is pruned to force lateral cane production at the base of the graft. These plants are ready for harvest the following fall. These are dug from the field when the plants are dormant, graded, and stored in cool storage rooms in damp shingletow.

Roses will root and grow on hybrid tea or floribunda canes. However, these roses are most often grafted to rose rootstocks, which stimulate better rose flower production. The most common understock used for commercial cut roses is *Rosa noisettiana* 'Manetti' (Manetti rootstock). Roses grafted to this rootstock do not go dormant during the cool cloudy weather of winter and remain in high production the year around.

Planting Roses

Rose plants are purchased from commercial rose nurseries for planting in greenhouses. These roses may be planted as soon as they are available. This is usually done from February through May. The root system on roses is quite large and since these roses will be in the production bench for 4 years or longer, the soils must be well prepared before planting. Roses are planted in deep ground benches filled with a soil mixture having good air and water drainage. Roses are planted at 12- × 12-inch spacings to a depth that leaves the graft union about 4 inches above the soil line. Before the plants are placed in the ground, the dead and broken roots are trimmed off. The plants are dipped in a fungicide solution (such as fermate at the rate of 1 pound/gallon of water) to aid in disease prevention. After the roses are planted in the production benches, a mulch material (such as composted manure, straw, or ground corncobs) is layered over the soil to a depth of 4 inches. This layer of mulch aids in conserving water and preventing the soil from drying too rapidly.

Pruning and Training Rose Plants

The first 3 months after roses are planted is used to build the plants for maximum flower production. New shoots that arise from larger canes near the base of the plant are selected. All weak growth and all blind shoots are kept pinched back. Blind shoots are those which arise near the base of the plant and fail to develop into flowering stems. These may be identified by the pointed bud on a very short stem.

The shoots that have been selected for later flower production are pinched by removing 1–2 inches of the terminal after at least 6 inches of growth has formed. These may be pinched again 4–6 weeks later. This practice prevents the formation of flowers and forces new growth from the base of the plant. During this building period, the plants are fertilized and watered heavily. The temperature is maintained at 60–62°F (15.5–16.6°C) during the night and high light intensity is provided during the daytime.

Timing Rose Production

The peak market demand for roses occurs at the major flower holidays and for weddings. Rose flower production is timed to provide more flowers at these sale periods than at other seasons. Roses normally produce more blooms during the summer months than at any other time of the year. The rose grower must time the crop to provide heavy production of flowers during the winter and spring months and reduce production during summer. This is done by selectively pinching the flowering stems to obtain flowers at the higher demand periods.

Flowers are produced from a soft pinched stem in 8 weeks for Christmas and Valentine's Day sales. An Easter crop is produced in 7 weeks following a pinch. Flower production is reduced to 5 weeks from a soft pinch for Mother's Day through summer. The stems pinched are those having pea-sized flower buds already formed. The pinch is made above the nearest node to the bud having a five-leaflet leaf (Figure 6–22). Stems are not pinched if they have a bud that will flower in less than a month or if no flower bud is visible. Stems that are to be flowered for a holiday crop are selected from plants growing on the outside rows of the benches, since environmental conditions are better in this location.

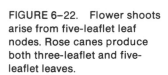
FIGURE 6–22.   Flower shoots arise from five-leaflet leaf nodes. Rose canes produce both three-leaflet and five-leaflet leaves.

Cutting Rose Flowers

Since rose plants will be in benches for 4 years or longer, the stems must be kept on a continuous flowering cycle. A rose flower stem should be cut at a location on the plant that will allow a new flower to form as fast as is possible. At the same time it is desired that the longest possible stem be left with the flower, because the value of the bloom is determined in part by the length of the stem.

Rose blooms are cut early in the morning and again in the afternoon so that the flowers will not open too far during the day. Roses cut during the late afternoon of a bright day will have an increased keeping quality. Blooms cut at this time have more carbohydrate food reserves for the flowers to use after they are cut. Roses are ready to be harvested when the sepals begin to curl downward or the first petal begins to loosen from the bud. Flowers that are cut too soon (in tight bud) may fail to open properly (Figure 6–23).

Rose stems are cut at a level that will leave at least one set of five-leaflet leaves on hybrid tea roses. A rose stem normally produces one or more three-leaflet leaves at nodes closest to the previously cut stem. Above these are found several nodes having five-leaflet leaves.

FIGURE 6–23. Rose flowers are harvested when the calyx begins to curl downward, revealing the petals of the bud.

When stems are cut above a three-leaflet leaf, it is more likely that a blind shoot will be produced. The blind shoot fails to form a flower. Cuts made above five-leaflet leaves are most likely to form flower stems.

During the fall and winter months, rose flower stems are cut at a level that will leave two five-leaflet leaves remaining on the stem. This is done to be certain that enough foilage remains on the plant for good flower production during this dark weather period. These cuts are made above the point where a previous stem was cut. This area is called a *hook*, because it includes a portion of the previous flower stem and the new lateral shoot. Stems cut during the spring and summer are cut increasingly lower on the stems. These cuts often include the hook so that the cut will be located at a five-leaflet leaf on a previously cut stem. This process of *undercutting* continues during the summer months as the plant height is gradually lowered. Reducing the height of the plant each year facilitates easier handling, care, and cutting of the rose plants.

Grading and Handling Cut Roses

Roses are harvested before the petals are open and taken directly to a cool room and placed in water. Standard roses are graded by color and stem-length increments of 3 inches, beginning with 12-inch stems. All hooks are removed from the stems as they are placed in bundles of 25 blooms of the same cultivar. Misshapen blooms and short or crooked stems are discarded. Roses may be spirally rolled or packed in bundles of 25 blooms in waxed tissue paper. Some markets also sell roses in packs of 13 blooms.

The graded roses are placed in warm water (preferably deionized) having a flower preservative and moved to a refrigerator as soon as possible. The cut flowers are held at 40°F (4.4°C) until they are shipped. Whenever possible the roses are stored by themselves in refrigerators. Roses held in a lighted refrigerator with cut carnations or orchids will open faster than they will when stored alone in the dark.

For information and sample schedules of some other potted plant crops, refer to *Appendices D–F.*

SELECTED REFERENCES

Ball, V. (ed.). 1975. *The Ball Red Book, 13th ed. George J. Ball, Inc., West Chicago, Ill.*

Ecke, Paul, Jr., and O. A. Matkin (ed.). 1976. *The Poinsettia Manual.* Paul Ecke Poinsettias, Encinitas, Calif.

Hartmann, H. T., and D. E. Kester. 1968. *Plant Propagation Principles and Practices,* 2nd ed. Prentice-Hall, Inc., Englewood Cliffs, N. J.

Kofranek, A. M., and R. A. Larson (ed.). 1975. *Growing Azaleas Commercially.* Division of Agricultural Sciences, University of California, Davis, Calif.

Langhans, R. W. (ed.). 1964. *Chrysanthemums: A Manual of the Culture, Diseases, Insects, and Economics of Chrysanthemums.* Department of Floriculture and Ornamental Horticulture, Cornell University, Ithaca, N.Y.

Langhans, R. W., and D. C. Kiplinger (eds.). 1967. *Easter Lilies: The Culture, Diseases, Insects, and Economics of Easter Lilies.* Department of Floriculture and Ornamental Horticulture, Cornell University, Ithaca, N.Y.

Laurie, A., D. C. Kiplinger, and K. S. Nelson. 1968. *Commercial Flower Forcing,* 7th ed. McGraw-Hill Book Company, New York.

Mahstead, J. P., and E. S. Haber. 1957. *Plant Propagation.* John Wiley & Sons, Inc., New York.

Mastalerz, J. W., and R. W. Langhans (eds.). 1969. *Roses: A Manual on the Culture, Management, Diseases, Insects, Economics and Breeding of Greenhouse Roses.* Department of Floriculture and Ornamental Horticulture, Cornell University, Ithaca, N. Y.

Nelson, K. S. 1967. *Flower and Plant Production in the Greenhouse,* 2nd ed. The Interstate Printers and Publishers, Inc., Danville, Ill.

Pirone, P. P., B. O. Dodge, and H. W. Rickett. 1960. *Disease and Pests of Ornamental Plants,* 3rd ed. The Ronald Press Company, New York.

TERMS TO KNOW

Bracts	Miniature Carnation	Standard Carnation
Continuous Cropping	Node	Terminal Shoot
Cultivar	Pinching	Thermoperiodic
Disbudding	Photoperiodic Response	Vascular Disease
Lateral Buds	Scion-grafted Bud	Vegetative Growth
Liner Production	Sports	Week Response

STUDY QUESTIONS

1. List the important cultural procedures in the production of the following crops: chrysanthemums, poinsettias, florists' azaleas, Easter lilies, carnations, and roses.
2. Describe how growth-regulating chemicals are used as tools in the production of florist crops.
3. List the major types of roses grown for cut flowers and describe each type.

SUGGESTED ACTIVITIES

1. Grow a few specialized crops in the school greenhouse.
2. Visit commercial greenhouses in your area to study specialized crop production.
3. Invite a greenhouse manager to the class to discuss the production of important commercial floriculture crops.

CHAPTER SEVEN

The Foliage Plant Business

The current popularity of tropical plants and flowering house plants has placed a tremendous demand on the floriculture industry to provide new plants and products to the buying public (Figure 7–1). With this increase in interest by customers in house plants has come a greater need for further knowledge of the plant needs and their uses in the home. The employees of a plant store or flower shop will be involved with the daily tasks of maintaining the plants in the shop, answering questions concerning the plant uses and care, and in making salable plantings for the shop.

An employee in both the foliage plant shop and a flower shop will benefit from training received in interior design classes. The use of plants in interior designs has become a primary requirement in modern home decorating. The plant store employee can be helpful to the customer by recommending plants that will fit each decorating requirement. This chapter will discuss the use of house plants for specific purposes. These include the construction, maintenance, and uses for the more common commercial houseplant plantings sold in retail plant stores. These are terrariums, planters and dish gardens, and hanging baskets. The plant store personnel will be required to be capable of designing, maintaining plantings, and aiding customers with problems that may arise with each of these houseplant specialty products.

FIGURE 7-1. The 'Red Princess' Philodendron planted in a decorative plant stand.

TERRARIUM CONSTRUCTION

Terrariums have enjoyed great popularity in the houseplant trade because they occupy small spaces in a home or office and require a minimum of maintenance. When properly planted with suitable species, these miniature gardens may remain healthy and beautiful for several years. Customers who normally avoid purchasing foliage plants are attracted to these unique miniature greenhouse plantings for adding a bit of nature to their rooms.

The greater population movement to apartments, condominiums, or even school dormatories has deprived many Americans of the opportunity to enjoy the larger potted houseplants. The terrarium is a natural answer to this dilemma, for they need not occupy much needed space in a room. Yet the growing plants, which maintain their own needs for water, change the personality of an interior design.

What Is a Terrarium?

The use of the first terrarium is credited to Nathaniel Ward, an English surgeon and amateur gardener, over 150 years ago. His hobby included the collection of wild plants from the wooded areas around his home. However, he found that some plants, such as the bog ferns, could not survive in the climate of his garden or home. He accidentaly found that these same ferns would germinate and grow beautifully when the same soil from the wooded area was placed in a bottle or glass case.

This discovery led Dr. Ward to experiment with other plant species and different styles of containers in his search for more ideal growth of plants in his home. His experiments showed that when plants were given proper light, humidity or moisture, and air temperatures, they might be maintained in a sealed container for many years without additional water or air from the outside environment.

These glass-enclosed containers later became known as "Wardian cases." They were so successful a century ago that for the first time plants could be transported over long distances (between continents) without damage to the growing plants. Exotic plants began to be exchanged between botanical gardens, and new industries were created in distant countries and in new continents. Plants did not survive well in Victorian homes heated by only a single fireplace. The Wardian case contained its own microclimate created by air heated by light and plant respiration, so were less affected by the chilly room air.

Once the value of the Wardian case caught on as a decorative item in Victorian homes, they became highly decorated and ornately designed pieces of furniture for growing plants. These beautiful gardens persisted during the early portion of this century. The popularity of these container gardens was slight until the late 1940s. Renewed interest came when the name *terrarium* was applied to the bottle gardens (*terra*, meaning "earth", and *arium*, meaning "of the home").

How a Terrarium Works

A terrarium regulates the plant processes within the enclosed space, provided that light is available (Figure 7–2). The bottle garden can be likened to a sealed greenhouse, where light enters to provide heat and energy for the manufacture of food by the plants. Carbon dioxide from the air around the plants and water from the soil are also ingredients used to produce plant sugars. The nutrients from the soil, along with the plant sugars, allow the plants to grow and maintain cellular functions.

Respiration, or the "burning" of the plant sugars, occurs both in the dark and in the light. During the process of respiration, oxygen is

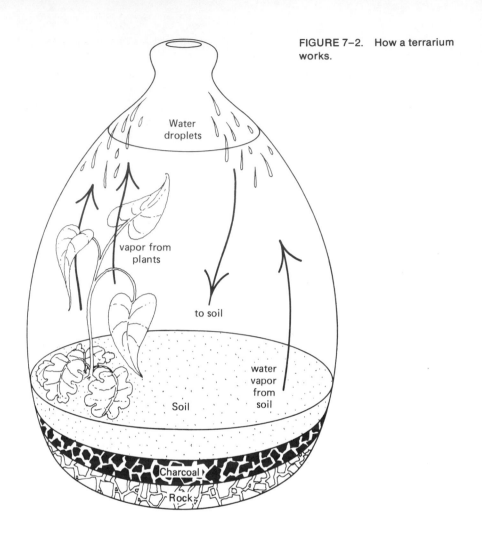

FIGURE 7–2. How a terrarium works.

Water droplets

vapor from plants

to soil

water vapor from soil

Soil

Charcoal

Rock

burned while carbon dioxide is given off by the plants. Water escapes from the leaves, forming a thin cloud of moisture within the terrarium. When the moisture content of the air becomes saturated, the water droplets that form fall to the soil. In this manner the oxygen, carbon dioxide, and water are constantly recycled within the confines of the terrarium. The terrarium provides the ideal low-maintenance garden for the amateur who is either too busy or lacks the skill to maintain houseplants. The properly constructed terrarium efficiently uses and reuses each of these growth ingredients to fulfill the needs of the plants. When moisture forms along the sides of the container, it runs down the sides until it reaches the soil. This cycle provides a constant moisture balance in the terrarium, which eliminates the need for watering the plants for long periods of time.

PLANNING THE TERRARIUM

Increased interest in the use of terrariums for home and office decorating has spawned the manufacture of many styles of containers and the introduction of new, more adaptable plants for the miniaturized gardens suitable for planting in terrariums. Before a terrarium is planted, the proper equipment must be on hand, a suitable container should be selected, plants should be selected for suitability in the planting and container, and a design for the terrarium planting must be determined.

Selecting a Container for the Terrarium Planting

A suitable container for use as a terrarium is one that (1) will admit light through either glass or transparent plastic, (2) will hold a suitable amount of soil to sustain the growth of the plants, (3) is large enough to allow some growth of the plants in the soil, (4) provides an adequate air volume when planted, and (5) may be covered easily to hold the moisture within the container.

Any container that fulfills these requirements can be made into a suitable terrarium. Department stores sell sophisticated and elaborate terrariums in the shapes of glass-covered coffee tables, free-standing plastic globes, as well as many other unusually designed containers. More common containers used for terrariums are large bottles, carboys, glass canisters, medicine jars, candy jars, aquariums, and display domes (Figure 7–3).

Theoretically, a true terrarium is a *sealed* container which is opened to the outside air only when maintenance of the planting is required. More recently, plant stores and flower shops have introduced the use of *open* terrariums. The two types differ primarily in the amount of light the terrarium planting can tolerate.

Closed terrariums cannot tolerate strong sunlight near a window. These are best placed in areas lighted only by diffused light or under fluorescent fixtures. Even large bottles having very narrow openings at the top will function as a closed terrarium. When placed in direct sunlight, the glass absorbs heat and collects it, resulting in temperatures inside the terrarium that far exceed the tolerance of the plants. The water vapor is not free to exchange with the outside environment to aid in cooling the plants.

An open terrarium will have an opening equal to nearly half the container's dimensions. These containers may be placed in direct sunlight because there will be a free air exchange between the atmosphere of the terrarium and the outside air. A brandy snifter or a glass globe (with the top removed) are examples of open terrarium containers. These differ from simple dish garden containers because the walls of these containers aid in retaining the moisture around the plant foliage. Even though these containers may be suitably placed in direct sunlight, a careful selection of high-light-tolerant plants is necessary when planting a terrarium for this purpose.

FIGURE 7–3. Some typical containers used for terrarium plantings.

Selecting Plants for a Terrarium

The plant department of any plant store should be well stocked with a selection of plants suitable for placing in terrariums. Plants to be used in a terrarium planting should be grouped according to their specific growth requirements for the type of terrarium container into which they will be placed. Plants that will thrive in most closed terrariums will prefer high humidity, moist soil, a constant temperature, and tolerate lower light. Another desirable characteristic is a slow rate of growth. Such plants will require little maintenance once they become established in the terrarium. A list of suitable plants is given in Table 7–1.

Each planting design should be carefully planned ahead of the actual selection of plants to be placed in the container. A variety of sizes, types, and colors of foliages will be desired to make each terrarium unique and more salable. A typical planting will consist of some tall plants, low-growing plants, and some plants that will provide a ground cover. Generally, some form of accent, whether a plant or simulated pool, will be desired. Ornate ceramic animals, miniature pagodas, and other unnatural items should be used with care in well-designed commercial terrariums. A more naturalistic miniature landscape is created when these artificial objects are completely left out of the planting design. A theme should be developed by selection of plants or the topography of the soil into a woodlands or tropical scene.

TABLE 7-1: *Selected List of Plants Suitable for Use in a Terrarium*

SCIENTIFIC NAME	COMMON NAME
Acoris gramineus 'pusillus'	Miniature Sweet Flag
Acoris gramineus 'Variegatus'	Variegated Sweet Flag
Achimenes sp.	Miniature forms of Achimenes
Actiniopteris australis	Australian Table Fern
Adiantum bellum	Bermuda Maidenhair Fern
Adiantum diaphanum	Maidenhair Fern
Adiantum reniforme	Maidenhair Fern
Aechmea miniata 'discolor'	Purple Coral Berry
Aechmea tillandsioides	Epiphytic Aechmea
Aeschynanthus marmoratus	Zebra Basket Vine
Aglaonema costatum	Spotted Chinese Evergreen
Allophyton sp.	Miniature Foxglove
Alternanthera bettzickiana	Dwarf Joseph's Coat
Anoectochilus roxburghii	Dwarf Jewel Orchid
Aphelandra squarrosa	Zebra Plant
Araucaria excelsa (seedlings)	Norfolk Island Pine
Ascocentrum miniatum	Miniature Orchid
Asplenium bulbiferum	Mother Fern
Asplenium nidus	Bird's Nest Fern
Begonia bowerae	Miniature Eyelash Begonia
Begonia imperialis	Carpet Begonia
Buxus microphylla (seedlings)	Boxwood
Calathea makoyana	Peacock Plant
Calathea micans	Miniature Maranta
Callisia elegans	Striped Inch Plant
Callopsis volkensis	Miniature Calla
Ceropegia woodii	Rosary Vine
Chamaedorea elegans 'Bella' (seedlings)	Neanthe Bella Palm
Chimaphila maculata	Pipsissewa, Princess Pine
Cissus antarctica 'Minima'	Miniature Kangaroo Vine
Columnea microphylla 'Minima'	Dwarf Goldfish Vine
Cordyline terminalis 'Minima'	Baby Ti Plant
Cryptanthus acaulis 'Ruber'	Miniature Red Earth Star
Cyanotis somaliensis	Pussy Ears
Davallia bullata 'Mariesii'	Ball Fern
Dizygotheca elegantissima (seedlings)	False Aralia

Dracaena godseffiana	Gold Dust Dracaena
Episcia dianthiglora	Flame Violet
Euonymus japonica 'Microphylla' (seedlings)	Box-leaf Euonymus
Exacum affine	Exacum
Festuca ovina 'Glauca'	Blue Fescue
Ficus pumila	Creeping Fig
Ficus pumila 'Minima'	Miniature Creeping Fig
Fittonia verschaffeltii	Nerve Plant Fittonia
Gesneria cuneifolia	Fire Cracker Plant
Goodyera repens	Rattlesnake Plantain
Helxine soleirolii	Baby's Tears
Hoya bella	Miniature Wax Plant
Hypoestes sanguinolenta	Polkadot Plant
Lagerstroemia indica 'Petite'	Dwarf Crape Myrtle
Maranta leuconeura 'Erythroneura'	Nerve Plant Maranta
Maranta leuconeura 'Kerchoviana'	Prayer Plant
Mitchella repens	Partridge Berry
Myrtus communis 'Microphylla'	Dwarf Myrtle
Nephrolepis exaltata 'Bostoniensis' Norwoodii	Norwood Fern
Nephrolepis exaltata 'Bostoniensis' Whitmanii	Feather Fern
Nertera granadensis	Coral Beads
Oxalis martiana 'aureo-reticulata'	Fire Fern
Paphiopedilum insigne	Ladyslipper Orchid
Pellaea rotundifolia	Button Fern
Pellionia daveauana	Watermelon Begonia
Peperomia caperata	Emerald Ripple
Peperomia sandersii	Watermelon Peperomia
Phalaenopsis amabilis	Miniature Moth Orchid
Philodendron micans	Velvet-leaf Philodendron
Philodendron sodiroi	Silver-leaf Philodendron
Pilea cadierei	Aluminum Plant
Pilea involucrata	Friendship Plant
Pilea microphylla	Artillery Plant
Pilea × 'Silver Tree'	Silver Tree
Platycerium bifurcatum	Staghorn Fern
Polypodium lycopodioides	Polypody Fern
Polypodium pisellodes	Polypody Fern
Polystichum tsus-simense	Holly Fern
Pteris cretica 'Albo-lineata'	Spider Fern
Pteris crética 'Wilsoni'	Fan Table Fern
Pteris ensiformis 'Victoriae'	Victory Fern
Saxifraga sarmentosa	Strawberry Begonia
Saxifraga sarmentosa 'Tricolor'	Magic Carpet
Selagenella emmeliana	Sweat Plant
Selagenella kraussiana	Trailing Irish Moss
Selagenella kraussiana 'Brownii'	Cushion Moss
Senecio jacobsenii	Weeping Notonia
Sinningia pusilla	Miniature Slipper Plant
Tillandsia fasciculata	Wild Pine
Tillandsia usneoides	Spanish Moss
Tolmiea menziesii	Piggyback Plant
Tradescantia fluminensis 'Minima'	Miniature Wandering Jew

Simple, miniature landscapes may be developed through the use of plants having different heights, shapes, foliage colors, and textures. When selecting the plants for the terrarium, have their basic positions planned before planting begins. Select plants that will be compatible with each other in growth rate, soil moisture, temperature, and light requirements.

Plants to be used in a terrarium should be free of existing diseases and insect pests. Employees should maintain a constant preventative pest control program with plants to be used for this purpose. Plants that show symptoms of any pest problem should be immediately isolated to a quarantine area away from the healthy plants. A foliar fungicide combined with a broad-spectrum insecticide is routinely sprayed on the foliage a day or more before planting the terrarium. As an added precaution, a soil fungicide is added to the first watering following planting to aid in the prevention of root-rotting infections until the plants become established in their new soil conditions.

In addition to the plants, other natural items might be collected for adding interest to these terrariums. Small lichen or moss-covered twigs or rocks may be desirable. Under the high-humidity conditions found in the terrarium environment, these small plants will soften and again grow. Shredded bark, very small leaves, and even snail shells will aid in simulating the naturalness of a forest floor.

Added interest may be given to the planned design of a terrarium when a small pool of water is added. This pool may be simulated by use of a mirror covered at the edges with rocks to give the impression of water. A plastic film covered with small rocks will also serve this purpose. The addition of a pool filled with a small quantity of water is even more attractive and is functional as well. An inverted jar lid or shallow saucer buried in the soil with small rocks used to camouflage the edges will provide a suitable container. The plastic film and small pebbles will also form a watertight basin for the pool. When additional water is required for maintenance of the terrarium, it may be added to the pool. By adding water in this manner, it is difficult to add too much, and the plantings will not be disturbed. When planning the designs for each terrarium, use your imagination to make each creation unique and attractive.

PLANTING THE TERRARIUM

The type and style of container selected for the terrarium will dictate to some degree the types of tools required for their planting. A commercial plant store or flower shop employee will be required to construct terrarium plantings in both open and closed containers, as previously described. The open containers are quite simple to plant because the designer can reach inside to manipulate the plants and soil. Closed containers are much more difficult for planting because specially constructed tools must be used instead of hands for doing the actual planting procedures.

Adding the Soil to the Terrarium

The growing medium in the bottom of the terrarium container will consist of three layers: drainage material, charcoal, and a potting soil mixture. The proper construction of the terrarium depends to a great extent on these elements (Figure 7–4).

Drainage material. Drainage material is added first to the bottom of the container. The purpose of this layer is to allow excess water to drain from the potting mixture and remain below the plant roots. The water is not allowed to stand at the root zone, which would promote root rotting and decay. Air is also allowed to penetrate the soil mix more readily for more ideal root growth. The drainage material may consist of sterilized pebbles, crushed rock, broken pieces of clay pots, or horticultural perlite. Before the drainage material is added to the container, a thin layer of sphagnum sheet moss may be lined around the sides of the container. This material will conceal the soil zones and make the terrarium more attractive.

Charcoal. Charcoal is thinly layered over the smoothed drainage material. The purpose of the charcoal is to absorb the noxious by-products from decaying plant material. Charcoal is highly absorbant and, when combined with the drainage material, will aid in the removal of the excess soluble fertilizer salts which accumulate at the bottom of the container. The charcoal material most used is aquarium filter charcoal that has been rinsed thoroughly to remove the fine dust found in this product.

Soil mixes for the terrarium. The commercial plant store will most likely maintain a supply of various types of potting soils for sale to customers. A suitable potting soil for use in terrariums will contain a high percentage of organic materials, such as leaf mold or humus. These soil mixtures should also contain high amounts of sand to ensure rapid drainage of excess water away from the root zone. The sterilized commercial potting soils have the advantage that they are already prepared for immediate uses. These should also contain a proper balance of fertilizers at low concentrations for maintaining plant growth.

The employee who designs terrariums should avoid the use of

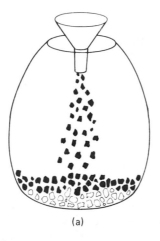

(a)

Step 1: Add drainage material
and charcoal to a clean container.

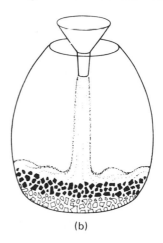

(b)

Step 2: Add soil
(not to exceed 20-25%
container volume) and
form landscape.

FIGURE 7-4.   Making a
terrarium.

(c)

Step 3: Place plants, rocks,
and accessories.

(d)

Step 4: Clean bottle and water lightly.

finely structured dark muck soils. These must be amended by the addition of liberal amounts of sand and peat before they can be used. A
suitable potting mixture will have small twigs and portions of leaves
visible and will feel gritty from the sand when rubbed between the
fingers.

Sand-sculptured large-mouthed terrariums require special skill in
preparation. The colored sands must be added to the sides of the

glass in layers as the soil medium is placed in the center. The addition of the ingredients is done in stages to include the drainage layer, charcoal layer, and soil mix as the colored sand is carefully shaped into designs around the perimeter of the container. Once the plants have been placed in the container, a layer of sand or pebbles is then added to the top of the soil to conceal the construction materials.

Preparing the Terrarium

Each component of the terrarium should be added to the container in the appropriate sequence until the soil has reached the proper level. Each component is leveled and firmed slightly before the next is added. The final soil volume, including drainage material and charcoal, should not exceed 20–25 percent of the volume of the container. When a narrow-mouthed container is used for the terrarium, a newspaper formed into a funnel will allow the materials to be poured through the opening and evenly distributed.

Special tools may be purchased or constructed for the purpose of planting narrow-necked bottle terrariums. Quarter-inch dowel rods may be prepared with various materials attached to the ends to form tools that will fit through the narrow necks of the bottles. A cork attached to one end will provide a suitable soil tamping device for firming the soil around the plant roots. A small measuring spoon attached to the dowel rod will allow the scooping of the soil to form variations in the topography of the miniature landscape. A spring-loaded pickup tool may be used to grasp the plants or small rocks as they are positioned in the container. Specially designed long-handled scissors can be purchased for the purpose of trimming plant foliage or a single-edged razor blade fastened to the end of a dowel rod will perform the same function. The pickup tool may be used to retrieve the unwanted debris from the container. A loop formed from a coathanger may be substituted for the pickup tool for handling plants as they are positioned in the container.

Often the soil ball on a plant will be too large to fit through the opening of a narrow-necked bottle. It may be easily inserted into the container, however, by reshaping the roots to fit the contour of the opening. The plant is removed from its pot and excess loose soil gently shaken from the roots. The remaining soil can then be shaped into a narrow cylinder by rolling it between a layer of newspaper or cloth. The plant should then easily fit through the top of the container.

The arrangement of plants in the design is best begun by placing the focal plant first. This will be the tallest or the most colorful plant in the design. The other plants are then placed around the focal area, being careful to provide adequate space for future growth, so the plantings should not be crowded. A small ground cover of baby tears, selaginella, or miniature creeping vine is then added to create a carpet in the terrarium.

Adding Water to the Terrarium

The terrarium will require cleaning once the planting operation is completed. This may be done most effectively using a squeeze bottle having a narrow orifice so that a small volume of water may be directed around the inside of the container to remove excess dust and dirt. A small sponge or cloth swab may be required for scrubbing stubborn dirt. A minimum quantity of water should be used during the cleaning process.

Water is added to the terrarium for two reasons. The water must saturate the soil to provide sufficient moisture to the plant roots for growth. The added water also initiates the rain cycle that maintains the terrarium environment. Whenever irrigation is applied to the terrarium, either deionized or rainwater is more suitable than tap water. The potting soil is capable of holding up to 40 times its weight in water, so the designer must be extremely careful when adding water to the terrarium for the first time. The drainage in the container is limited and any excess moisture is difficult to remove. It is much more desirable to add the water in small portions until a balance is reached rather than to overwater the first time. Overwatering of the terrarium will most surely cause the roots to rot, requiring complete renovation of the terrarium planting.

Water may be applied to a terrarium at the location of the small pool, if one has been included. The first watering is more difficult, because splashing should be avoided to prevent disturbance of the planting and soiling of the glass. Following the initial addition of water, the top of the closed container should remain off for a few days to allow excess water to dissipate. This will allow the terrarium atmosphere to reach the proper balance. After being closed, the top should be removed or opened slightly if moisture formation on the glass is observed and does not dissipate each morning. Successive watering of a terrarium should only be done when the plants appear to wilt slightly and moisture on the glass is no longer observed. Usually, watering is required only after several months for a closed terrarium. These maintenance waterings may be done with a small plant mister or a long-necked watering can. The amount of water required will be much less than for the first watering.

Locating the Terrarium

A terrarium is a miniature greenhouse, so it must be treated in the same manner. A terrarium must never be placed in a location where it will receive direct sunlight and raise the temperature to a level that may kill the plants. Open-topped containers are not as seriously affected by sunlight, since air is free to exchange between the container and outside environment.

Most terrariums will be located in areas away from a bright window. Plants selected for use in a single terrarium should tolerate the same amount of light. Some species may tolerate very low light levels in corners across the room from a bright window. Others will grow best under higher light regimes (approximately 6–8 feet from a bright window).

Lighted plant stands are excellent light sources for both potted plants and terrariums. Miniature flowering plants that will tolerate the high humidity of a terrarium require approximately 400 foot-candles of light for 14 hours each day, if flowering is desired. This is the amount of light that causes a slight shadow over the leaves when a hand is placed between the plant and the light source. Fluorescent light fixtures containing one warm white and one cool white tube provide adequate light for terrariums when placed 15–18 inches above them. Incandescent light sources should be avoided, since they provide too much heat and will raise the temperature of the air within the terrarium.

The temperature of the average home is quite satisfactory for most terrarium plantings. The average home remains between 65 and 75°F (18.5–24°C), which is the ideal range for these tropical plants. In addition, terrarium plantings are not affected by the lower house humidities during the heating and air-conditioning seasons, since they maintain their own constant environment. Temperature changes are much slower within the closed terrarium because a separate microclimate is established.

Terrarium plantings will eventually require some maintenance once the plants have become established. Plants may become too large for the container. These should be trimmed to bring them back into scale with the container. Fertilizer is added to the terrarium to provide the nutrients that are vital for good plant growth. However, fertilizer is added at only one-tenth strength (commercial plant food) every 3–4 months. The plants should not be stimulated to grow rapidly, merely to maintain their original beauty and stature.

When diseases or insects are observed in a terrarium planting, a pesticide treatment may be required for proper control. An insecticide may be placed in a small bottle cap on the soil surface. This material will evaporate into the terrarium atmosphere and kill the insects. In severe cases, the foliage may have to be sprayed with a contact insecticide to provide complete control of the pest. A soil fungicide may be applied by mixing it with the irrigation water. When this is done, it is important that the material does not splash onto the walls of the container or an unsightly residue will result.

A terrarium will require an occasional exchanging of air between the container and the outside environment. Small, closed terrariums utilize the carbon dioxide and oxygen content of the air quickly in the growth processes of the plants. The lid of the container should be opened slightly to allow the air exchange once every month for proper growth of the plants. Large terrarium containers provide a greater volume for air and humidity, so will require opening much less frequently. When the correct plants are selected, watering is properly applied following planting, and the container is matched in size to the planting, the terrarium may not require opening for maintenance for several years.

Terrariums for Other Purposes

The high-humidity atmosphere created by a terrarium container is useful for many purposes in plant propagation and rejuvenation. When a container is covered with a sheet of glass or plastic it becomes an excellent germination chamber for seeds. A simple plastic freezer bag placed around a pot filled with peat moss or soil mix provides an ideal environment for rooting cuttings from plants that have become too large. An unused aquarium may be converted into a small greenhouse by simply adding a transparent cover. This will provide adequate space for transplanting and other work which may be required later.

Smaller plants may be placed in a closed aquarium with a small amount of water to keep the plants fresh while on vacation. When plants are infested with insects, they may be treated with an insecticide and placed in the enclosed terrarium. The pesticide will treat the foliage more effectively in the high humidity created by the terrarium.

PLANTERS AND DISH GARDENS

A typical planter or dish garden found on the commercial market is all too often a haphazard arrangement of unrelated plant types having no artistic design. These planters are created by placing plants together so closely that they have no room for further growth, only to appear "full" at the point of sale. These planters also lack sufficient soil to support the growth of the plants. Little, if any, drainage space is provided in these containers. The result of these poorly planned and constructed planters is a distraught customer who may never again choose to purchase plants from the shop.

Planters and dish gardens that contain tropical plants, cacti, or succulent plants are favored for gift items and for decorating dwellings. These miniature gardens may be moved from one location to another or be left in a window sill to create a more attractive window treatment. Planters may grace a bookcase, coffee table, fireplace mantel, or a dressing table, because the compact containers occupy only a small space. Cut flower blooms are sometimes added to the finished planter to add allure when used for a gift.

Planning the Dish Garden

The dish garden or tropical planter differs drastically from the terrarium, since no controlled "microclimate" is present. The miniature garden is placed in a shallow dish or container which allows the upward growth of the plant to any height, restricted only by the root growth. A well-designed dish garden will convey a central theme, be it a desert scene or an alpine garden. The selection of a suitable container, a composition of compatible plants chosen from Table 7–2 or 7–3, and a blending of design elements are required for a well-constructed planter.

TABLE 7–2: *Selected Plant List for a Tropical Dish Garden*

SCIENTIFIC NAME	COMMON NAME
Acorus gramineus 'Pusillus'	Miniature Sweet Flag
Acorus gramineus 'Variegatus'	Variegated Sweet Flag
Adiantum capillus-veneris	Maidenhair Fern
Aglaonema commutatum	Variegated Chinese Evergreen
Aglaonema simplex	Chinese Evergreen
Alternanthera bettzickiana	Dwarf Joseph's Coat
Anthurium scherzerianum	Flamingo Flower
Aphelandra squarrosa	Zebra Plant
Asparagus densiflorus 'Meyeri'	Foxtail Fern
Asparagus densiflorus 'Sprengeri'	Asparagus Fern
Asparagus plumosis	Plumosis Fern
Asparagus plumosis 'Nanus'	Dwarf Plumosis Fern
Aspidistra elatior	Cast Iron Plant
Asplenium bulbiferum	Mother Fern
Asplenium nidus	Bird's Nest Fern
Begonia species	Fancy-leaf Begonias

Buxus microphylla 'Japonica'	Japanese Boxwood
Buxus microphylla 'Koreana'	Korean Boxwood
Calathea makoyana	Peacock Plant
Callisia elegans	Striped Inch Plant
Camaedorea elegans 'Bella'	Neanthe Bella Palm
Cissus antarctica	Kangaroo Vine
Cissus rhombifolia	Grape Ivy
Codiaeum variegatum	Croton
Cyperus alternifolius 'Nanus'	Dwarf Umbrella Plant
Dizygotheca elegantissima	False Aralia
Dracaena godseffiana	Gold Dust Plant
Dracaena sanderiana	Ribbon Plant
Episcia cupreata	Flame Violet
Euonymus japonica 'Microphylla'	Box-leaf Euonymus
Fatsia japonica	Japanese Aralia
Ficus pumila	Creeping Fig
Fittonia verschaffeltii	Nerve Plant Fittonia
Grevillea robusta	Silk Oak
Gynura aurantiaca	Velvet Plant
Hedera helix	English Ivy
Hoya carnosa	Wax Plant
Hypoestes sanguinolenta	Polkadot Plant
Iresine herbstii	Bloodleaf
Maranta leuconeura 'Kerchoveana'	Prayer Plant
Maranta leuconeura 'Erythroneura'	Nerve Plant Maranta
Nephrolepis exaltata 'Bostoniensis'	Boston Fern (many variations)
Pachysandra terminalis	Japanese Spurge
Pandanus veitchii	Screw Pine
Pedilanthus tithymaloides cucullatus	Devil's Backbone
Peperomia caperata	Emerald Ripple
Peperomia sandersii	Watermelon Peperomia
Philodendron oxycardium	Heartleaf Philodendron
Pilea cadierei	Aluminum Plant
Pilea involucrata	Friendship Plant
Pilea microphylla	Artillery Plant
Pilea × 'Silver Tree'	Silver Tree
Plectranthus australis	Swedish Ivy
Podocarpus macrophyllus	Podocarpus
Polystichum tsus-simense	Holly Fern
Pteris cretica 'Albo-lineata'	Spider Fern
Pteris cretica 'Wilsoni'	Fan Table Fern
Pteris ensiformis 'Victoria'	Victory Fern
Rhaphidophora aureus (Scindapsus)	Devil's Ivy
Sansevieria trifasciata 'Hahnii'	Bird's Nest Sansevieria
Sansevieria trifasciata 'Laurentii'	Striped Sansevieria
Sansevieria zeylanica	Snake Plant
Saxifraga sarmentosa	Strawberry Geranium
Syngonium podophyllum	Arrowhead Plant
Syngonium podophyllum 'Tri-leaf Wonder'	Nephthytus
Tolmiea menziesii	Piggy-back Plant
Tradescantia fluminensis	Wandering Jew
Tradescantia fluminensis 'Minima'	Miniature Wandering Jew
Zebrina pendula	Purple Wandering Jew

TABLE 7–3: *Selected List of Plants Suitable for a Desert Dish Garden*

SCIENTIFIC NAME	COMMON NAME
Aloe vera	Medicine Aloe
Astrophytum myriostigma	Bishop's Cap
Crassula argentea	Jade Plant
Echeveria derenbergii	Painted Lady
Echeveria elegans	Mexican Snowball
Echinocactus grusonii	Golden Barrel Cactus
Echinocereus dasycanthus	Rainbow Cactus
Echinopsis multiplex	Easter Lily Cactus
Euphorbia lactea	Candleabra Cactus
Gasteria × hybrida	Ox Tongue
Gymnocalycium mihanovichii 'Friedrichii'	Rose Plaid Cactus
Hawarthia cuspidata	Star Window Plant
Kalanchoe tomentosa	Panda Plant
Lithops bella	Stone Face Living Stones
Mammillaria elongata	Golden Stars
Mammillaria magnimamma	Mexican Pincushion
Notocactus haselbergii	Ball Cactus
Opuntia microdasys	Bunny Ears
Sedum morganianum	Burro Tail
Sempervivum tectorum	Hen-and-Chickens
Senecio serpens	Blue Chalk Sticks
Stapelia hirsuta	Hairy Toad Plant

Selecting the Correct Container

The plants and container should complement each other. Containers used for most planters are manufactured from glass, ceramic, plastic, and fired clay. Shapes of these planting containers vary from simple bonsai dishes to ornate ceramic and metal planters. The container should provide a depth of at least 3 inches (7.6 centimeters) to allow suitable root penetration while permitting adequate drainage of the soil. A more suitable container will be of a natural color so as not to contrast strongly with the planting. Containers having a black, white, brown, green, or gray color are most desirable. Planters that are deeper than 3 inches may provide adequate drainage without the need for drainage holes at the base. Containers less than 3 inches in depth must have drainage holes. Containers sold for the purpose of creating bonsai plantings are excellent for tropical planters, since drainage holes are provided and they are available in an assortment of shapes and sizes.

Planting the Dish Garden

A dish garden is prepared in a similar manner as a terrarium, especially when a drainage hole is not provided in the base of the container (Figure 7–5). Containers having drainage holes are unsuitable for many locations in the home. These containers require the use of a saucer or tray placed under them to prevent water damage to tabletops. Since the free runoff of this excess water is so desirable, a container having a built-in saucer beneath the container may be most practical.

(a)   Charcoal, drainage material, and soil are first added to the base of the container.

(b)   Plants are located within the container and soil added until the roots are covered and firmed.

(c)   The foliage is cleaned lightly with a leaf polish and the planter watered lightly.

FIGURE 7–5.   Planting the dish garden.

Containers that do not provide drainage holes must have drainage space added during their preparation. A layer of drainage material must be added to a depth of about 1 inch to the bottom of the container. Crushed rock, perlite, or broken clay pots may be used for this purpose. A thin layer of washed aquarium charcoal is layered over this drainage material. The charcoal is used to "sweeten" the drainage water by pulling out excess fertilizer salts and to neutralize rotting materials. The soil layer may then be placed over the charcoal to a depth suitable for anchoring the plants.

Planters having drainage holes are capable of removing the excess water rapidly through the base of the container. The holes are often large enough, however, to allow washing of the soil through them when the container is watered. A small broken clay pot chip or bottle cap may be placed over the holes to allow the soil to remain in the container. Generally, a thin layer of horticultural perlite is placed on the bottom of the planter to aid in the collection of excess water.

The soil mix recommended for a tropical dish garden should be a commercially prepared, sterilized mixture. These generally contain peat, humus, and vermiculite. They should have a high organic content, yet provide adequate drainage and aeration for the roots. It is often desirable to add a small amount of sterilized sharp builder's sand to these packaged soil mixes to increase their drainage capabilities. Slow-release fertilizer capsules, found in many plant stores, may be mixed with the potting soil during its preparation. The soil depth in the planter should be at least 2 inches, enough to support the plant roots while growing. It may be necessary to add a slightly raised mound to the center of the planter to accommodate large plants. Adequate space must be provided at the container's edges, however, to allow water to be absorbed into the soil.

Bromeliad Planters

Soil mixes for epiphytic bromeliads may best be planted in media that contain no soil at all. These plants receive their water and nutrients through the "vase" formed by the foliage. The plants produce anchorage roots which have a limited capacity for water absorption. Bromeliads are placed in a mixture of sphagnum peat and sand. The dry peat is thoroughly soaked in water, wrung dry, and then mixed with an equal volume of clean sand. The peat must be kept slightly damp to prevent it from drying out to a point where rewetting is difficult. Bromeliads may also be wired to a piece of osmunda fiber (from palm stalks) for anchorage. Watering is done through the funnel-shaped leaves. The organic fertilizers, such as fish emulsion, may be mixed with the irrigation water occasionally to feed the plants.

Desert Gardens

Planters constructed with arid-tolerant plants, such as cacti and desert succulents, are grown in a fast-draining "dry" soil mix. Sand and peat moss are also combined to provide a suitable planting medium. The peat is again dampened to make it easier to mix. The mixture will consist of three parts of sand and one part of the peat moss. Cacti are particularly intolerant of wet soils and will rot easily if the soil does not drain rapidly. The small amount of peat in the planting medium provides some nutrients and an ample amount of water retention for growth of cacti.

Selection of Plants

When a plant shop employee is selecting plants for the various containers to be prepared, it is important that they be grouped according to moisture and light requirements. Some plants which are more suitable for terrariums because they require high humidity should be avoided in flat dish gardens. Plants having extremely fast growth rates or those which will quickly outgrow the confines of a restricted root system are best used for larger specimen potted plants. Planters that are to be sold by a commercial plant store must be well designed and reflect the quality of the other products sold in the shop. It is imperative that the plants be free of pests and disorders and that they have been selected for growth in a dish garden. Cacti and succulent plants should only be used in arid dish gardens, and tropical plants are also used together. First select the plant that will form the focal point. Add companion plants which will complement the main plant in the container. A special rock or piece of driftwood may be selected as the focal point in the container. The other plants or accessories must then be subordinated in color, height, and foliar texture. The planning of the dish garden should also include adequate space for future growth of the plants. Use of fully grown specimens in a dish garden will require their replacement much sooner than is desirable.

Succulent plants thrive best in a soil medium consisting of equal parts of sand, peat or leaf mold, and loam soil. Both cacti and succulent plantings should not be watered for several days following the construction of the dish garden. When the containers are watered, the medium should be thoroughly soaked. Excess water may need to be carefully run from the sides of the container if too much is added. It is best to allow the soil mix to dry completely and then add only small amounts of water, to prevent the waterlogging of the roots.

Cacti and succulent plants grow best when placed in full sun. Under high light they develop striking color patterns. This is in sharp contrast to the tropical foliage plants, which generally grow better in filtered light. During the low-light and cooler-windowsill temperatures of winter, cactus and succulent dish gardens should be allowed to dry more thoroughly. No fertilizer should be used in late fall and winter, as the plants become dormant during this period. Any attempt to force the plants during this time with high temperatures and water may cause them to rot in the soil mixture.

HANGING BASKETS

The popularity of plants suspended at eye level in decorative containers has stimulated an entirely new phase of the foliage and houseplant business. New species of tropical and flowering plants which are suitable for use in hanging baskets have been introduced to the retail market, making the list of plants available filled with unusual and beautiful specimens (Figure 7–6, Table 7–4). These hanging basket plantings add a new dimension to indoor and outdoor living areas. Plants suspended from the ceiling, patio, trees, or porch create a means for adding plants in otherwise undecorated areas. Plants

FIGURE 7–6. Hanging basket planters have become one of the most popular items purchased at plant stores.

may be viewed from all sides at eye level. Those plant species which might otherwise require training on totems may be allowed to grow gracefully from their suspended position.

TABLE 7–4:   *Selected List of Plants for Hanging Baskets*

SCIENTIFIC NAME	COMMON NAME
Adiantum hispidulum	Australian Maidenhair Fern
Adiantum raddianum 'Pacific Maid'	Delta Maidenhair Fern
Adiantum tenerum 'Wrightii'	Fan Maidenhair Fern
Aeschynanthus lobbianus	Lipstick Vine
Aeschynanthus marmoratus	Zebra Basket Vine
Asparagus densiflorus 'Meyeri'	Foxtail Fern
Asparagus densiflorus 'Sprengeri'	Asparagus Fern
Asparagus plumosis	Plumosis Fern
Asplenium nidus	Bird's Nest Fern
Asplenium viviparum	Mother Fern
Begonia acetosa	Hanging Basket Begonia
Begonia foliosa	Fern Begonia
Begonia fuchsioides	Fuchsia Begonia
Begonia × hiemalis 'Rieger's Schwabenland'	Reiger Begonia
Begonia rex	Rex Begonia
Begonia semperflorens	Wax Begonia
Begonia tuberhybrida 'pendula'	Pendulous Begonia
Browallia speciosa 'major'	Browallia
Calathea makoyana	Peacock Plant
Calathea roseo-picta	Redleaf Zebra Plant
Calathea zebrina	Zebra Plant
Ceropegia woodii	Rosary Vine
Chamaecereus silvestri	Peanut Cactus
Chlorophytum comosum 'Picturatum'	Variegated Chlorophytum
Chlorophytum comosum 'Vittatum'	Variegated Spider Plant
Chlorophytum elatum vittatum	Spider Plant
Cissus antarctica	Kangaroo Vine
Cissus rhombifolia	Grape Ivy
Columnea linearis	Goldfish Plant
Columnea microphylla	Small-leaved Goldfish Vine
Cuphea platycentra	Cigar Flower
Cyanotis kewensis	Teddy Bear Vine
Davallia fejeensis	Rabbit's Foot Fern
Duchesnea indica	Mock Strawberry
Episcia cupreata	Flame Violet
Fatsia japonica	Japanese Aralia
Ficus pumila	Creeping Fig
Ficus radicans 'Variegata'	Variegated Rooting Fig
Fittonia verschaffeltii	Nerve Plant Fittonia
Gibasis geniculata	Tahitian Bridal Veil
Graptopetalum paraguayense	Ghost Plant
Gynura 'Sarmentosa'	Purple Pashion Vine
Hatiora salicornioides	Drunkard's Dream
Hedra helix	English Ivy
Hedra helix 'Needlepoint'	Needlepoint Ivy

Hedra helix 'Scutifolia'	Sweetheart Ivy
Hemigraphis colorata	Red Ivy
Hemigraphis 'Exotica'	Waffle Plant
Hoya carnosa	Wax Plant
Hoya compacta	Dwarf Wax Plant
Hoya compacta 'Mauna Loa'	Lura-Lei
Hoya compacta 'regalis'	Hindu Rope Plant
Hoya minima	Miniature Wax Plant
Hypoestes sanguinolenta	Polkadot Plant
Impatians sultanii	Impatiens
Lobelia erinus	Lobelia
Lygodium japonicum	Climbing Fern
Lysimachia nummularia	Creeping Jenny
Maranta leuconeura 'erythroneura'	Nerve Plant Maranta
Maranta leuconeura 'Kerchoveana'	Prayer Plant
Nephrolepis exaltata 'Bostoniensis'	Boston Fern
Nephrolepis exaltata 'Bostoniensis compacta'	Dwarf Boston Fern
Nephrolepis exaltata 'Bostoniensis Petticoat'	Boston Petticoat Fern
Nephrolepis exaltata 'Bostoniensis Verona'	Verona Lace Fern
Nephrolepis exaltata 'Florida Ruffles'	Florida Ruffles Fern
Nephrolepis exaltata 'Fluffy Ruffles'	Fluffy Ruffles Fern
Oxalis braziliensis	Brazil Oxalis
Pelargonium peltatum	Ivy Geranium
Pellaea viridis	Green Cliffbrake
Pellionia daveauana	Watermelon Begonia
Pellionia pulchra	Satin Pellionia
Peperomia obtusifolia	Pepper Face Peperomia
Peperomia orba	Princess Astrid
Peperomia prostrata	Hanging Basket Peperomia
Peperomia scandens 'Variegata'	Variegated Philodendron Leaf
Peperomia trinervis	Nerved Peperomia
Peperomia viridis	Green Peperomia
Petunia × *hybrida*	Cascade Varieties
Philodendron × *'Florida'*	Florida Philodendron
Philodendron micans	Velvet-leaf Philodendron
Philodendron oxycardium	Heart-leaf Philodendron
Philodendron × *'Red Princess'*	Red Princess Philodendron
Piea cadierei	Aluminum Plant
Pilea depressa	Creeping Pilea
Pilea involucrata	Friendship Plant
Piper crocatum	Saffron Pepper
Platycerium bifurcatum	Staghorn Fern
Plectranthus australis	Swedish Ivy
Plectranthus oertendahlii 'Variegatus'	Variegated Prostrate Coleus
Plectranthus purpuratus	Moth King
Polypodium scalopendria	Polypody Fern
Pyrosia macrocarpa	Pyrrosia Fern
Rhaphidophora aureus (Scindapsus)	Devil's Ivy
Rhipsalis cassutha	Mistletoe Cactus
Rhipsalis clavata	Clavata Cactus
Rhipsalis rhombea	Rhombea Cactus
Rhoeo spathacea (discolor)	Moses-in-the-Cradle
Saintpaulia ionantha	African Violet

Saxifraga sarmentosa	Strawberry Geranium
Scindapsus aureus 'Marble Queen'	Marble Queen
(also *Rhaphidophora aureus 'Marble Queen'*)	
Sedum morganianum	Burrow Tail
Sedum rubrotinctum	Christmas Cheers
Senecio macroglossus 'Variegatus'	Variegated Wax Ivy
Senecio mikanioides	German Ivy
Setcresea purpurea	Purple Heart
Streptocarpus saxorum	Dauphin Violet
Tolmiea menziesii	Variegated Wandering Jew
Tradescantia albiflora 'Albo-vittata'	Piggyback Plant
Tradescantia fluminensis 'Minima'	White Inch Plant
Tradescantia fluminensis 'Variegata'	Miniature Wandering Jew
Vinca major 'Variegata'	Band Plant
Zebrina pendula	Silvery Wandering Jew
Zebrina pendula 'Discolor'	Tricolor Wandering Jew
Zebrina purpusii	Bronze Wandering Jew

Locating Hanging Basket Plantings in the Home

Hanging plants provide an excellent window treatment for a personalized interior design. Sun-loving plants may be suspended at various levels to create a hanging garden in any room with a window. These plants may aid in blocking an undesirable view or to create privacy from an adjoining neighbor's home. Flowering plants especially will grow best in a window area. Foliage plants, which grow best in the more shaded areas of the home, may best be used to create a division of spaces, such as between a stairwell and the living room or hall (Figure 7–7). Plants that tolerate darker areas may be

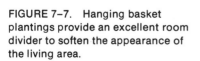

FIGURE 7–7. Hanging basket plantings provide an excellent room divider to soften the appearance of the living area.

suspended in corners to soften the lines where walls meet. Hanging baskets placed in the corners of a room break the space where other room decorations are seldom found.

Hanging basket plantings may be used effectively in an outdoor living area as well. These planters may be suspended from the ceiling of a patio deck to create an outdoor living space. Hanging basket plants provide a limited amount of privacy to the patio while adding the interest of growing plants (Figure 7–8). When these planters are suspended by chains from trees in the yard to be viewed from the home or patio, they create an added attraction to the landscape. These hanging plantings may be exchanged frequently to vary the areas that provide accent in the yard.

FIGURE 7–8.   Hanging basket planters may be used in outdoor living areas also to create privacy, divide spaces, and provide a living screen.

Plants should be grouped according to their requirements for temperature and light when located in a room or patio. Plants requiring full sun are best placed in sunny windows, in well-lighted porches or sunrooms, or from trees that provide very light shade. During the summer months, even sun-loving plants should be provided some protection from the intense sunlight through windows.

Flowering hanging basket plants and some foliage plants grow best under the high-light conditions of a sunny window area. The variegated foliage on tropical plants will be more brightly colored when placed in a bright location. Although many of these plants may grow in low-light areas, the foliage colors may be absent, with only a deep green color being evident.

Most tropical plants are best grown under diffused light conditions. The zone 4–8 feet from a sunny window provides adequate light for satisfactory growth of most plants. The air and leaf temperatures are not as great in this location as they would be closer to the lighted window. A north-facing window does not receive direct sunlight at any time of the day, so is adequate for these lower-light-requiring plants. Keep in mind, too, that plants suspended at eye level or above will be in a warmer environment than those at table height. The additional air movement around all sides of the containers and the increased temperature of the air around the plants makes hanging basket gardens more demanding on water. The air temperature around the leaves of a plant in the sunny window may cause the plants to lose water more rapidly than the roots may provide it. If plants wilt repeatedly, even when the container soil is moist, the plants should be moved to diffusely lighted areas.

Hanging baskets should be located in areas where they will not restrict foot traffic in the home or patio. The containers should be placed at a level where the plants may be viewed to their greatest advantage. For most pendulous ivy plants, this means that the container will be at the same level or above the head of the average person. The elevated position of these plant containers may prove a difficulty when watering. The baskets may have to be removed from their hooks to add water, or long-spouted watering cans may be required for those containers placed too high to be reached conveniently. Commercial pulleys are also available which will conveniently raise or lower the baskets for watering with the simple pull of a string.

Most of the plants sold in plant stores for use in hanging baskets are normally used for indoor decoration. Many of the annual bedding plants often used in outdoor gardens may also be used for hanging baskets around the patio. The cascading petunias and other sun-loving landscape plants having a suitable growth habit may be grown outdoors in hanging baskets. The restricted root systems in these baskets dry out rapidly, however, in the increased air movement of the outdoor environment. All hanging basket plants placed outdoors should be provided with adequate shade from direct sunlight to reduce elevated leaf temperatures and rapid drying of the soil in the containers.

Planting Hanging Basket Containers

A wide array of hanging basket containers is available commercially. Most containers used for hanging basket plants are designed with attached hangers and drainage saucers. These are the most suitable containers for use as hanging baskets in interior locations. The saucers aid in collecting drainage water, while allowing the excess fertilizer salts to be removed from the roots. These saucers are not effective in removing the soluble fertilizer salts contained in the drainage water, however, unless they are drained an hour or so after each watering. If the saucers are not drained of the runoff water, the fertilizer salts will again be absorbed and accumulated in the soil. Unless the containers are watered carefully, more drainage water flows from the base of the container than can be held by the saucer.

Containers without drainage holes (such as glazed pottery, glass, and metal) require more special care in watering, since water will easily accumulate in the root zone and exclude the soil-air. Drainage material must be placed in these containers as an aid in removing excess water from the root zone. This drainage material (gravel and charcoal) is added to the base of the pot in the same manner as was explained earlier for planters and dish gardens.

Plastic, clay, metal, and pottery baskets having drainage holes may be planted in the same manner as any flower pot. A layer of broken clay pot, pebbles, or gravel is added to the lower portion of the pot over the drainage hole. Then the soil mixture is added to the rim of the pot. As the cuttings or potted plants are added to the basket, the soil is pressed lightly to firm it around the plant roots. When the basket planting is completed, the soil level should be at least 1 inch below the pot rim to allow a suitable watering space.

Wire mesh baskets are also common. These may be lined with either a specially designed cardboard liner or with sheet moss. When sheet moss is used as a liner, plants may be placed in the sides as well as in the top of the container, for a more uniform planting. These are especially attractive when such plants as bear's paw and rabbit's foot ferns are placed in them.

Wire mesh baskets are prepared by first soaking sheet moss in water until it is thoroughly saturated. The excess water is drained off and the moss lined around the inside walls of the container. Soil mix is then added to the container and the plants inserted. Some plants may be placed in the sides of the wire container by opening spaces in the moss to allow their insertion. The soil in the wire baskets will dry out more rapidly than in any other container. The moss acts as a wick to draw water from the soil as the air moves around the base of the container.

Hanging basket containers range in width from 4 inches to about 2 feet. The most commonly sold baskets are either 6, 8, or 10 inches wide. A 6-inch basket will suitably hold only a single plant or only a few rooted cuttings and still allow adequate rooting into the soil. The 8- and 10-inch baskets eventually develop into the most attractive plantings, since several potted plants may be placed in them initially. The containers need not be planted to only a single species.

Many of the more desirable hanging basket plants grow rapidly in a trailing manner to hang from the pot. These plants do not grow upward or fill in the center of the pot. A mixture of climbing or flowering plants may be placed in a container with the trailing species to add special interest to the basket garden. When combining different plant species in the same container, be certain that the plants require the same environmental conditions.

Fertilizing Hanging Basket Plantings

Since hanging basket plants utilize more water than comparable plants in pots in the same room, fertilizer is leached from the soil more quickly also. Newly planted hanging basket plants should not require additional fertilization for at least 1 month in interior locations, since they are still carrying excess fertilizer from the nursery for growth in the lower light of the home.

Fertilizer applications should ideally be matched with the rate of plant growth, rather than being added on a set schedule. Rapidly growing plants in ideal conditions may require a dilute fertilizer application every month, while slower-growing plants will require less frequent fertilization. A typical fertilizer for these plants will be of a 1–2–1 (N–P–K) ratio, such as 5–10–5 or 6–12–6 formulations. Most water-soluble fertilizers contain a dye which aids the grower in determining when fertilizer is being applied through a hose proportioner. This dye may stain carpets and table cloths when spilled. All hanging basket containers will drip water, even when saucers are attached, so special care must be taken when watering. Hanging baskets should be placed so that water will not drip onto furniture. A large pitcher may be placed under the basket when watering to catch any drainage water that cannot be held by the saucer.

Some species of hanging basket plants are particularly sensitive to the fluoride added to most city water systems. Chlorophytum, dracaena, and cordyline are sometimes observed with browned leaf tips. The fluoride content of most city water systems is high enough to cause this leaf symptom. When fluoride is suspected of causing the leaf tip browning, either distilled water or rainwater should be used when watering these species. Chlorine in city water rarely causes this symptom when adequate soil leaching is allowed during each irrigation. Softened water is safe for irrigating house plants. It should not be used, however, on terrarium plants in closed containers.

THE PLANT SHOP EMPLOYEE

The vast array of tropical, foliage, and flowering house plants available in retail stores places a great responsibility on the employees who must maintain, sell, and answer questions concerning their care. The employees should be capable of identifying each plant that is sold. This is aided by the addition of plant labels in each pot for easy identification when a customer is present. The employee should also be familiar with the general culture of each type of plant when making a sales talk to a customer. The advantages and special-interest characteristics should be emphasized for each plant, especially when selecting plants for a particular decorating purpose.

Most plant stores have copies of plant dictionaries or encyclopedias which list each plant by common name and scientific name, and include a picture to aid in identifying unusual or unknown plants and their cultural requirements. A successful plant store employee will, however, be capable of identifying the more common plants on sight and will be able to speak intelligently of their care and environmental requirements. The names (both scientific and common) should be memorized along with the general growth requirements for each plant. Much of this information may be obtained from the many houseplant books available in school or public libraries. The greatest training in house plant container design, identification, and care will be obtained from experience in the plant shop and constant contact with plants.

SELECTED REFERENCES

Arthurs, K. 1974. *How to Grow House Plants*, 2nd ed. Lane Publishing Company, Menlo Park, Calif.

Arthurs, K. 1973. *Terrariums and Miniature Gardens.* Lane Publishing Company, Menlo Park, Calif.

Bachmann, R. 1974. *Caring for Indoor Flowers and Plants.* The Pillsbury Company, Minneapolis, Minn.

Coleman, M. J. (ed.). 1974. *Foliage Plants for Modern Living.* Merchants Publishing Company, Kalamazoo, Mich.

Crandall, C. 1973. *Success with Houseplants.* Chronicle Books, San Francisco, Calif.

Elbert, G., and E. Hyams. 1968. *House Plants.* Funk & Wagnalls, Inc., Reader's Digest Books, Inc., London, England.

Florists' Transworld Delivery Association. 1976. *Professional Guide to Green Plants.* Southfield, Mich.

Graf, A. B. 1968. *Exotica 3,* 4th ed. Roehrs Company—Publishers, East Rutherford, N.J.

Graf, A. B. 1976. *Exotic House Plants,* 10th ed. Roehrs Company—Publishers, East Rutherford, N.J.

Kayatta, K., and S. Schmidt. 1975. *Successful Terrariums.* Houghton Mifflin Company, Boston, Mass.

Kramer, J. 1975. *Miniature Gardens in Bowl, Dish, and Tray.* Charles Scribner's Sons, New York.

Laurie, A., D. C. Kiplinger, and K. S. Nelson. 1968. *Commercial Flower Forcing,* 7th ed. McGraw-Hill Book Company, New York.

McDonald, E. 1975. *Miniature Gardens.* Grosset & Dunlap, Inc., New York.

Wallach, C. 1976. *Interior Decorating with Plants.* Macmillan Publishing Co. Inc. (Collier Books), New York.

TERMS TO KNOW

Bonsai	Photosynthesis	Succulent Plants
Bromeliad	Plant Respiration	Terrarium

STUDY QUESTIONS

1. Describe the duties of a plant store employee.
2. List the steps involved in making a terrarium.
3. Develop basic guidelines for the planting of a dish garden.
4. Describe the ideal location for a hanging basket planting.
5. Give reasons for a plant store employee to be well trained in plant identification and plant care.

SUGGESTED ACTIVITIES

1. Visit a plant store and report your observations to the class.
2. Make terrariums, planters, dish gardens, and hanging baskets to be displayed in the classroom.

CHAPTER EIGHT

House Plant Care

Foliage plant production and sales have increased tremendously during the past decade (Figure 8–1). Growing plants indoors is becoming more of a family effort each year. The reasons for this increased interest in foliage and flowering houseplants are many. The primary reason for this interest is that America is becoming highly urbanized. People are being crowded together, whether within the inner city (nearly 80 percent of the population lives in essentially urban conditions) or in a suburban environment. More people are living in apartments or condominiums, which provide little area for gardening and the enjoyment of natural surroundings. As our population is provided with more leisure time, gardening and an interest in plants is increased.

Technological improvements have also increased the homeowners' opportunity to enjoy plants in the home. Improvements in home heating and cooling units have only recently been appreciated for their compatibility for growing plants. Gardening under lights in homes has developed to such an extent that interior designers routinely include plant lighting in home decoration. Horticulture suppliers have developed improved soil mixes, plant stands, pots, and hanging baskets for the market. Among the greatest changes taking place in the florist business has been the tremendous interest among young people in learning about tropical plants and growing them in their rooms, apartments, or campus dormitories.

The rapid increase in foliage plant sales has created a greater demand for plants than what suppliers can occasionally produce. Growers and nurserymen are constantly seeking new and more unusual plants for the market. With so many plants available, each with its own specific growth requirements, the employees of a retail plant store or flower shop are constantly asked for cultural information. A primary requirement for all sales personnel employed at a retail store where flowering and foliage plants are sold is to obtain sufficient knowledge to answer these questions intelligently. The employees must be able to identify each plant sold and be familiar with their proper care in the shop and in the home.

FIGURE 8–1.   The popularity of foliage house plants has created a demand for experienced and knowledgeable plant store employees.

CARING FOR PLANTS IN THE SHOP

When a shipment of foliage plants arrives at the shop, each plant should be carefully inspected for damage. The foliage may harbor insects or disease organisms that may go undetected unless inspected carefully. The newly arrived plants should then be set aside in an isolated location for observation. These plants may first be watered thoroughly and then treated with a plant insecticide and fungicide as a precaution against the spread of pests to other plants in the showroom.

The environmental conditions of most retail plant stores are less than is desirable for the proper growth of foliage plants for extended periods of time. The temperature, light, soil, humidity, and water are all important factors in the culture of foliage plants. Most foliage plants cannot take direct sunlight from the front window of the store. Plants can be grown in dark corners of the store only when additional light is provided for up to 12 hours each day.

During the winter heating season, the atmosphere of the average plant store is too dry and too warm for best plant growth. Plants should be kept in the coolest location of the shop under artificial growth lamps. Specially designed plant growth spot lighting will greatly improve the appearance of plants. Humidifiers may be used in the store to increase the moisture content of the air around the plants. When these are not available, the potted plants may be placed in trays lined with fine gravel. The gravel is kept damp, but water is not allowed to stand in the trays. The water that evaporates from the gravel will raise the humidity around the plant foliage and prevent excessive drying of the leaves. Periodic misting of the plants from a spray bottle will also freshen the foliage.

The foliage plants in the retail store should have their foliage cleaned periodically to remove dust and dirt. Newly arrived plants may have a residue of fertilizer and pesticide sprays remaining on the foliage. This residue should be removed as soon as the plants are unpacked. Many florists prefer to apply a foliage polish to their green plants. This is done for the purpose of cleaning the leaves and to add a deeper luster to the foliage. Many commercial products are available for this purpose, but light mineral oil may also be used. When applying a plant polish, a soft rag should be used to wipe the upper surface of the leaves only. The major portion of the plant stomata are located on the underside of the leaves and might be sealed by the oily polishes. These plant polishes should be used only infrequently, as their constant application will cause harm to the foliage.

Watering and fertilization of foliage and flowering plants is critical while they are in the retail store. The potted plants will carry a residual amount of fertilizer salts from the production nursery when they arrive at the shop. These plants have been grown under ideal conditions which favored rapid production of foliage. High amounts of fertilizer are generally used by growers to achieve rapid plant production. When these plants arrive at the shop, the soil should be leached of excess fertilizer salts. This is done by watering the plants heavily to allow at least 10 percent of the water to drain from the pots. The plants should not be allowed to remain in this drainage water or the fertilizer salts will be reabsorbed by the plant roots. This water should be discarded. If plants are placed in saucers, the water should be discarded about 1 hour following the leaching.

Foliage plants do not generally remain in the retail store for extended periods of time. Since they are sold within several weeks of their arrival, the potted plants should not require additional applications of fertilizer while in the shop. If, however, the plants should remain for a month or longer, additional fertilization is necessary. The common foliage plant fertilizers may be used for this purpose. These may either be used as a soluble liquid fertilizer (mixed with the water), as a tablet applied to the soil (dissolves when wet), or as a slow-release fertilizer capsule. The soluble liquid fertilizers are mixed in water at the rate recommended by the manufacturer. During the darker months of winter this rate may be cut in half, since most plants are not growing actively at this time. The analysis of these fertilizers is generally 6–12–6 or 5–10–5. These figures correspond to 6 percent nitrogen, 12 percent phosphoric acid, and 6 percent potash (for 6–12–6). A lower analysis for nitrogen than for phosphorus is recommended for best plant growth in interior locations.

The plants should not be watered daily simply because it fits the shop routine. Foliage and flowering plants should be watered only when required. Overwatering of the plants in the darker areas of the store may cause the plants to lose their foliage or rot. The plants should be watered thoroughly with water no colder than room temperature. The excess water is then allowed to drain from the pots and is discarded. Water is not required again until the soil is dry to the touch. Several models of plant moisture meters are available for aiding in the determination of plant water needs. These are simply plunged into the soil and will indicate the water status of the root zone (Figure 8–2).

Each plant to be sold in the retail store should be clearly marked

FIGURE 8–2. A soil moisture meter will indicate when the planting medium requires added water.

with a tag bearing the plant name. Such tags are available which also give recommendations for the proper care of that plant. The price may also be attached to this plant care tag. The employees should become familiar with the proper growth requirements and the correct names for each of the plants sold in the store. When aiding a customer in the selection of the proper house plants for his/her home, the sales personnel should be able to provide accurate information concerning the proper environment, fertilization, watering, and light-level needs of each plant. The tendency among many plant shop employees is to oversell the performance of the plants. When the correct growth requirements and expected problems are fully explained, the customer will be much more satisfied with the purchase. The more knowledge expressed by the sales clerk concerning these plant needs, the greater the potential for sales. A customer will readily purchase house plants from a store where the plants are given proper care and the employees are knowledgeable of their requirements for growth.

HOUSE PLANTS IN THE HOME

Foliage plants may be grown successfully in almost any area of a home when given proper care and growing conditions. However, no plants are naturally found in interior gardens, so all plants must be acclimatized to their new environment. House plants are grown commercially in greenhouses, in shade structures, or in tropical fields under conditions of full sunlight. These plants are grown as rapidly as possible using high rates of fertilizer, high light, water, and humidity. Each type of plant grown has different growth requirements, so the degree of acclimatization necessary for adaptation to the home environment will also differ.

Since the plants are grown under different conditions by the nurserymen, these must be modified during the acclimatization process. These environmental conditions requiring modification are *nutrient levels, light intensity*, and *humidity*. All plants received from commercial growers and retail stores require some degree of acclimatization, but this is especially true for those grown in Florida and Central America. Light intensities in the tropics may be in excess of 14,000 foot-candles, while in the home the maximum light may be no more than 150–200 foot-candles. The relative humidity in these tropical nurseries may exceed 70–80 percent, yet in homes heated during the winter the moisture content of the air may be no greater than 20 percent. Under these high-light and high-humidity conditions, commercial foliage growers may use 10–20 times the amount of fertilizer required for houseplants in the home environment.

Light Acclimatization

Under high-light conditions, plants form smaller, thicker foliage. The anatomy of the leaves is compact, with the chlorophyl-bearing cells stacked upon each other. This protects the leaves from cell damage caused by the high light intensities. These leaves are adapted to the high light available in the production field. When moved to an interior location, however, these leaves are unable to manufacture sufficient carbohydrates for proper growth.

Plants are adaptable to these changes in light intensity. They may be made to grow new, more light-efficient leaves under shaded conditions. Generally, plants grown under lower light regimes will develop larger, darker green foliage. These same plants, however, also develop softer, less sturdy stems. Under very low light conditions in most homes it is impractical to maintain large specimen plants, such as schefflera or *Ficus benjamina*. If these plants are grown under lower light conditions continuously, their trunks and stems would not be capable of supporting the foliage. Therefore, they must be grown under high light to achieve their size and support and then adapted to the lower light of the home environment.

Plants grown under high light and then acclimatized for the low light in homes will possess two types of leaves. The lower leaves which developed under high light will be smaller, harder, and less able to conduct photosynthesis for food production. The upper leaves, which are acclimatized, are broader, larger, and have a deeper green coloration. Foliage plants produced in Florida and other subtropical areas of the world often lose the lower sun-grown leaves when moved into darker interior plantings. This leaf drop may be reduced by allowing the plants to acclimate in shaded areas for several weeks before being sold to the consumer.

Placing house plants directly outdoors in the spring after a long dark winter season might cause as much damage to them as bringing them from high light to a darkened room. The plant foliage which has adapted to the low-light growing conditions is unprotected from the intense rays of the sun. Plants that have been indoors all winter require at least 3 weeks to adapt to the increased sunlight before being moved into direct sunlight. Moving plants directly from a room having 200 foot-candles of light to a patio which receives 8000 foot-candles of light in early May will have a devastating effect on the foliage. The intense sunlight will burn the plant cells so that the plant will be stunted and will not grow or flower properly.

In preparation for acclimatizing plants for outdoor summer conditions, first place them in the brightest windows of sunlit rooms. If this is not possible, they should be placed under artificial lights for long periods during the day. After about 1 week of this treatment and when the outdoor temperatures remain above 45°F (7°C) during the evenings, the plants may be moved to the outdoor environment. The plants should first be moved to a shaded carport, garage, or porch, where they will receive indirect sunlight. If a shaded deck or patio is selected, many of the green foliage plants may remain in this location for the entire summer (Figure 8–3). The plants may be fertilized with a commercial plant food at this time.

FIGURE 8–3.   Most foliage plants may remain in a protected location outdoors during the summer months after being provided with a suitable acclimatization period.

The plants may be moved once again following 1 week in the heavily shaded location. Hanging baskets may be suspended from the branches of trees or the overhang of the north side of the roof to increase the light on the plants. Some plants may require the added protection of a light cloth or shade to prevent excessive foliage burning at this time. The plants that are adapted to full sunlight for flowering may then be moved to a sunny location. If any of the leaves show signs of yellowing or burning along the margins, they should be immediately moved back to a location that provides heavier shading for a longer period. If these steps are taken by the homeowner to properly acclimatize the plants, they will be more satisfied with their house plants and will experience fewer problems in their care.

The customer can do much to acclimate newly purchased plants for their proper adaptation to lower light regimes of the home. If plants are desired for a darkened wall or corner opposite a window, the plants should be brought to that lower lighted area gradually. The plants may first be placed near the window, where they will receive diffused, but not direct, sunlight. They should remain in this location until new growth is apparent. The plants may then be placed in the darker areas of the home. Even though many types of foliage plants are adapted to the darker areas of a home, they will produce healthier foliage when regularly rotated with other plants from these darkened areas to locations providing better light.

Plants may be selected according to their light requirements to fit any decorating need. Plants will not produce flowers unless they receive adequate sunlight of the proper intensity and duration. Under the natural lighting conditions of most homes, this requirement is met best by an east or north windowsill. Flowering plants in a west or south window may have to be placed away from the window to prevent foliage burning from direct sunlight. Plants grown only for their attractive foliage will not require fully sunlighted areas for adequate growth in the home. Foliage that produces colored variegations (striping or colored islands of pigmentation) are more attractive when grown in higher light areas of a room. Only plants that produce dark green foliage and are not expected to form flowers may be grown best in the darker corners of a room.

Supplemental artificial lighting may be added to a room to enhance the growth of plants and to allow a greater latitude in the types of plants that may be grown. Specially designed growth spotlighting or fluorescent fixtures may be situated for providing proper lighting in a room. Lighted plant stands also provide areas for rejuvinating plants which may be exchanged with plants placed in other parts of the home. Plant-growth wide-spectrum spotlights or fluorescent lights are most often used to provide supplemental light in a home or office building. These are placed approximately 15–18 inches above the tops of the plants. This artificial lighting may also be provided from fluorescent fixtures with reflectors containing one warm white and one cool white or daylight tube, to approximate natural sunlight in the home.

Humidity Acclimatization

Plants grown under the high humidity of a tropical nursery do not experience the water stresses often encountered in home interiors. For this reason their root systems are often not as extensive and are not developed adequately to support the plants under drier conditions. The result of this poor root-to-shoot balance is that the plants transpire (lose water) faster than the root system can resupply it.

The plants can compensate for the lower humidity conditions in most homes if the root system is allowed to develop properly. This may be accomplished by paying special attention to the root zone of the plants to be grown in the home. The roots must "breathe," just as do leaves, stems, and flowers. Every cell of the plant uses oxygen for metabolism and growth. To provide this necessary oxygen, the plants must be grown in a soil mixture that allows adequate aeration.

Further consideration should be given to the type of pot used for growing the plants. The root systems of most greenhouse plants are highly restricted in space. A normal root system will grow wider and deeper than it is capable of doing in a pot. This means that there are more roots competing for the available air and moisture of potted plants. During acclimatization, the root system can be improved by reducing nutrition levels and lengthening the watering-frequency intervals. This will reduce foliage growth while stimulating the growth of the roots. This will also harden the foliage so that it will be less likely to wilt under low-humidity conditions.

Nutritional Acclimatization

Plants purchased from a plant store or flower shop often carry a residual effect from the high levels of slow-release fertilizer added by the grower. Once these plants are placed in the less favorable environment of a home or office, the need for fertilizer is reduced manyfold. Even though the excess fertilizer salts should have been leached from the soil by the plant store employees, it is a good idea to do this once again when taken home. Too much fertilizer in combination with low light will make the growth soft or spindly.

Foliage plants should not require fertilization more than once every 3–4 months for adequate growth in interior plantings. Commercial fertilizers may be used at the recommended rates during the summer months and at half strength during the dark weather of winter in northern climates. Fertilizers should be applied to previously dampened soil. The plant container is first watered lightly before the soluble fertilizer is added. Application of a soluble fertilizer to dry soils may cause root burning and damage to the plant. Dry fertilizers or slow-release fertilizers may be placed on dry soil, followed by a thorough watering to dilute the fertilizer.

Plant nutrient deficiencies can, however, occur even under the slower growing environments of homes or offices. This is particularly true when plants are taken to sunny rooms for rejuvenation or when fertilizer is withheld for fear of accumulating excessive salts in the soil. Many of these nutrient-deficiency symptoms may appear to be those caused by other environmental factors, so a wise plant shop employee will want to investigate all possible causes for plant failure during the diagnosis. Some of the more common nutritional-deficiency symptoms on plants and their causes are as follows:

Nitrogen. Plants are stunted in growth. Young leaves are green, while the lower leaves are yellow and may fall from the plant.

Phosphorus. Growth of plants is retarded, with leaves being darker green and smaller than normal foliage. Younger leaves may roll unusually, while older leaves may drop from the plant while still green.

Calcium. The plant growth is stunted. The affected parts of the plant are the root tips and stem tips. Addition of gypsum or agricultural lime to the soil will remedy this condition.

Potassium. The older, more mature leaves are affected first, with symptoms of marginal yellowing and browning. This symptom progresses upward from the older to younger leaves as the nutrient is translocated to the growing tip.

Iron. Interveinal chlorosis on newer leaves first. The chlorosis may be identified by the dark green skeleton outline of the veins of the leaves, while the tissue between the veins is yellow.

Magnesium. Interveinal chlorosis appears on older, more mature leaves.

Minor Nutrients. These deficiencies are rare when a complete fertilizer or trace elements are added to a soil mixture.

DIAGNOSING HOUSE PLANT PROBLEMS

House plant problems can often be avoided by maintaining the proper environment in which a plant is grown. House plants are still subject to insect, disease, or other disorders which are not easily avoided and which require immediate attention. Plant store and flower shop employees receive numerous requests daily concerning information relative to correcting plant problems. The employees must be familiar with the various maladies that afflict house plants and be capable of recommending corrective measures.

The most effective way to control house plant disorders is, of course, to practice preventive measures. However, even when a customer may give each plant personal attention, maladies will occur. When the plants do become "sick," the customer will either bring them to the plant store or call the shop for advice. The employees must learn to identify the problems and their possible causes, often over the telephone. Diagnosing foliage plant disorders over the telephone is difficult, but by asking the proper questions the employee can often determine the problem and recommend a cure.

The more important causes for plant disorders are related to improper watering and light regimes. These are followed by temperature, fertilization, and plant pests. The employee at a plant store or flower shop should develop a "checklist" of questions that will aid in the diagnosis by the process of eliminating each possible cause for a disorder. The first question that should be asked, of course, is the type of plant that has the malady. The correct identification of the plant will often greatly aid in determining problems related to the species.

The plant should be inspected visually, if the customer brings it to the store, for the presence of insects or diseases. If this is not possible, the symptoms must be assessed before plant pests can be ruled out. Diagnosing plant troubles can be difficult due to the variety of growing conditions in homes, apartments, or offices. Through the correct questioning of the customers or observation of the plant, the sales clerk may determine the problem from the following list of plant symptoms:

1. *Chlorosis of leaves (yellowing of foliage)*
   (a) *On mature leaves*

   Possibly caused by nutrient deficiencies in nitrogen, magnesium, phosphorus, or potassium.

   A root or stem disease may be affecting transport of food and water.

   Occasionally, a toxicity develops these symptoms (such as excesses of boron or sodium).

   (b) *On young leaves*

   Waterlogging of the soil caused by overwatering is the most common cause.

   Root diseases (again caused by overwatering) may create this symptom.

   Mineral deficiencies may also affect young tissue. Interveinal chlorosis may be caused by deficiencies of iron, zinc, or manganese. A uniform chlorosis of only young tissue may be caused by a deficiency of sulfur.

   (c) *On any leaves*

   Low light intensity caused by improper acclimitization.

   Fertilizer salt accumulation in the soil.

   Underwatering, which causes root death from drought conditions.

   Herbicidal contamination may cause this symptom.

2. *Chlorosis and foliage drop*

    Chilling temperatures from drafts may cause leaves to fall.

    Poor drainage and poor aeration of the soil, compounded by overwatering, is often the cause of leaf drop.

    Gas fumes from poorly ventilated burners may damage plants at low concentrations.

3. *Leaf burn*

    (a) *Scattered leaf burn*

    Pesticide phytotoxicity burn from insecticide application. Either the wrong chemical was used or it was applied when temperatures were too high.

    Fungal disease may be present. Foliar leaf spots may be caused by some diseases.

    Nutritional deficiencies of phosphorus or copper may cause these symptoms.

    (b) *Edge burn on mature leaves*

    This is most often caused by soluble fertilizer salts accumulation in the potting soil. Cause may be over-fertilization or improper watering. Cure is by thorough leaching of soil and withholding of additional fertilizer until new healthy growth is resumed.

    (c) *Edge burn of immature leaves*

    Caused by the death of cells along the edge of the leaves. Plant has become overly wilted from being watered too late.

    Extremely high temperatures from too-high light levels may burn the leaves.

    Pesticide residues may burn some plants.

    Diseases of roots or stem which restrict water uptake may kill cells in the leaves. Irreversible foliage wilting usually precedes this symptom.

4. *Stunted plant growth, small leaves, and poor foliage color*

    Poor watering practices, which restrict root growth.

    Badly restricted root growth or mechanical injury which necessitates repotting.

    Poor temperature control. Excesses in heat or cold will affect the growth rate of plants.

    Poor lighting for the plant needs.

5. *Rapid death of the plant*

    Although no cure is available in this case, it is necessary to determine the cause of death to avoid its repetition with other plants.

    Freezing of plant tissue.

Excessive soluble salts accumulation in the soil.

Root- or stem-rotting diseases which were undetected earlier.

Insect damage severe enough to kill the plant.

Herbicides accidently used on the plant or drifted onto them from the lawn.

Watering Practices

One of the first questions to be asked of the customer should be to determine the size of container in which the plant is growing. The soil mixture in a small pot will dry more quickly than when in a larger container. Some customers believe that when a small plant is placed in a large container, the watering practices will be the same as with a small pot of soil. Since the soil dries out slowly in a large pot, overwatering often occurs. Generally, as the size of the container increases in relation to the plant, the frequency of watering should be decreased.

Most people have the tendency to "kill their plants with kindness." They often water the plants on a frequent basis to maintain the soil in a constantly moist state. When the soil is maintained in a constantly moist condition, little air will penetrate to the roots. The result of this is ultimately the death of the fine feeder roots or the death of the entire plant. When water is applied, adequate irrigation water should be provided to allow drainage from the bottom of the container. The soil should then be allowed to dry to the touch before additional irrigation water is added. Soil in large containers should be allowed to become nearly dusty dry before water is added. When the top of the soil in a large container is dry, water is still available to the roots at the bottom of the container to maintain plant growth.

The type of pot or container within which a plant is growing will determine to some extent the frequency of watering required. The soil mixture in a clay pot will dry out more rapidly than when in a plastic or ceramic container. The clay pot is porous enough to allow passage of excess water through the side walls. Air is also free to pass into the soil medium through the fine holes in the pot. Plastic pots do not allow this passage of air and water. Plants grown in plastic pots must be watered more infrequently than those in clay pots.

Plants grown in containers having no drainage holes present a difficult problem for watering. The tendency for most customers is to overwater these containers, resulting in a pool of stagnant water at the base of the pot within the root zone of the plant. No knowledgeable plant store employee will ever sell a dishgarden or foliage planter constructed with insufficient drainage. These planters in solid bottom containers should be planted by double potting. A suitable plant is placed in a pot that is smaller than the more decorative container to be used. Pea gravel may be placed below the pot and peat moss placed around the sides to make the pot secure in the larger, more decorative pot. Decorative rock or sheet moss may then be placed over the rim of the inner pot as a top dressing to conceal the added container. Water will be able to drain freely from the drainage hole in the bottom of the inner pot without running out onto the floor or furniture.

The most common result of overwatering house plants is the incidence of root-rotting diseases. Excessive watering causes an oxygen deficiency in the soil, resulting in root death and subsequent rotting. Some soil-borne fungi and bacteria will invade root systems that are injured or weakened. When poorly drained soils are used in the potting of the plants, the problem of root rotting is aggrevated. When root rot is suspected by the presence of foliar symptoms, the plant should be removed from its container for inspection of the roots. A healthy root system will appear fibrous, with many fine white root hairs. When the rotting organisms are present, the roots will appear blackened, with various degrees of slimy, black decay material present. Mild cases of root rotting may be corrected by repotting the plant in clean, sterilized soil. A mild application of a suitable soil fungicide will aid in elimination of further damage to the roots. The drainage and watering factors that led to the root rotting must also be corrected. When severe root rotting has caused the collapse of the lower portion of the stem, the upper portion of the plant must be severed and rerooted to prevent the total death of the plant.

DISEASES AND INSECT PESTS ON HOUSE PLANTS

Foliage and flowering plants may become infected with diseases or infested with insects under proper conditions at any time. The luxurious growth promoted by nurseries during rapid production favors diseases and insects. Nurserymen spray their plants regularly to combat these pests, but some invariably escape the pesticides and are transported to retail stores and to customers' homes.

Both retail shop employees and customers should inspect newly purchased plants. These should be isolated for a short time from other plants in the home or shop. During this time the plants should be inspected periodically to determine whether microscopic eggs have hatched or hidden insects have appeared since the purchase of the plants. The plants may require a preventive insecticide application to avoid the spread of pests to healthy specimen plants in other parts of the home or store. Often insects will be evident within the period of a week if they were present when the plant was purchased. When the plants have been isolated for a period suitable to convince the customer that no pests are present, they may be placed with the other plants in the home to proceed through the acclimatization procedures.

Foliar Diseases

Aside from the root- and stem-rotting diseases brought about by overwatering, various other diseases may infect foliage plants when environmental conditions are favorable. Bacterial and fungal leaf spots may affect foliage and flowering house plants. The more common foliar leaf spots are caused by fungal organisms. These leaf spots vary in size, shape, and color. Often the spots will have a distinctive margin. The infected foliage may wither and die. Fallen leaves may contain blotched areas covered with dark dots (fruiting bodies of the fungus organism).

Bacterial leaf spots will appear as sunken lesions (spots) having concentric ring patterns around them. These rings may progress until they run together. The bacterial organism may enter the vascular tissue of the plant and progress upward to the growing tip. Bacterial diseases leave the stem shrunken and wrinkled. The growing tip of the stem becomes drooped and soft in appearance.

Both types of leaf spot diseases may be prevented by keeping the foliage dry. The organisms cannot invade healthy leaf tissue unless water droplets are present for their germination. When the plants are watered, water should not be allowed to splash onto the leaves. Watering is best done in early morning to allow the leaves to dry before evening. In the dry atmosphere of most homes, these foliar diseases are rarely found. However, the practice of misting foliage plant leaves frequently stimulates invasion of these disease organisms. Too-frequent misting of the foliage may actually do more harm than good. The major advantage in favor of misting plants routinely is that it causes the customer to observe the plants frequently. This routine inspection may allow them to discover potential problems before they become serious.

Plants that become infected with leaf spot diseases should be immediately isolated from other plants in the home or office. If a light infection is observed, the infected leaves may be removed and destroyed. Dense foliage should be thinned and the air circulation around the plant increased to keep the foliage dry. Severely infected plants may be treated with a suitable foliar fungicide until a cure is obtained. Remedies for severe bacterial leaf spot diseases are costly, but effective. A bactericide spray is required for this malady. These bactericides may damage some foliage plants, however, so labeled directions should be followed explicitly.

Insect Pests of House Plants

Insects found on a house plant may have been present at the time the plant arrived at the plant store or they may have spread from another plant in the home which was already infested. Scales, mealybugs, spider mites, cyclamen mites, aphids, whiteflies, and a small black fungus gnat may at one time or another be observed on house plants. A plant store employee must become familiar with this insect population in order to make a quick and accurate diagnosis.

Aphids.   Aphids are small, soft-bodied insects which usually do not fly, but flying stages are present. These may be green, brown, or black in color. Aphids usually feed on the soft stems at the growing tip of the plant, causing it to wilt and become distorted in growth. Light infestations may be washed from the plant with soapy water, and rinsed. Systemic pesticides or foliar applications of an insecticide give best control of aphids.

Spider mites.   The most common mite of foliage plants is the two-spotted spider mite, but the red spider mite may also be found occasionally. These are extremely small in size and may be identified easier from the webbing they spin over the leaves and stem tips. Spider mites suck the plant juices from the undersides of the leaves, giving a bronzed, speckled appearance to the foliage. Control of spider mites is by washing the foliage, then applying insecticide (miticide).

Whiteflies.   Whitefly adults are tiny, white-winged insects. Immature forms crawl on the undersides of the leaves, where both forms suck the plant sap. As they feed, a sticky honeydew material is excreted, which supports a black sooty mold (breadmold) growth. Infested plants will have yellow leaves with a moldy appearance on the undersides and will drop from the plant. Insecticides are effective control measures following thorough washing of the plant.

Cyclamen mites.   Cyclamen mites are so small that they cannot be seen with the unaided eye. They suck the plant juices at the young growing tips and lower buds. The leaves and buds become deformed on infested plants. This pest is most often seen on aralias, false aralias, fatsia, and English ivy. Control of cyclamen mites is by rinsing the leaves followed by an insecticide application.

Mealybugs. Mealybugs may be identified by the cottony material found on the stems and along the leaf veins of the plant. This cottony material serves as a protective covering for the small, soft-bodied insects. These insects also suck plant juices and excrete a honeydew which supports the growth of sooty mold. The adults that are covered by the cottony mass are protected from the insecticide applications sprayed on the plant. The immature mealybugs are not protected when they move to a new feeding site, so are vulnerable to insecticides at this time. Regular use of a foliar insecticide is required for complete control. Light infestations of mealybugs may be controlled by dipping a cotton swab in alcohol and applying it to the insects. This must be done routinely to eliminate the pests as they hatch.

Scale. Various types of plant scale may be found on houseplants. Each type has a hard shell that covers and protects the soft-bodied insect beneath (Figure 8–4). Scale insects suck plant juices and cause generalized yellowing of the stems and leaves, or death of the plant when heavily infested. Scale insects may be treated in the same fashion as mealybugs, since the protective shell is impermeable to foliar-applied insecticides. Systemic insecticides kill all stages (except eggs) of the feeding insects. However, the shell will remain on the plant, making it appear to be still infested. These may be re-

FIGURE 8–4. Scale insects on a fern frond; a noninfested frond is on the left.

moved manually with a cotton swab dipped in alcohol. The alcohol must be used carefully. Too much applied to the plant may burn the plant cells.

Making Pesticide Recommendations to Customers

The Federal Environmental Protection Agency (E.P.A.) requires that the owner of any business that sells pesticides be certified by the state and instructed in proper use and distribution (Figure 8–5). All pesticides are potentially hazardous and should be treated with care and respect. Each employee who makes recommendations for pesticide applications to customers should be well informed on their proper application and labeled instructions for each product sold. A conscientious employee may also wish to attend the certification lectures and be examined for the purpose of receiving the proper training in handling pesticides. When making pesticide recommendations to customers it is imperative that the pesticide mentioned be registered for this use. The plants and insects or diseases that may be treated and the proper spraying concentrations and conditions are listed on the label of the product. The customer should also be instructed as to what protective clothing or breathing aid must be worn when using the pesticide.

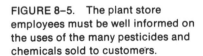

FIGURE 8–5.   The plant store employees must be well informed on the uses of the many pesticides and chemicals sold to customers.

Because the laws are strict and rather exacting concerning the use of pesticides, a plant store employee must be careful in making recommendations to customers. An inexperienced employee might be wise to copy each pesticide label from the containers onto separate pages. These instructions may then be quickly referred to when discussing the plant pest and the correct control measures. It is wise,

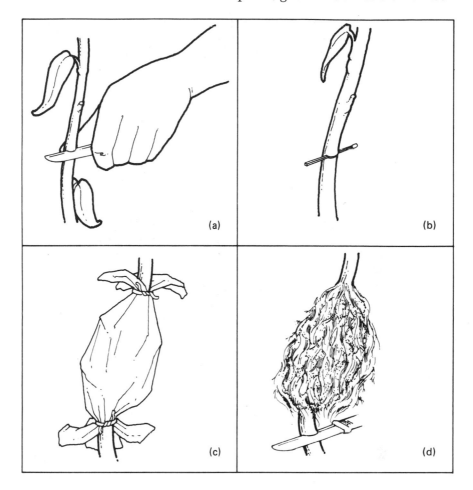

FIGURE 8–6. Air layerage of plants. When houseplant specimens become too large for their location or require drastic action to preserve their natural beauty, this technique may be used.

(a)   Make an incision through the bark of the stem at the point where rooting is desired.

(b)   Place a match stick between the folds of tissue created by the incision.

(c)   Pack moist sphagnum moss around the stem at the location of the cut stem and wrap this with plastic.

(d)   When the moss is suitably filled with new roots, the stem may be completely severed from the original plant below the new root zone. This new plant may then be placed in a pot.

also, that store employees accept for treatment no plant that is to be returned to the customer upon payment of a fee for treatment. That would classify the store manager as a commercial pesticide applicator, which requires licensing.

PLANTS IN INTERIOR DESIGN

Plants are considered to be a very important factor in modern interior designs. As a salesperson in a plant store or flower shop, it is helpful to have a knowledge of how plants can be assets in an interior design. A generally recognized rule among interior designers is that every room in the home will be more pleasant to live in and attractive to guests when it contains healthy, living plants. Foliage and flowering plants bring life into an otherwise empty dwelling. Plants aid in breaking the starkness of architectural lines. They create a warmth that helps the home environment become closer to nature through growth and change, which create interest in life.

An important design guideline dictates that good items should be functional. Plants serve many functions in a home setting. One obvious function is to add a dramatic element or aid in creating a mood when a specimen plant is used as a focal point (accent). Lighting may be used on such a plant to help emphasize the center of interest in the grouping.

Plants are sometimes used as room dividers. They are quite effective in creating the feeling of a defined space and also in aiding in the reduction of noise levels between the spaces defined by means of strategic placement of plants. Good choices for plants used as room dividers are large foliage specimens, hanging basket planters, small potted plants on shelves, or a combination. The pots selected should suit the room's decor.

Plants may also be used effectively to camouflage architectural "eyesores." Exposed water pipes or an old radiator may go almost unnoticed when a beautiful plant is vying for attention. A fireplace that is unused during the summer months, with its sooty and unused central location in the room, is not too attractive. However, when decorated with living plants, its prime location can provide a fresh and pleasant focal point for the decor.

Plants are especially effective when used as part of a window treatment. They can help to break the plainness of a large expanse of glass. Often, plants are used to create more interest at a window or to aid in making the outside appear to be an extension of the indoor living area. Plants may also be used to screen undesirable views, while allowing light to enter the room.

House plants can be an important asset where a problem area or empty, uninteresting space exists in a room. They can provide a personal touch to decorating, especially when someone has as a hobby the collection or propagation of interesting plants. Plants can be an extension of a color scheme. Flowers, fruits, and foliage can accent a decor because they possess interesting colors, textures, and shapes. Examples of house plants that lend special color are African violets, orange and lemon trees, begonias, crotons, and plants with colorfully variegated foliage.

REVIEW OF THE DIAGNOSIS OF PLANT DISORDERS

Whether sales personnel must diagnose plant problems over the telephone or at the store, a checklist of possible causes for these disorders and their corrective measures is helpful. As an employee becomes more experienced at diagnosing plant disorders, a simple examination of the plant may be all that is necessary to make an intelligent recommendation for a cure. Such a checklist is as follows:

1. Determine the identity of the plant. Some typical problems are commonly found on only certain types of plants. For example, scale insects might be expected on an ailing spider plant (*Chlorophytum*).

2. Next determine whether insects are present. These should be easily identifiable either through direct observation of the pests or by the typical damage symptoms they create.

3. If insect damage is not found, determine whether foliar diseases are present. Foliar leaf spots indicate disease of the leaves. Determine whether the disease is caused by a fungus or a bacterial organism.

4. If leaves are dropping from the plant after becoming chlorotic, the cause may be any one of several factors. The employee should inspect the roots personally, if practical, to determine whether they are dying. This root deterioration may be corrected by transplanting into a coarse potting soil mixture with addition of a soil fungicide to the irrigation water.

5. The sales clerk should also determine the location where the plants have been growing, how often and how much water is applied, and the frequency and amount of fertilization of the plant. Any of these factors may aid the employee in making a sound recommendation to the customer.

6.   If a pesticide application appears to be the best remedy, instruct the customer to read and follow labeled directions explicitly. Under no circumstances is it wise for the employee to offer to treat the plant in the store at an additional fee. Always be doubly certain that the correct pesticide is selected for the particular plant and pest problem to be treated.

SELECTED LIST OF PLANT REQUIREMENTS FOR INTERIOR PLANTINGS

Light Requirements

High Light.   Plants tolerate bright light or full sun during most of the day.

SCIENTIFIC NAME	COMMON NAME
Acalypha wilkesiana	Copper Leaf
Aloe vera	Medicine Plant
Beaucarnia recurvata	Pony Tail Palm
Brassaia actinophylla	Shefflera
Cissus rhombifolia	Grape Ivy
Chamaecereus silvestri	Peanut Cactus
Codiaeum variegatum	Croton
Crassula argentea	Jade Plant
Euphorbia lactea	Candleabra cactus
Euphorbia splendens	Crown-of-Thorns
Fatshedera lizei	Ivy Tree
Gynura aurantiaca	Velvet Plant
Hedera canariensis	Algerian Ivy
Hedera helix	English Ivy
Homocladium platycladum	Tapeworm Plant
Hoya australis	Porcelain Flower
Hoya carnosa	Wax Plant
Kalanchoe tomentosa	Panda Plant
Mammillaria sp.	Cactus
Pedilanthus tithymaloides	Slipper Flower
Rhoeo spathacea	Moses-in-the-Cradle
Sansevieria cylindricus	Spear Sansevieria
Sansevieria trifaciata 'hahni'	Birdsnest Sansevieria
Sansevieria trifaciata 'Laurenti'	Striped Sansevieria
Sansevieria zeylanica	Snake Plant

Medium Light. Plants grow best in partial shaded, diffused, or filtered light.

SCIENTIFIC NAME	COMMON NAME
Acanthus montanus	Mountain Thistle
Adiantum capillus-veneris	Maidenhair Fern
Aphelandra squarrosa	Zebra Plant
Asparagus sp.	Asparagus Ferns
Araucaria excelsa	Norfolk Island Pine
Begonia rex	Rex Begonia
Brassaia actinophylla	Shefflera
Chlorophytum sp.	Spider Plants
Chrysalidocarpus lutescens	Areca Palm
Cissus antarctica	Kangaroo Vine
Cissus rhombifolia	Grape Ivy
Crassula arborescens	Silver Dollar
Dieffenbachia amoena	Giant Dumbcane
Dieffenbachia picta	Spotted Dumbcane
Dizygotheca elegantissima	False Aralia
Dracaena deremensis	Striped Dracaena
Dracaena godseffiana	Gold Dust Dracaena
Dracaena marginata	Madagascar Dragan Tree
Dracaena sanderiana	Ribbon Plant
Ficus benjamina	Weeping Fig
Ficus elastica	India Rubber Plant
Ficus elastica 'decora'	Wideleaf Rubber Plant
Ficus lyrata	Fiddleleaf Fig
Ficus pumila	Creeping Fig
Hedera helix	English Ivy
Monstera deliciosa	Split-leaf Philodendron
Nephrolepis exaltata	Sword and Boston Ferns
Pandanus utilis	Screw Pine
Pandanus veitchii	Variegated Screw Pine
Peperomia caperata	Emerald Ripple
Peperomia minima	Miniature Peperomia
Peperomia obtusifolia	Baby Rubber Plant
Peperomia sandersi	Watermelon Peperomia
Philodendron oxycardium	Heart-leaf Philodendron
Philodendron selloum	Selloum Philodendron
Pilea cadierei	Aluminum Plant
Pilea microphylla	Artillery Plant
Polyscias balfouriana	Dinner Plate Aralia
Pteris ensiformis	Victory Fern
Rhaphidophora aureus (Scindapsus)	Devil's Ivy
Sansevieria sp.	Sansevierias
Syngonium podophyllum	Arrowhead Vine
Tradescantia fluminensis	Variegated Wandering Jew
Zebrina pendula	Wandering Jew

Low light. Plants grow slowly, yet survive in a dimly lighted area of a room.

SCIENTIFIC NAME	COMMON NAME
Aglaonema commutatum	Chinese Evergreen
Asplenium nidus	Birdsnest Fern
Hedera helix	English Ivy
Monstera deliciosa	Split-leaf Philodendron
Pellionia daveauana	Watermelon Begonia
Philodendron oxycardium	Heart-leaf Philodendron
Philodendron selloum	Selloum Philodendron
Sansevieria sp.	Sansevierias

SELECTED REFERENCES

Arthurs, K. (ed.). 1974. *How to Grow House Plants,* 2nd ed. Lane Publishing Company, Menlo Park, Calif.

Bachman, R. 1974. *Caring for Indoor Flowers and Plants.* The Pillsbury Company, Minneapolis, Minn.

Cornell University. 1974. *Cornell Recommendations for Commercial Floriculture Crops. Part II: Disease, Pest, and Weed Control.* Ithaca, N.Y.

Crandall, C. 1973. *Success with Houseplants.* Chronicle Books, San Francisco, Calif.

Elbert, G., and V. Elbert. 1974. *Plants That Really Bloom Indoors.* Simon and Schuster, New York.

Elbert, G., and E. Hyams. 1968. *House Plants.* Funk & Wagnalls, Inc., Reader's Digest Books, Inc., London, England.

Florists' Transworld Delivery Association. 1976. *Professional Guide to Green Plants.* Southfield, Mich.

Graf, A. B. 1976. *Exotic House Plants,* 10th ed. Roehrs Company—Publishers, East Rutherford, N.J.

Laurie, A., D. C. Kiplinger, and K. S. Nelson. 1968. *Commercial Flower Forcing,* 7th ed. McGraw-Hill Book Company, New York.

Nelson, K. S. 1967. *Flower and Plant Production in the Greenhouse*, 2nd ed. The Interstate Printers and Publishers, Inc., Danville, Ill.

Pirone, P. P., B. D. Dodge, and H. W. Rickett. 1960. *Diseases and Pests of Ornamental Plants*, 3rd ed. The Ronald Press Company, New York.

Taylor, J. L. 1977. *Growing Plants Indoors*. Burgess Publishing Company, Minneapolis, Minn.

Wallach, C. 1976. *Interior Decorating with Plants*. Macmillan Publishing Co., Inc. (Collier Books), New York.

Widmer, R. E. 1970. *Care of House Plants*. Agricultural Extension Service Bulletin No. 274. University of Minnesota, St. Paul, Minn.

TERMS TO KNOW

Acclimatization	Humidifier	Scale
Aphids	Interveinal	Systemic Pesticide
Chlorosis	Mealybugs	Toxicity
E.P.A.	Photosynthesis	Variegations
Foliar Disease	Residual	

STUDY QUESTIONS

1. Discuss the importance of carefully inspecting a new shipment of plants before placing them in the greenhouse or plant store.
2. What equipment can be utilized in the plant store to provide a more natural climate for tropical plants?
3. How can overwatering damage plant?
4. Why can stagnant water at the base of a pot cause problems in house plant growth?
5. Discuss how plants can be acclimatized in the transition from the greenhouse to the home.
6. Make a chart describing some common nutritional deficiency symptoms of plants.

7. List some insect pests typical to house plants and briefly describe how they can be identified.

8. Tell how plants can be used in interior designs.

SUGGESTED ACTIVITIES

1. Make plant-care tags to be placed on the foliage plants for sale in your department.

2. Develop a plant-care pamphlet which can be distributed in your school and community.

3. Set up a display showing house plant problems and suggestions for good care practices.

4. Invite an interior designer to class to discuss how plants can be used in home decoration.

5. Conduct a sick-plant clinic for students or community members, to diagnose house plant problems and recommend good methods for plant care.

PART THREE

The Florist Business

CHAPTER NINE

The Retail Flower Shop

Retail and wholesale florists play an important role in the marketing of flowers and plants produced in commercial greenhouses. The wholesale florist purchases flowers from many different growers throughout the country and sells the various items to the many retail florists in the area. The retail florists then use these flowers in arrangements that are sold to the consumers.

The florist business is highly seasonal, the greatest demand for flowers being on Christmas, Valentine's Day, Easter, and Mother's Day. Flowers are also in great demand during the wedding months of June and August. Most of the profits for a shop and most of the work for the employees occur at these periods. During the months between the flower holidays, florist shop employees are kept busy designing for funerals, weddings, hospital gifts, and other occasions. A large portion of retail flower shop sales are for special occasions or sentimental designs.

TYPES OF FLOWER SHOPS

The majority of flower shops found in the United States are *full-service* shops. These florists provide personalized service for their customers. They design floral arrangements for funerals, weddings, parties, and other occasions. They may design arrangements to suit the decor of the customer's home. Each design is constructed once it is ordered, rather than being mass-produced for later sale. Full-service shops also provide delivery service for their customers. The person who makes these deliveries is responsible for setting up wedding decorations, displaying funeral pieces properly, and assisting customers at locations away from the flower shop.

A full-service shop will often sell other items besides the cut flower arrangements. These products are generally sold to complement flower sales. Pottery, containers, and accessories are sold for the purpose of adding cut flowers later. The florist may make additional sales in this manner. Permanent flowers are often sold in flower shops as well. These flowers may be arranged and sold to customers who desire the added touch of flowers in their homes, yet cannot afford to purchase live flowers on a regular basis.

Living potted plants are sold in nearly all flower shops. These are specially decorated with paper or foil to cover the pot, and ribbon is added to serve as an accent for the flowers or foliage. Florists also create living arrangements using foliage plants. The designers arrange various small tropical plants in containers so that they may be enjoyed for many months thereafter. The employees of a full-service flower shop are well trained in the care of plants and will assist customers with their plant problems.

In addition to full-service flower shops, many *specialty flower shops* may be found. These shops are usually located in areas where there is more traffic and a better opportunity for sales. The specialty flower shop is designed for customers who want to purchase flowers or gifts on impulse. These shops are located in areas where such potential customers are most likely to be found. A typical specialty flower shop might be located at a large hospital. A customer can find there cut flower designs that are specially arranged for hospital patients. Potted plants and other gift items might also be sold there.

Many flower shops located at shopping centers and in hotel lobbies are also specialty shops. The designs sold at these shops are most often arranged at another location and brought to these shops for sale. Specialty flower shops provide little service, since they are generally very limited in space. These shops are often operated by a full-service flower shop, which provides all the plants and floral arrangements.

Many of the larger mass merchandising firms in the United States are presently selling florist products. These chain stores purchase potted plants directly from growers for sales at prices that are lower than at a full-service flower shop. Cut flowers are also sold in packages ready to be arranged. These items may be purchased by customers as they shop for groceries, clothes, or hardware items. Since these floral products are sold in high-traffic areas, large numbers of plants are sold. A full-service florist may sell prearranged designs to these chain stores for direct sales. These floral products are sold on a cash-and-carry basis without additional services (such as delivery).

EMPLOYEES OF A FLOWER SHOP

Most of the flower shops in the United States are operated by individual owners or as family-operated establishments. These businesses are managed by their owners. In a small shop, all the functions required in the business might be handled by the owner, his/her spouse, and one or more of their children. As the business grows, additional employees may be hired to assist with the work.

The larger flower shops may employ several designers, delivery truck drivers, sales clerks, and designers' assistants. These shops are often operated as part of a chain of retail shops, with each store being run by a store manager. Large flower shops provide the greatest opportunity for advancement. After a period of training and experience, a highly motivated designer's assistant may be promoted to a position as a designer. This person could advance to become head designer or shop manager.

Employee positions may be categorized as to specific duties only when the staff is large. Smaller shops divide the duties in a manner that requires each employee to be involved with several aspects of the business. In large shops, however, an individual employee will follow a more rigid job description. These job descriptions might be: delivery truck driver, sales clerk, assistant designer, and store manager.

Delivery Truck Driver

Providing delivery service is one of the most important functions of a full-service flower shop. Although the person who performs the duties of delivering flower arrangements may not necessarily be a skilled designer, he is one of the most important employees at a flower shop. The person who makes deliveries for the shop is often the only person making direct contact with the customers. For this reason, the delivery person must make a good impression on the public and promote goodwill in the community for the flower shop.

Any person who is licensed to drive an automobile is capable of driving a delivery truck. In some states, however, it may be necessary to obtain a chauffeur's license to drive the florist's vehicles. Although drivers must primarily be competent in the operation of delivery trucks, they must have other qualifications as well. They must be well groomed, well mannered, and have a good knowledge of the flower shop. Some of the specific qualifications required of delivery truck drivers are:

1.   Drivers must be familiar with the delivery area. They should know the streets and the manner in which houses are numbered. It is important that drivers make deliveries as quickly as possible, with the least amount of backtracking.

2.   Drivers must be courteous and safety-conscious. They must make certain that the floral arrangements arrive at their destinations in the same condition in which they left the shop. Also, the driving habits of the delivery personnel will affect the opinions of the customers toward the flower shop.

3.   Delivery personnel should always be well groomed and well mannered. If a delivery jacket or uniform is provided by the shop, it should be worn at all times. Again, the impression that is made on the customers by the drivers will directly affect the public image of the flower shop.

4.   Drivers will also be responsible for the upkeep of the delivery vehicle. They should see to it that the trucks are serviced regularly and kept filled with fuel. A routine check should be made periodically to determine that the trucks are in good operating condition. When the tires are worn seriously or other potential problems develop, it is their responsibility to inform the shop manager that these are in need of immediate attention.

5.   Drivers are responsible for the proper handling of the merchandise during delivery. Each item to be delivered should be wrapped properly for best display and protection. During cold or inclement weather, potted plants and arrangements should be enclosed in a plastic or paper wrap to prevent damage during delivery. Each address should be double-checked before leaving the shop to avoid the return of merchandise.

The person who makes deliveries for the flower shop will occasionally be required to make collections for items that are delivered. These C.O.D. (cash-on-delivery) sales require that a certain amount of cash be kept in the truck for making change. The driver must be able to make change quickly and must also protect the cash on hand. The money that is collected should be returned to the shop at the end of each delivery run.

Many deliveries are made to hospitals, funeral homes, and other locations on a frequent basis. These deliveries often must be made at rather precise hours or on regular schedules. It is the responsibility of the driver to see that all the deliveries that are scheduled to go to one location are ready at the same time. Drivers should check the order board and the delivery table after each run to minimize the number of trips made to the same area.

When wedding flowers are delivered to a church, a certain amount of decorating is often required. The sanctuary is often prepared by placing candelabra bouquets, and other equipment at the front. An aisle runner (carpet) may require special care in its placement down the aisle. The reception area is also decorated with arrangements, foliage, and other floral garnishments. In a large church these decorations may require a great amount of effort and designing skill to set up correctly. It is often of benefit for drivers to have some designing skills for this type of work. They should clean up all materials used in designing the church before guests arrive. Since most of the wedding-setup work is done just before the wedding is scheduled to begin, delivery personnel must work quickly and precisely to avoid delaying the wedding.

Funeral deliveries also require that the designs are in place at the proper time for the funeral or wake. Since funerals are never scheduled more than 2–3 days in advance, the florist must arrange all the sprays, bouquets, and the casket spray on a very short notice. This means that an entire truckload of floral pieces may be delivered at the same time. These floral designs must be taken to the appropriate chapel and placed on display racks before the scheduled visitation period. The pieces should be arranged for best effect and all litter cleaned up, as with all deliveries.

When the driver is not making deliveries, he should be assisting with other duties at the shop. Often flowers that have arrived in shipping boxes must be unwrapped and placed in water for refrigeration. When no other jobs are required, the driver should always assist with cleaning the design and storage areas. When the designers are busy, they often do not have time to keep the design room cleaned.

Sales Clerk

The position of sales clerk is an excellent place from which to learn the operation of a retail flower shop. Although no specific design experience is required, sales clerks are very important employees in the flower shop. They will be in constant contact with customers, so must have pleasant personalities, be well groomed, and be capable of making sales.

One of the most important duties of sales clerks is to determine what customers wish to purchase and how much they are willing to spend. Even though sales clerks may not be skilled designers, their constant contact with the designers will help them in assisting customers with orders. When helping customers with a purchase, sales clerks should always be as helpful as possible. They should not merely wait for the customer to decide on a purchase and then take the order. Often a larger sale can be made by simply suggesting related items to complement the desired purchase. For example, when a customer wishes to purchase a plant, the sales clerk might suggest that a pottery container be purchased to place it in.

Sales clerks must also be skillful in making change and correctly operating a cash register. Each cash register is operated differently, but the sales clerks are responsible for making all register-tape entries in the appropriate manner. They will be held accountable for any missing funds at the end of the day, so all transactions involving cash must be entered on the register tape.

Whenever purchased items are to be taken from the store by customers, sales clerks should either box or wrap each item. When a flower arrangement is purchased, it should be placed in a box to prevent it from tipping in the customer's car. During inclement weather, floral arrangements should be placed in plastic wrap to protect the flowers. Since many of the purchases made in a flower shop are to be used as gifts, the sales clerk may assist the customer in selecting an appropriate card. This is tied to the arrangement or gift before the customer leaves the store.

The sales clerk will take many orders for flowers over the phone. Some of the orders may be from local customers desiring flowers for a special occasion. Other calls may be from florists in another city or state for local delivery. These orders must be taken quickly and accurately, since often a long-distance call is being made. The sales clerk writes the order on a standard order form and places it on the designer's board.

When the sales clerks are not busy taking phone orders or waiting on customers, they may be required to keep the sales room properly arranged. They should keep all potted plants watered, groomed, and decorated, and see that a *care tag* is placed on each plant. A care tag tells the name of the plant and instructs the customer on its proper care. The shelves, pottery, and display cases will need to be dusted or cleaned on a regular basis. This is very important, because customers will not purchase any item that appears to be "shopworn" or dirty. Candles that are displayed must constantly be rotated or exchanged with fresh merchandise to prevent sale of discolored goods.

One of the most important duties of sales clerks is to see that all items on display are correctly priced. They must keep the display shelves neatly arranged and well stocked at all times. This may require unpacking the merchandise from the storage room and marking each item as it is placed in the sales room. Sales clerks must also be aware of the current prices being paid for the flowers used in designs. Since the prices fluctuate during holiday seasons, they must know what prices should be charged for the arrangements being sold. Generally, arrangements are sold in various sizes, each having the same type of flowers but for different prices. The larger designs will contain more flowers for a higher price. Sales clerks must be able to explain to customers what might be expected for each range of prices quoted.

In addition to changes in prices, which occur periodically, the availability of flowers varies with the time of year. Most of the flowers sold by florists are available year-round. However, many specialty flowers are available only during specific months. The spring bulb crops (such as tulips, daffodils, and iris) are only sold from late January through Easter. These could not be ordered or sold at any other time, unless they are forced in a local greenhouse.

Sales clerks, together with delivery truck drivers, have the most contact with the public. It is very important that sales clerks are always helpful and cheerful around the sales room. The attitude and interest in their jobs will be reflected in their work and will result in more sales for the flower shop. This attitude should also be extended toward the other employees. Any friction between employees is immediately felt by the customers and may result in a lost sale.

Designer's Assistant

The designer's assistant is a person being trained as a floral designer. The assistant is not responsible for arranging large specialty floral pieces, but is expected to quickly learn to make smaller designs. The trainee will work directly with the more experienced designers and will have ample opportunity to observe a wide variety of arrangements being constructed. The assistant will be responsible for keeping the designers supplied with materials. The assistant will be taught to place wires on stems, make bows, corsages, and do various duties required for the preparation of arrangements (Figure 9–1).

Designers' assistants will often be instructed in design principles through use of dried and permanent flowers. When assistants are not working directly with designers, they may be able to make up permanent designs to be sold. This gives assistants the necessary practice before they handle live cut flowers.

FIGURE 9–1. Designer's assistant. Wiring stems and making bows are part of the duties involved in supplying the designer with materials.

An important duty of designers' assistants is to coordinate orders from the designer's board. Assistants may be responsible for selecting the orders that require the most immediate attention for the designers. As the arrangements are being constructed by the designers, the assistant will fill out a card or delivery tag to be placed on the design. The assistants will then organize all the arrangements for delivery. Each cut flower arrangement or potted plant will have a care tag attached, and the shop card or delivery tag will be added.

Designer

The duties of the designer and the designer's assistant are similar, but the designer is more experienced in flower arranging. All designers and assistants are responsible for filling orders from the order board. The head designer will assign specific work to each of the other designers. The assistant must prepare all of the containers, bows, and other accessories that will be used by the designers. In the case where wedding designs are arranged, each designer and assistant will be given a specific portion of the work. An assistant designer may become a designer after 1 or 2 years of training. To become a designer, however, the assistant must be able to make any arrangement that is asked for and must have worked during all the major flower holidays. An assistant designer could begin work in the flower shop during holidays and on weekends while still in high school.

The head designer will usually be responsible for making sales that involve special treatment. The selection of floral pieces and decorations for weddings and large parties requires the training and experience of a skilled designer. The ideas and requests of the bride or hostess must be translated by the designer into the finished arrangements. The designer's assistant will learn to do this also after a period of observation, training, and practice.

The head designer also has the responsibility for ordering flowers and supplies that are required for the shop. The manager of most flower shops is also the head designer, so all materials and stock to be sold are ordered by the manager. Purchasing supplies and flowers for a flower shop is not an easy job. The buyer must be able to predict what items or quantities of flowers will be needed in advance. The flower and supply needs can be determined to some extent by comparing advanced orders. However, many arrangements and plants are sold on impulse or for sentimental reasons, so advanced orders are not made.

A designer's assistant will not be required to purchase flowers or supplies. During the first few months of training, however, the assistant will be able to aid in decisions concerning these purchases. As they become more qualified as a designer, designers' assistants will be more familiar with pricing and with the types of flowers available.

Both designers and assistants hold extremely important positions in the flower shop. The arrangements that are constructed should always be of the best possible workmanship. The flowers should be fresh and neatly arranged before any floral design leaves the shop. At the same time, the arrangements should contain no more value than was purchased. The flower shop would quickly lose its profits if each designer added a few more flowers to each design just to make it prettier.

A designer in a flower shop must be familiar with the principles of good floral design. Each employee in the shop must be able to recognize each flower, plant, and type of greenery used in the design. They must also be familiar with the cost and the selling price of each item or flower sold. These designs must be constructed quickly and yet be arranged properly so the customer will return.

HANDLING CUT FLOWERS AND POTTED PLANTS

Cut flowers and potted plants are received regularly from the wholesale florist or greenhouse. These must be unwrapped and cared for as soon as they arrive if the flowers and plants are to remain fresh and healthy. Since live flowers and plants are the primary items sold from a flower shop, all personnel will be responsible for their proper handling and care.

Receiving Cut Flower Shipments

Cut flowers are received from wholesale florists at least once a week. Most flower shops order their cut flowers from wholesalers located in another city, so these flowers are packed in special boxes. The cut flower shipping boxes or cartons are constructed of corrugated cardboard to provide protection from weather and handling. When the shipping carton arrives, the bundled flowers should be carefully removed and inspected for damage. If damage has occurred, this should be reported to the shop manager. He will then file a claim for the damages with the shipping agency (Figure 9–2).

As the flowers are sorted onto tables, the packing invoice should be checked to determine whether the order has been accurately filled. It is possible to receive fewer flowers than were ordered or listed on the invoice. Any discrepancies should be reported to the manager. The flowers are then unwrapped or untied and placed in containers of water. They should be placed in a refrigerated cooler while the rest of the flowers are unpacked. The flowers should not be allowed to remain at room temperature for any extended period of time. When all the flowers are removed from the box, the newspapers used for insulation are neatly folded and the box is saved. Most wholesale florists will purchase the boxes back from retail florists.

Before cut flowers are placed in the refrigerated cooler for storage or display, they must receive further treatment. Roses must have the bases of their stems recut with a sharp knife to allow greater water uptake. At this time the thorns are also removed from the stems with a knife. Carnation stems are recut with a knife or sharp scissors as they are placed in containers. Cut greenery is stored under refrigeration in their original shipping cartons.

FIGURE 9–2.   A cut flower shipment; prompt unpacking, inspection, and refrigeration of fresh cut flowers is important.

The stems of standard and pom-pon chrysanthemums are first cut with a scissors about 1 inch above the base of the stem. The lower 2–3 inches of the recut stem is then crushed by pounding the stems with a hammer. This opens the stem for rapid water absorption (Figure 9–3).

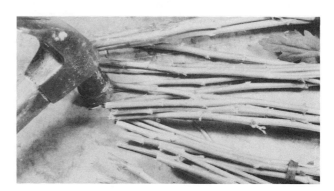

FIGURE 9–3. Crushing stems; cut chrysanthemum stems one inch above the bottom and pound with a hammer to help them absorb water faster.

Gladiolus are shipped in special cartons called "hampers." These flowers are shipped in a tight-bud condition and must be held at room temperature for a short period to allow the blooms to open. The bunches of gladiolus are untied and the stems cut at the base with a scissors. The gladiolus are then placed in large containers in 4–6 inches of water and held at room temperature for a few hours to allow the blooms to open slightly.

All containers used for storing flowers should be clean and watertight. The water used for previously stored flowers should be discarded. The containers should then be cleaned with a solution of detergent, household bleach, and water. The containers are then refilled to one-third of their depth (except with gladiolus) with tepid water. The water should be warm to the touch, but not too hot. Warm water is taken into the flower stems faster than is cold water.

A flower preservative is added to the water in each container to aid in prolonging the life of the cut flowers. Most commercial flower preservative compounds contain ingredients that reduce the growth of bacteria and provide a food source for the flowers. Sugar is included in these preservatives to provide a source of carbohydrates to the still-growing flower stems. Bacteria growth must be controlled in holding solutions for cut flowers. The bacteria will foul the water where stems and leaves are in contact with liquid. Bacteria also may plug the base of each stem, thus preventing water uptake by the cut flower (Figure 9–4).

FIGURE 9–4.   Adding flower preservative; these compounds reduce the growth of bacteria and provide a food source for the cut flowers.

Handling Potted Plants in the Shop

Potted flowering plants are delivered regularly by a wholesale greenhouse or are picked up at the greenhouse by the delivery truck driver. Plants are generally wrapped in a paper sleeve when they are shipped. Upon arrival at the flower shop, the wrap is removed and each plant is inspected for damage. The pots are dressed or decorated to make them more attractive to the customers. A wrap of foil, cloth, or heavy paper is used to hide the pot, since it often has become stained and soiled while the plant was growing in the greenhouse (Figure 9–5). The pot decorations may consist of the pot wrap only, or a bow may be added. The plants are watered thoroughly after the pots are decorated. A care tag is attached to the plant, which gives the purchaser special instructions about the care of that particular plant, and then it is priced.

Potted plants must be cared-for properly while still in the flower shop. Flowering plants on display will grow very slowly in the low light of the shop. They should be placed out of the main traffic lanes to protect them from damage. Plants should not be placed in areas where they will receive drafts from air conditioners, doorways, or heaters. The flowers will also need protection from the hot sunlight that enters the store windows. The decorated plants will require a light watering daily to prevent excessive drying of the soil. Some florists use a small moisture meter to determine when the plants require additional water.

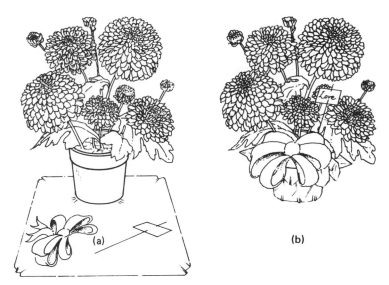

FIGURE 9–5. Decorating a flowering pot plant: (a) This plant is to be wrapped with decorative foil and decorated with a picked bow and flower care card. (b) The finished product.

WHOLESALE FLORISTS

The flowers and products used by retail florists are obtained from wholesale marketing firms. A greenhouse operator may sell his cut flowers and potted plants directly to florists in a local area. The retail florist may also obtain cut flowers and supplies from the wholesale florist in his locality. The wholesale florist acts as a middleman between the growers and the retail florists. A wholesaler provides a large warehouse for storing cut flowers and staple supplies received from growers and manufacturers all over the country. These items include permanent or dried flowers, ribbon and corsage materials, and other hardware supplies.

Cut flowers are purchased from both local growers and from cut flower distributors in other states. The wholesaler must supply materials to retail florists which are not grown in the local area. Often the cut flowers grown by local greenhouses are sold on consignment. This means that the wholesaler does not actually purchase the flowers, but receives a commission for their sale. The cut flower materials supplied by a wholesale florist include cut greenery, roses, carnations, chrysanthemums, orchids, gladiolus, and various other seasonal flowers. Wholesale florist firms do not usually supply potted plants to retail flower shops.

The wholesale florist is a very important part of the marketing channel for flowers. The wholesaler obtains flowers from distant suppliers where they may be grown more economically. For example, some orchid blooms are purchased from growers in the tropical climates of Australia and Hawaii. Carnations grown in Colorado are well known for their quality, so wholesalers purchase these primarily to fill the orders of florists. Cut greens, such as ferns and palm leaves, are obtained from the northwestern United States. Gladiolus are supplied by growers in Florida during the winter months and from northern growers during the summer. The wholesaler can obtain large quantities of flowers from growers at any season and distribute them to various retail flower shops.

Since the demand for cut flowers is seasonal, the prices and quantities of flowers available at any time is determined by the demand for them. The wholesale florist must be able to provide the best possible prices to the grower for his flowers, yet offer the same flowers to retail florists at a fair market price. To provide these services to the industry, the wholesale florist firm must hire many salespeople. Each salesperson develops a list of retail florists whose orders he will handle personally. The salesperson must keep informed of the current list of flowers, supplies, and their prices. Whenever a retail florist makes a special request for a particular type or color of flower, the salesperson is responsible for carefully preparing the order.

The wholesale florist firm employs truck drivers who supply the retail florists in the area. These drivers may be required to drive great distances to make deliveries to flower shops in other cities or states. Since the delivery truck drivers are the persons making personal contact with the retail florists, they must always attempt to be pleasant and provide assistance whenever necessary.

Floral designers may be hired by the wholesale florist to prepare made-up designs (Figure 9–6). A floral designer employed at a wholesale florist firm will generally be responsible for designing permanent or dried flower arrangements. These designers provide readymade arrangements or funeral pieces to retail florists, using materials that might not otherwise have been sold by the wholesaler. These designers will generally have received their training at a retail flower shop or school of design before beginning work for a wholesale florist.

The wholesale florist's warehouse provides coolers for cut flowers, storage for dry goods and hardware, and packing areas for deliveries. The shipping cartons that are received are unpacked quickly upon arrival. The cut flowers are sorted and placed in buckets of water with a floral preservative. The containers are placed in various coolers according to the storage temperature requirements of the flowers being stored. The sales personnel may then prepare each order for shipping from these refrigerated coolers.

FIGURE 9-6. The floral designers at a wholesale florist firm prepare permanent arrangements to be sold to retail florists.

FIGURE 9-7. Flowers are shipped in insulated boxes packed with ice and newspaper insulation. Local florists may take their cut flower purchases wrapped only in paper.

Cut flowers are packed in insulated boxes for shipping to retail florists (Figure 9-7). Each box is lined with newspaper to protect the flowers. Bundled cut flowers are placed in boxes with the stems pointing toward the middle. Ice is added to the center of each box to keep the cut flowers cool during summer deliveries. Each box is then covered with a lid and tied with a rope or steel strap. The delivery personnel are responsible for correctly labeling each box for delivery.

Potted plants and foliage plants are often delivered to retail flower shops in trucks owned by the growers. Retail florists may order specific plants for delivery or purchase plants directly from the truck. Wholesale growers provide regular delivery service to retail flower shops in their area. This ensures that the freshest and most long-lasting flowers will be available to customers of a retail shop at all times of the year.

A Rewarding Career

The employees or managers of a retail flower shop or wholesale florist firm will enjoy a promising career. These employees will be in constant contact with flowers and plants. They will be able to work with the public while enjoying the floral arrangements they create and sell. A skilled designer might be capable of managing or purchasing his/her own flower shop after a suitable period of training. The possibility of owning one's own shop is a strong incentive to a person entering this business as an employee.

SELECTED REFERENCES

Florists' Transworld Delivery Association. 1963. *Flower Shop Operation as a Career.* Southfield, Mich.

Hoover, N. K. 1963. *Handbook of Agricultural Occupations.* The Interstate Printers and Publishers, Inc., Danville, Ill.

Liesveld, J. H. 1951. *The Retail Florist.* Macmillan Publishing Co., Inc., New York.

Pfahl, P. B. 1968. *The Retail Florist Business.* The Interstate Printers and Publishers, Inc., Danville, Ill.

Richert, G. H., W. G. Meyer, and P. G. Haines. 1962. *Retailing Principles and Practices.* Gregg Publishing Division, McGraw-Hill Book Company, New York.

The Society of American Florists. 1969. *Opportunities for You in the Florist Industry,* SAF, 901 North Washington Street, Alexandria, Va.

U.S. Department of Agriculture. 1968. *A Graphic View of the Florist Industry.* Marketing Research Report No. 788, Economic Research Service. U.S. Government Printing Office, Washington, D. C.

U.S. Department of Agriculture. 1964. *Profile of the Retail Florist Industry*. Marketing Research Report No. 741, Economic Research Service, U.S. Government Printing Office, Washington, D.C.

Weyant, J. T., N. K. Hoover, and D. R. McClay. 1965. *An Introduction to Agricultural Business and Industry*. The Interstate Printers and Publishers, Inc., Danville, Ill.

TERMS TO KNOW

Aisle Runner	Cash and Carry	Floral Preservative
Candelabra	Cash on Delivery	Full-Service Shop
Care Tag	Chain Store	Wholesale Florist

STUDY QUESTIONS

1. Contrast the services offered and types of products sold in a full-service shop with those in a specialty flower shop.
2. List the qualifications required of delivery truck drivers working for a retail shop.
3. Describe the duties of a sales clerk in a retail flower shop.
4. Discuss how the seasonal holiday sales periods place greater demands on flower shop employees.
5. Describe the qualifications and duties of a designer in a retail flower shop.
6. Describe the methods used to prepare cut flowers when they are received from a wholesale florist.
7. Describe how the wholesale florist's services are of benefit to retail florists.

SUGGESTED ACTIVITIES

1. Invite the owner of a retail flower shop to visit your class to describe the duties of the shop's employees.
2. Visit a wholesale florist business in your area to observe how flowers and hardware products are handled.
3. Visit various types of flower shops in your community and discuss their differences.

CHAPTER TEN

Floral Designs —
Arrangements

The modern American florists' designs have evolved from the styles of flower arrangements used throughout history. These modern arrangements include the designer's imagination and originality in the creation of an artistic design. A successful floral designer must receive ample training and practice before being able to arrange flowers and foliage in a container in a pleasing manner for the commercial market. This chapter will explain the basic fundamentals of commercial floral design and will assist in the training of the floral designer's assistant.

HISTORY OF DESIGN

Floral designs as we arrange them today are a blending of two styles, oriental and European. The early *European* floral arrangements were based on mass and color. These designs were often influenced by the artists and rulers of the era. The Greek and Roman periods were characterized by flowers worn as garlands or wreaths. Fruit bowls were used to adorn banquet tables. The classical bouquets were first found during the Renaissance and Dutch–Flemish eras of floral history. The famous artists of these periods set the styles of floral art in containers. Since they were more interested in a total composition for their paintings, their designs were created with large masses of brightly colored flowers.

The favored containers and flower colors or styles changed often during the succeeding years of floral design history. The basic designs were dictated primarily by the rulers in power at the time. The first written design principles (rules) were established by the French. These rules were used and modified by floral designers in the creation of new floral patterns in more recent history. The basic European floral designs have become known as *mass* arrangements. These designs were all similar in that the large, closely spaced flowers were located at the perimeters of the design. Floral artists used lavish quantities of flower types and colors to create a massive display.

The *oriental* art of floral design began in ancient China and was later refined by the Japanese. These designs were created primarily for the observance of various religious ceremonies. The art of oriental design emphasizes a type of simplicity that has become known as *line* design. A simple design is created from natural materials to signify an event or tell a story. A typical oriental arrangement might contain three elements: a fully open flower, a half-opened bloom, and a tight bud. These would signify the past, the present, and the future. The historic placement of the three flowers in an oriental design signify *heaven, man,* and *earth.* The component that signifies heaven is always the tallest and earth is the shortest. Each twig or flower is placed at a precise angle. Although there are many variations in oriental floral designs, ranging from formal or classical to modern free style, they are all based on the basic placement of plant materials.

Modern florists' designs incorporate the simplistic line from the oriental art and the massed arrangements of European designs. Today's American floral pieces are called *line–mass* arrangements. Larger flowers are used at the center to create interest and balance in the designs. Colors are used more harmoniously for best floral display. The modern designer uses these skills to create designs that have more depth, simplicity, and originality than were present in the past.

PRINCIPLES OF DESIGN

A successful floral designer must become familiar with the basic tools of design before attempting to create a unique arrangement. Each floral design (whether a bouquet, centerpiece, corsage, or funeral spray) is a planned composition of container, flowers, and foliage. Before one attempts to create a floral arrangement it is essential to have a basic knowledge of the principles of design. These design principles are then used to determine the placement of components in an arrangement. In all design work, the floral designer is striving for emphasis, balance, proportion, rhythm, harmony, and unity.

Emphasis. *Emphasis* is achieved in an arrangement by creating an accent or focal point. The focal point will draw the most attention and cause the viewer's eye to be brought to one location in the floral design. The accent area of most floral designs will be located immediately in front of and above the lip of the container. This will be at the geometric center of a well-balanced design.

Accents may be created by the proper selection of flower types and colors. Bright colors against a neutral or dark background will draw attention. Large flower types and those that have opened most create emphasis because of their size. Accent is best created in a design when the smallest, least-open flowers are placed at the perimeter of the arrangement and the large, fully developed flowers are located at the focal point. Often a single flower will create an accent, but more often a floral designer will use three flowers to emphasize a larger arrangement. The purpose of the focal flowers is to visually draw all the elements of the design to a single location, the *center of interest*.

Balance.   A well-designed arrangement will appear to be stable and self-supporting—will be in *balance*. It will give the impression that the visual center of the design (focal point) bears the weight of the arrangement. This is achieved by placing the heaviest flowers (deepest colors, largest blooms, or massed effect) at the focal point. An arrangement may be composed of symmetrical or asymmetrical balance (Figure 10–1). A symmetrical design will appear to be the same on each side of a vertical center line. Most floral designs are arranged with asymmetrical balance. This gives the appearance of a more natural design and allows the arranger more originality in his/her work. Balance is also best achieved when the design is ar-

FIGURE 10–1.   Balance in design.

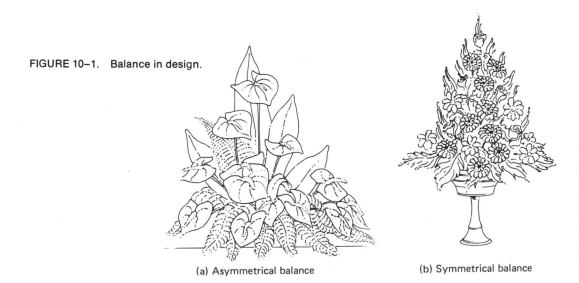

(a) Asymmetrical balance   (b) Symmetrical balance

ranged from the back of the container toward the front. This gives the arrangement a feeling of visual depth and support.

Proportion. A floral arrangement appears best when all the component flowers are of the proper sizes. *Proportion* in an arrangement is accomplished by the scaling of flowers toward the focal point. This means that the smallest buds are placed farthest from the center of the design. The flowers are then scaled by the placement of increasingly larger flowers in the arrangement until the center of interest is reached. Here are placed the largest or most showey flowers in the arrangement.

Proportion is also achieved by the use of flowers of varying size. The use of large orchids with tiny violets may appear out of place. At the same time, a monotonous design would be created with the use of flowers of exactly the same size throughout the arrangement. The use of scaled flower sizes allows the creation of a pleasing design (Figure 10–2).

FIGURE 10–2. Scaled flower sizes; use of flowers that are similar in size helps to achieve a feeling of good proportion within a design.

Rhythm. The viewer of a floral arrangement should be able to see all areas of the design, yet be led to the visual center of interest. This visual movement is created through the use of *rhythm* in the design. Motion is created most easily in a design with a curved line. Here the eye must travel from the top of the arrangement, through the focal point, and out the other end. In a triangular design the points of the arrangement must be made to appear to radiate from the center.

Rhythm may be achieved in an arrangement in several ways: through radiation, gradation, or color. When flowers are placed at opposing points away from the focal point, the center of interest is emphasized. The gradation of scaled flowers also leads the eye to the focal point. Color can be used to easily create motion in a design. The darkest or brightest color may be placed at the focal point. From the center, the intensity or tone of the color is graded to be more subdued. This will create motion within the design which leads the eye to the center of interest in the arrangement.

Harmony. When an arrangement possesses *harmony*, all the parts of the design fit together. This could be interpreted to mean that the idea or theme of the design has been successfully created. In order for harmony to be achieved, all of the component parts of the design must blend together. This will include the proper use of color, texture, and design in a pleasant relationship.

Unity. *Unity* is created when all the parts of the design blend together to form a single idea or impression. Although each flower or area of the arrangement may be distinctive, all elements must be pulled together. Unity is best achieved by repeating the same flower types or colors throughout the design. This helps to pull the components of the arrangement together, with the focal point as the center of interest. When one color or type of flower is used at the perimeter of the arrangement and another at the center, no unity is felt by the viewer of the design (Figure 10–3).

FIGURE 10–3. Lack of unity; unity is lacking when the arrangement can be divided into separate parts.

Lack of unity

DESIGN ELEMENTS

The principles of design are used to create a successful flower arrangement. The elements of good designs are used in achieving those principles, much like building blocks are used to make a structure.

The principles of design are similar to a recipe, with the elements of design being the ingredients. In order that we might create successful arrangements through the use of good design principles, the following design elements must be understood: line, form, texture, and color.

Line. *Line* in a flower arrangement is the visual path that the eye travels as it passes through the arrangement. The line establishes the skeleton of the design, particularly when linear flowers or foliages are used. Line may also be created by repetition of similar flower colors, textures, or shapes. Certain emotional qualities are seen in the line of the arrangement. A vertical line gives the arrangement a feeling of strength and is considered a masculine design. A curved line adds gentleness or restfulness and has the feeling of motion. The curved designs convey a feminine feeling. A horizontal line is more relaxing or informal, so is used most often for table arrangements.

Form. The flowers, foliages, and containers used in flower arranging have various shapes, or *form*. Flower and foliage shapes add a visual quality that is important in developing harmony, creating rhythm, and establishing a focal point.

Texture. *Texture* is a term that is most easily understood. It refers to such surface qualities of plant materials as smoothness, glossiness, and roughness. Texture may also be used to describe leaf or stem patterns. The texture of each part of an arrangement should blend pleasingly with its neighbor. A bold contrast in textures will create an accent (focal point).

Color. Pleasing *color* schemes may be seen all around us. The colors in the wings of a butterfly or a bird are combined in pleasing patterns. The colors found in the petals of the flowers themselves are attractive. The floral designer must be able to combine these flower and foliage colors in a pleasing combination. The *color wheel* has been developed to aid the arranger in the development of successful color schemes (Figure 10–4).

Primary colors are the fundamental pigments from which all others are derived. The primary colors are *yellow, red,* and *blue.* Another word that is used in place of color is *hue.*

Secondary colors are those that are created by mixing equal proportions of any two of the primary colors. These are *orange, green,* and *violet.*

Tertiary colors are made when a primary color is mixed with a secondary color. These carry the names of both colors used when creating them. For example, when equal parts of red and orange are combined, they create red-orange.

Colors also have other characteristics that are important in the design of flowers. A flower color possesses an *intensity*. This is a visual quality of brightness or dullness created by adding different proportions of *gray*. The more gray that is added, the less intense or bright is the color.

A color also will exhibit various degrees of *value*, depending on the amount of black or white pigment added. When black is added to a color, it is darkened, and a *shade* of the color is created. The addition of white creates a *tint* of a color. Pink would be a tint of red.

Colors are combined in a floral arrangement to create pleasing color schemes. Some of the most common color schemes used by florists when designing are: accented neutral, monochromatic, analogous, complementary, and split-complementary.

The *accented neutral* color scheme is composed of flowers having any hues used with a neutral background. A single accent color is selected for emphasis. The background flowers and foliage may be white, gray, brown, or black. The accented color then becomes predominant.

A *monochromatic* color scheme is created from the tints and shades of a single hue. A truly monochromatic design must be composed of flowers, foliage, and a container having only a single hue.

An *analogous* color scheme is created by combining any three hues found next to each other on the color wheel. This color scheme is successful because any three colors that lie next to each other on the color wheel have one primary color which has been used in the development of each of the other hues. For example, yellow is a part of the mixing of yellow, yellow-green, and green in an analogous color scheme.

Any two colors that are located directly across from one another on the color wheel are complements. When the two colors are used together in an arrangement, a *complementary* color scheme is created. Since such colors both attract attention, it is best to use the tints or shades of one hue, while using a small amount of the opposite color in a bright and intense tone. Examples of complementary colors are red and green, yellow and violet, or blue and orange used together.

The *split-complementary* color scheme resembles a complementary color scheme except that three hues are used. In this color scheme a hue located on one side of the color wheel is used with each of the hues located on either side of its complement. An example of a split complementary color scheme is red used with blue-green and yellow-green. This color scheme is more attractive and is easier to construct than is the complementary color scheme.

Colors are used in arrangements for creating emotional impressions. *Warm* colors are those in which yellow and red tones predominate. Warm colors suggest such feelings as fire and sun. These colors appear best in daylight or under incandescent lights. *Cool* col-

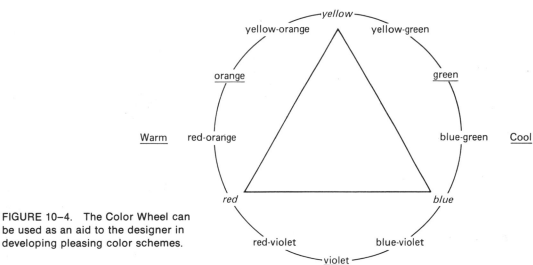

FIGURE 10–4. The Color Wheel can be used as an aid to the designer in developing pleasing color schemes.

ors are composed of green and blue tones. These colors suggest such feelings as sky and grass. The cool colors may be used with candle-light or fluorescent lighting.

Red gives the feeling of excitement. It should be used sparingly with large quantities of other colors (preferably white) for balance. *Yellow* creates a mood of cheerfulness. It is a good choice for use in a corsage to be worn by a girl having a blond complexion. Yellow is not always a good color to be used with red or pink. *Green* is a natu-ral color and is used best to provide a background for other flower colors. *Blue* adds darkness to an arrangement. It is intensified when used with brown, silver, or gold. *Violet* symbolizes dignity and roy-alty. This color is often used in religious designs. Violet-colored flowers are most attractive when combined with yellow. Brown pro-vides an excellent background for violet-colored blooms.

SELECTION OF FLORAL CONTAINERS

The container is an important part of any floral arrangement. The container is the starting point or foundation of a design, so it should be suited to the particular design that will be arranged. The container selected should be suitable for arranging flowers. It must be capable of holding water, be stable enough to support the weight of the

FIGURE 10–5. Floral containers; a selection of suitable containers in various sizes is helpful to the designer.

flowers, and be deep enough to allow the stems to reach the water (Figure 10–5).

A container must be suitable for the design to be constructed. A low, flat container is used for low table arrangements. Tall and narrow containers are best used for vertical designs. Stemmed compotes are the best choices for curved designs. The shape of the container may suggest the line to be used in a particular design.

The container should not dominate the floral composition. The container, flowers, and room decor should complement each other. A simple, unadorned container is often more effective than an ornate or highly decorative one. The container should provide visual balance as well as mechanical balance. This implies that the container will be heavy enough to hold the arrangement, will have a large enough opening at the top for proper arranging of flowers, and will be deep enough to support the stems. A properly selected container will offer these qualities without attracting undue attention from the arrangement.

The color of a container will affect its suitability for flower arrangements. A dark container appears heavier than one which is light in color. Dark containers suggest masculine lines and larger arrangements. Light-colored containers suggest dainty, feminine lines in the design. The color of the container should not pull too much attention away from the flowers. Containers having earth colors (brown and green) or neutral colors (white, gray, or black) are the most versatile.

The texture of the container should not contrast strongly with the texture of the flowers being used. A delicate texture of flowers indicates a need for a smooth textured container having a light visual appearance. These might be constructed of fine china, crystal, or stemmed glassware. Flowers having a coarse texture require a coarse-textured container. These might be arranged in pottery or pewter containers. It is not necessary to use an elegant container to design a beautiful arrangement. If a container is too fancy it may fight for attention and detract from an otherwise well-designed arrangement.

SELECTION OF FLOWERS AND FOLIAGE

Flower and foliage shapes influence the line and form of an arrangement. The designer creates a style of arrangement by properly selecting those flower and foliage types that will best express the desired effect. Flowers and foliage types may be classified for best use by their shapes. The shapes of flowers and foliages used by florists are classified into four groups: line, form, mass, and filler.

Line flowers. *Line flowers* help to create the skeleton or outline of a design. These flowers are composed of a tall, erect spire of blossoms which add height to an arrangement. Some examples of line flowers are gladiolus, snapdragons, stocks, delphiniums, and branches from spring-flowering shrubs. Foliage types that have a linear shape are used to complement linear flowers. Some examples of line foliage are gladiolus leaves, eucalyptus, sansevieria, and scotch broom.

Form flowers. *Form flowers* can easily be used as accents or to create a center of interest. Form flowers have distinctive shapes and may be used to create a visual path for the eye to follow by pointing the blooms in a line. These flowers should not be massed together. For best effect, space should be left between them. Some form flowers are best used to establish the outline of an arrangement. Since these flowers point in one direction, they require a skilled designer to use them properly. Examples of form flowers are lilies, anthurium, orchids, bird-of-paradise (*Strelitzia*), and daffodils.

Mass flowers. *Mass flowers* are composed of a single stem with one solid flower at the top. These are used to add depth to an arrangement. Larger blooms may be used to create a focal point, because they will add weight to the arrangement. The flowers are graded in size from the top of the arrangement to the focal point. The height of each bloom is varied so that each flower is revealed. Some examples of mass flowers are rose, chrysanthemum (standard), daisy, carnation, and aster. Mass foliage types that are used with these flowers might include lemon leaf (salal), caladium, and leather leaf fern (Baker's fern).

Filler flowers and foliages. *Filler flowers* are used in an arrangement to add a finishing touch. Foliage having a light and airy form aid in concealing holes and spaces in a design. Asparagus ferns and huckleberry are most often used by florists for this purpose. Filler flowers are found in two types: bunch and feather. *Bunch*-type filler flowers have many stems with small, mass-type heads. Examples of these are pom-pon chrysanthemums, button mums, aster, and sweet pea. *Feather*-type filler flowers add a misty or delicate appearance. These give the arrangement a more feminine or softened impression. Some examples of feather-type filler flowers are statice, gypsophilla (baby's breath), and heather.

MECHANICAL AIDS FOR ARRANGEMENTS

The various types of florist's containers used in a flower shop require some method for anchoring flower stems in arrangements. Several styles of flower stem holding devices are available for use in the containers used by florists. Stem support for weak-stemmed flowers is also provided by wires inserted in the stems by several methods. A bow may be added to a design to create a subtle accent. The beginning floral designer will need to master the techniques required for handling these floral aids before progressing to the arranging of flowers in containers.

Flower stem supports. Large baskets and bouquets require a firm support for the tall stems used in their arranging. The stems must be submerged to a depth of 4–6 inches and must not slip out of place during delivery or handling. The support aid used by florists for these large arrangements usually consists of *chicken wire* and a shredded *filler* material. The filler material may be of shredded styrofoam plastic or chipped floral foam. After the shredded filler material is pressed into the cavity of the container, a layer of chicken wire is placed over the top. This is done by bending and molding the wire to fit the container opening. The wire will prevent the filler from floating out of the container when water is added. It will also serve as a strong anchor for the large stems that are inserted into the arrangement (Figure 10–6).

Floral foams. Floral foams are now quite popular with florists for most design work. Such products as Oasis,® Quickee® foam, and Hydrofoam are available in round or square blocks. These products are extremely porous, so they provide water to the inserted stems used in the design. A flower preservative is also included in these foams to aid in extending cut flower life.

Some of the containers used by florists are specially designed for anchoring floral foam blocks. When these are used, the foam block is simply pressed into the holder provided at the base of the container. When these foams are used with other styles of containers, other an-

FIGURE 10-6. Large flower stem supports; chicken wire with a filler material is sometimes used for large arrangements in papier-mache containers.

choring methods are required. The foam blocks are cut to fit small-mouthed containers and then pressed into the opening. A small wedge-shaped opening is cut from the block to provide a space for adding water to the container. The block should be saturated with water at all times for the best cut flower life. The foam block is inserted to a depth that leaves 1½–2 inches of the block extended above the lip of the container for a more attractive arrangement.

When large containers are used for arranging, the foam blocks may not be anchored securely by merely pressing them into the cavity. A water-resistant florist tape is required for holding the foam block in place. The lip of the container must be thoroughly cleaned and dried before the tape is applied. A single strip of tape is placed over the foam block and secured over the lip of the container. In cases where large foam blocks are required, more than one strip of tape may be necessary for firm anchorage. The tape that extends over the lip of the container may be easily concealed with foliage or ribbon when the flowers are arranged.

The foam blocks must be thoroughly saturated with water before being used. The quick-filling floral foams may be simply placed in a sink or bucket filled with water. After a few minutes the foam block will sink to the water surface. These floral foam blocks should not be used until they are completely saturated with water, or severe flower wilting may occur. These foam products are available in various grades and sizes. The designer should select the foam that is best suited for the arrangement and the flowers to be used.

Needlepoint holders. Needlepoint holders are rarely used by commercial florists, but are popular aids in home designing. They are also called pin holders or "frogs." The needlepoint holder is secured to a container with florist's clay. The bottom of the container is cleaned and thoroughly dried before the clay and holder are secured firmly. A needlepoint holder consists of many sharp-pointed nails which hold the stems firmly when inserted. When a small stem is inserted between the points, a larger stem may be wedged against it to anchor it securely.

Styrofoam. Styrofoam is used for many purposes by florists. This product is generally used in blocks cut from sheets 2 inches in thickness. Although fresh flowers may be picked into these blocks, the greatest use is for anchoring permanent and dried flowers in containers. The styrofoam sheets may be obtained in several colors or they may be colored with floral spray paints.

A properly sized styrofoam block may be easily cut with either a knife or a fine-bladed saw. The block may be secured to the container in several ways. A small amount of specially designed clay may be used for this purpose. Some florists prefer the use of heated wax or "hot glue" for creating a strong seal between the container and the styrofoam block. Both permanent and fresh flower arrangements may be constructed in styrofoam through the use of picks or wires applied to the stems. Fresh flowers are used with styrofoam holders only for special pieces, such as for funeral or wedding work. Since the styrofoam will not hold water in the manner of the floral foams, the flowers will not remain fresh for extended periods.

Floral picks. Picks used for securing flowers in arrangements are available in a wide variety of styles. The introduction of the mechanical picking machine has nearly replaced other types of picking methods once popular with florists. This machine applies a metal pick to a stem by the depression of a handle. These picks are available in several sizes and secure the stems more tightly than conventional picks (Figure 10–7).

Wooden floral picks are still used by florists, especially for securing large flower stems in styrofoam blocks for funeral sprays or wreaths (Figure 10–8). These wooden picks may be supplied with a wire at the top to aid in securing a flower stem to the pick. Sizes of these picks range from 2½ to 6 inches in length. The pick is placed at the base of the stem so that two-thirds of the pick extends below the cut edge of the stem. The wire is tightly wrapped around the stem and then to only the pick for added support. When dried or permanent flowers are picked, the pick and wire may be wrapped with an appropriately colored floral tape to cover the area. Large, fleshy stems (such as those of gladiolus) are picked by inserting the top one-third of the pick into the center of the stem before the wire is wrapped around it.

When leaves or other dressings are applied to sprays, wreaths, or easel pieces, greening pins are used. These are hairpin-shaped, double-pointed pins having an S-curved top. The foliage or paper is

FIGURE 10–7. Metal pick machine; a metal pick is applied to a rose stem.

FIGURE 10–8. Wooden floral picks; this technique is generally used for funeral sprays and wreaths.

placed in a flattened position on a styrofoam block or frame. The greening pin is inserted through the base of the foliage or at a fold in the dressing. The pins are concealed by the next layer of leaves or folds.

Wiring flower stems. Florist wire is available either in enameled green or as unpainted wire strands. The wire is sold in 18-inch lengths in boxes containing 12 pounds of wire. The wire thicknesses used by florists range from the larger diameter of 18 gauge to the very thin wire of 36 gauge. Most florists will use two gauges of heavy wire (Nos. 18 to 24) and several thicknesses of light wire (Nos. 26 to 36). The heavier wires are used primarily for reinforcing large stems in designs. The lighter gauges of wire are used most often in the construction of corsages and in wedding work.

Flower stems are wired by the florist for several important reasons. First, the addition of a stiff wire to a stem may prevent a blossom from being inadvertently broken during delivery and handling. As flowers age, they become weak in the stems. The addition of a support wire will hold the bloom erect after the stem can no longer provide adequate strength. Wire is used to combine several blooms together to create a massed effect. One of the most important reasons for wiring flowers is for the purpose of straightening a crooked stem, or to add a curve to an otherwise straight stem. The use of wire allows the designer to create the desired curves and lines in an arrangement.

Fresh flowers are wired by several methods, depending on the flower type and the preferences of the designer. Roses, carnations, and other similar mass-type flowers are supported by the insertion of the wire into the base of the calyx with the remaining wire loosely wound around the flower stem (Figure 10–9). Larger, fleshy-stemmed flowers (such as gladiolus, tulips, and daffodils) are supported by inserting the wire through the center of the stem to the base of the flower head. Flowers having flattened heads on weak stems (such as pom-pon chrysanthemums and daisies may be supported by first running the end of the wire completely through the flower at a point near the stem attachment to the calyx. A small hook is shaped at the end of the wire, which is then pulled back into the center of the petals until it is securely fastened. Whenever flowers are wired for use in arrangements, the wires are to be concealed by foliage or flowers.

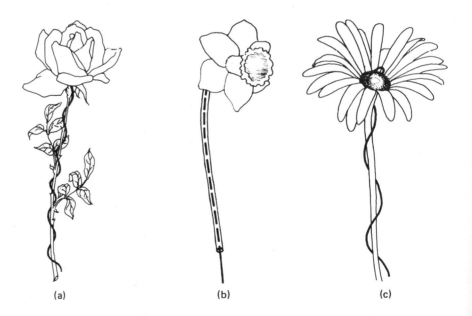

(a)   (b)   (c)

FIGURE 10–9.   Wiring for arrangements: (a) Loosely wrapped wire is used for roses. (b) Insert a wire through the center of fleshy, stemmed flowers. (c) The hook method is used for flat headed flowers.

FLORAL ARRANGEMENTS

The beginning designer's assistant will most likely be responsible for preparing the containers and cut flowers to be used by the designer for arrangements. The assistant's duties will include wiring the flowers. When necessary, picks may have to be applied to a large number of flower stems. Floral containers should be kept in stock and, when desired, should be filled with shredded filler and chicken wire. The designer will appreciate having containers ready for use when they are needed.

As the assistant designers gain the skills and knowledge necessary for design work, they may be given the responsibility of arranging some small pieces. The basic techniques required for constructing these arrangements are easily learned. However, before more complicated arrangements may be attempted, the assistant designers must be able to create pleasing designs both rapidly and accurately. This period of training is both challenging and rewarding for the ambitious and creative employee.

Shapes of Floral Designs

Flower arrangements are generally more pleasing to the eye when their outline creates a geometric pattern. The basic shape of the arrangement establishes the line and creates unity of the design, whether it is traditional or modernistic in mood. The outline of the arrangement is the framework from which the floral composition is constructed. The proper establishment of the floral framework makes the difference between a planned floral design and flowers that are merely placed in a container.

The most common shapes used in florists' designs include the following (Figure 10–10):

- Circular
- Triangular
- Radiating
- Crescent
- Horizontal
- Hogarthian curve

Other floral design patterns that may be constructed by a florist are the *right triangle, inverted T, vertical,* and *free-form* arrangements. Although these designs can be beautiful, they are not as marketable as

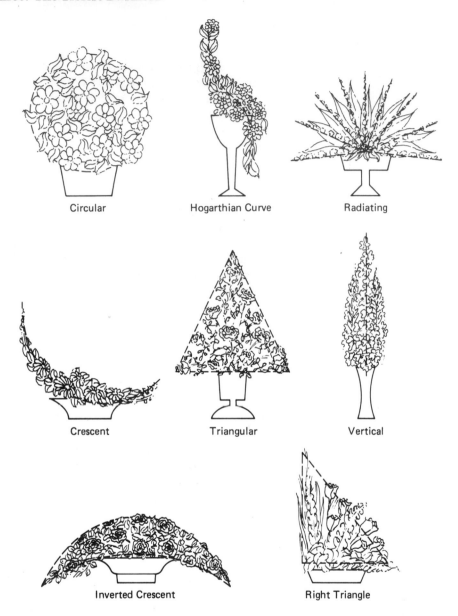

FIGURE 10–10.   Shapes in floral designs; these are the traditional shapes for flower arrangements.

the more conventional designs. Most customers prefer the "bouquet" look in designs, because such arrangements appear to have more flowers. This makes the customers feel they are receiving their money's worth in flowers. Free-form styles of arrangements result from the use of symbolism and an expression of the designer's own personality. This self-expression may not be consistent with the customer's own feelings.

Circular designs. *Circular* designs are widely used because they are relatively simple to construct and serve many purposes for the customer. Circular designs may be constructed so as to be viewed from only one side. However, they are usually designed to be seen from all sides (all-around designs). The circular design will make an excellent centerpiece for a low table and for dinner or reception parties. These arrangements, when designed to be seen from all sides, are especially attractive when placed in front of a mirror.

Circular designs are constructed in low, round containers or they may be arranged in baskets that hold water. The shape of the design is determined by making each of the "skeleton" flowers of equal lengths and placing these in position at the top, front, back, and both sides of the arrangement. Each successive flower is then positioned so that the stem lengths will be the same as the "skeleton" flowers. The foliage is placed between the flowers to give a natural appearance and to hide the flower stems and holder. The circular design will not have a focal point when it is arranged to be viewed from all sides. Since the design appears best from the top, the same sizes of flowers may be used in the construction of a circular centerpiece arrangement (Figure 10–11).

FIGURE 10–11. Circular arrangement; the lines of the lamp are repeated in the arrangement for this setting.

Triangular designs. *Triangular* designs may be of two basic types: symmetrical and asymmetrical (Figure 10–12). The symmetrical design creates a mirror image between the left and right sides of the arrangement. The asymmetrical design creates a more informally balanced appeal.

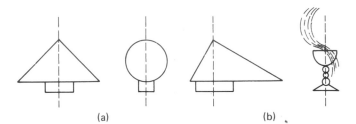

FIGURE 10–12.   Balance: (a) formal balance and (b) informal balance.

(a)

(b)

Triangular designs are constructed by first determining the height and width of the arrangement. An equilateral triangle-shaped arrangement will be equally as tall as it will be wide. The tallest flower is placed exactly in the center of the container and as far to the rear of the holder as is possible. The two "skeleton" flowers are then placed at each side. These should then have a total length equal to the height of the vertical flower. A short-stemmed flower is placed at a horizontal angle with the rim of the container at the front of the arrangement. The focal point flower will be positioned immediately above and in front of the rim of the container. The triangular arrangement is completed by filling in with the remaining flowers and foliage.

An asymmetrical triangle will be constructed in a similar manner, except that the height and width of the arrangement will be altered. Although these arrangements are designed to be viewed from the front only, a professional designer will make certain that the back of the arrangement is finished also. The stems and floral foams or other holder may be covered attractively with a few sprigs of foliage which remain after the arrangement is completed (Figure 10–13).

Radiating designs. *Radiating* designs have the general outline of a fan. This design is used extensively for the larger arrangements at large group affairs. This is a traditional design for wedding, commencement, and funeral bouquets. Line flowers or foliages are used to form the outline. Florists use gladiolus, snapdragons, flat fern, and palm fronds most often for this purpose. The height of the arrangement is established first. After the width of the design is determined by the placement of flowers at each side, the fan shape is created by placing flowers or foliage across the back to give a rounded appear-

FIGURE 10–13.   Preparing a gladiolus; unopened buds that will bend away from gravity and create excess length are broken off the top of a gladiolus stem.

ance (Figure 10–14). The designer must be especially careful to avoid the lack of unity with this particular design. The flowers selected to form the fan-shaped background must be repeated throughout the entire design. A lack of unity will result, for example, when gladiolus are used to create the skeleton of the arrangement and carnations or chrysanthemums are used to fill out the center of the design.

Crescent designs.   The designer who creates a *crescent*-shaped arrangement needs to be skillful in achieving visual balance. The crescent design has a naturally free-flowing outline which has a formal appeal. The designer will select materials having a natural curve when making this style of design. Florists usually select easily obtained foliages such as scotch broom, eucalyptus, forsythia, cattails, and other easily curved materials for this purpose, when they are in season. The crescent design is constructed in a "half-moon" shape having a height at the left side being two-thirds the length of the entire line of the arrangement. The curved foliage is placed to the side (usually left of the center) so that its tip nearly reaches the visual center line of the design. The focal point is located directly beneath this point at the base of the arrangement. The flowers used in this design will be smallest at the points and largest at the center of interest of the arrangement. Flowers having a heavy visual weight placed at the focal point create balance in the design.

Horizontal designs.   The *horizontal* design makes an excellent centerpiece because it is beautiful when viewed from either the front or the back. This design is arranged with a low profile so it will not block the view of seated dinner guests. The long and narrow shape of

(a)　Establish the height and width first.

(b)　Create the overall shape.

(c)　Place additional flowers and the bow within the shape.

(d)　Add the enframement foliage, if desired, and finish the back of the bouquet.

FIGURE 10–14.　Constructing an arrangement.

this style of arrangement repeats the long horizontal line of a rectangular table. This repetition of lines aids in establishing unity in design.

The horizontal design is constructed in the same manner as a triangular arrangement, but with two notable changes. The height of the arrangement is reduced so that the horizontal length becomes 1½–2 times the length of the container. The horizontal line appears best when the tips of the flowers placed in line with the container droop slightly downward. This gives the arrangement the appearance of being nearly like an inverted crescent design. Since this design is to be viewed from both sides, it is arranged from both sides of the container. The skeleton flowers are placed midway between the back and front of the container. A focal point may then be established on each side to attract attention to the design. This style of arrangement may easily be used with candles for an evening dinner party. When this is done, three candles are placed in the arrangement. The center candle is tallest, with the two other candles placed to the sides at the center line of the arrangement. All foliage and flowers located near the candles should be low enough so they will not be burned as the candle is shortened by the flame.

Hogarthian curve. The *Hogarthian curve* is a sophisticated asymmetrical design named after William Hogarth, an English artist. This design has a graceful appearance which lends formality to a room decor.

The Hogarthian curve design has the outline of an S, hence its nickname. Tall, stemmed compotes or raised containers are used for this design, because a portion of the floral line extends below the rim of the container. The S outline is constructed in exactly the same manner as the crescent design, with the exception that the line is reversed on one side of the focal point. The S shape is separated into two elements, with the upper curve consisting of two-thirds the height of the total design. The lower portion of the curve may be constructed with either curved material or dainty, dangling flowers. The focal point is often depicted (particularly in permanent designs) by a cluster of grapes gracefully dangling over the rim of the container. This is a rather attractive design, but is usually sold only for special occasions.

DESIGN RULES AND AIDS

1. Keep the balance and proportion of each design in the proper perspective. The overall height of the arrangement is usually 1½ times the height or width of the container (Figure 10–15).

2. Before you begin construction, know what size and shape your arrangement will have upon completion.

3. Work quickly and precisely when constructing an arrangement. Flowers soon wilt when left out of water. Also be sure that the container is amply filled with water before the flowers are added.

4. The focal point, when present, should be clearly apparent, yet it should not be created by a massing of closely spaced

FIGURE 10–15. Height of an arrangement, the overall height of an arrangement is usually 1½ times the height or width of the container.

flowers. This technique would give the effect of a circle drawn around the center. Carefully placed flowers having a larger size or unusual shape will be most effective in creating a center of interest. Use an odd number of flowers when creating the design and especially at the focal point. Odd numbers of flowers (1, 3, 5, 7, and 9) are more pleasing to the eye and aid in avoiding complete symmetry in designs.

5.   Foliage should be placed in such a manner as to give a natural appearance. Foliage placed between, and extending through, the flowers will soften the arrangement. The foliage should also be used to conceal the stems and holder at the base of the container. Keep in mind that the foliage used in florist's arrangements is often as expensive as the flowers. Even the small and broken branches or leaves may be used in the completion of the arrangement.

6.   Flowers are placed so space is allowed between each bloom. Generally, the petals from one flower will not touch those of another. Foliage is inserted in these spaces.

7.   Flowers as they grow in the wild do not all grow at the same height or in straight lines. When creating line or rhythm in a design, avoid the placement of flowers in lines or rows. A staggered row will create the desired affect, without allowing flowers to be placed at the same exact heights. A vertical staggered row is constructed by placing a flower slightly below and to the right of the first. The third flower is then placed slightly below the second and on the left side of the first. This pattern is continued throughout the design until the desired effect is created.

8.   When flowers of the same size, color, texture, or form are used in the creation of an arrangement, it is best to use an odd number of blooms. The flower numbers are apparent up to nine blooms. When flowers of the same type (but different colors) are used, the more dramatic color will create the accent in the design.

9. Create unity in your designs by using each color, texture, or type of flower throughout the entire arrangement. Smaller blooms may be used at the outside perimeters of the arrangement, while the larger blooms are used toward the center. A lack of unity exists when one type of flower is used to form the skeleton or backbone of an arrangement and the center of the design contains another.

10. A successful design will have three dimensions. This is created by constructing each arrangement from the back of the container toward the front. When done properly, the focal flower (accent) will be the most prominent and placed in front of all other flowers in the arrangement.

11. Finish the back of all one-sided arrangements. Even though the arrangement is intended to be viewed only from one side, the addition of greenery or a few left-over flowers will conceal the holder and flower stems at the rear of the container. This step in the construction of any floral arrangement tells customers that designers are conscientious and take pride in their work.

12. Once designers become skilled in the principles and elements of design, they may use their creative imagination when arranging flowers. Often advanced designers will not follow these rules directly, but will be able to create designs that are subtly attractive and appealing to customers. Such designers have reached the highest level of the profession.

SELECTED REFERENCES

Benz, M. 1966. *Flowers: Geometric Form.* 3rd ed. San Jacinto Publishing Company, Houston, Tex.

Bode, F. 1968. *New Structures in Flower Arrangement.* Hearthside Press, New York.

Cutler, K. N. 1967. *How to Arrange Flowers for all Occasions.* Doubleday & Company, Inc., Garden City, N.Y.

Davis, Margaret. 1971. *Living Flower Arrangements.* Henry Regnery Company, Chicago.

Florists' Transworld Delivery Association. (Current). *FTD Floral Selection Guide.* Detroit, Mich.

Fort, V. P. 1962. *A Complete Guide to Flower Arrangement.* The Viking Press, New York.

Hawkes, F. A. 1969. *The Gracious Art of Flower Arrangement.* Doubleday & Company, Inc., Garden City, N.Y.

Hillier, F. B. 1974. *Basic Guide to Flower Arranging.* McGraw-Hill Book Company, New York.

Klamkin, M. 1968. *Flower Arrangements That Last.* Macmillan Publishing Co., Inc., New York.

Pfahl, P. B. 1968. *The Retail Florist Business.* The Interstate Printers and Publishers, Inc., Danville, Ill.

Rockwell, F. F., and E. C. Grayson. 1960. *The Rockwells' New Complete Book of Flower Arrangement.* Doubleday & Company, Inc., Garden City, N.Y.

Rutt, A. H. 1958. *The Art of Flower and Foliage Arrangement.* Macmillan Publishing Co., Inc., New York.

Soules, K. 1957. *Modern Florist Designing.* Florists' Publishing Company, Chicago.

Tolle, L. J. 1969. *Floral Art for Religious Events.* Hearthside Press, New York.

TERMS TO KNOW

Accented Neutral	Harmony	Rhythm
Analogous	Intensity	Secondary Colors
Balance	Line Arrangement	Split-Complementary
Color Wheel	Mass Arrangement	Tertiary Colors
Complementary	Monochromatic	Unity
Compote	Primary Colors	Value
Emphasis	Proportion	

STUDY QUESTIONS

1. Explain how the oriental and European floral design styles have influenced the types of flower shop arrangements sold today.
2. List the design principles and give an explanation of how each is used in a floral design.
3. Explain how each of the design elements is used to create the principles of design.
4. Discuss how each of the various floral design shapes may be used in decorating for best display.

SUGGESTED ACTIVITIES

1. Make a color wheel to be used in creating various design color schemes.
2. Conduct a class survey by having the students write down the first word they think about when various colors are mentioned. Discuss how these colors convey specific moods as a summary of this activity.
3. Invite a retail floral designer to demonstrate the construction of the various commonly sold flower arrangements for the class.
4. Practice the construction of the various flower arrangements discussed in this chapter.

CHAPTER ELEVEN

Floral Designs — Corsages and Specialty Pieces

The florist business is largely one of seasonal sales. The major sales periods occur during the special florist holidays. The business depends heavily upon other sentimental or special occasions for flower purchases during the periods between major flower holidays. The holiday season flower sales also require some specialized floral arranging techniques by the designers. Although the traditional wedding months are June and August, a wedding may occur during any week of the year. A florist may have weddings booked months in advance of the actual marriage date.

Funeral business is much less easily predicted, with sales of funeral pieces occurring at all times of the year. The demand for flowers at funerals is waning in some areas but remains steady in others. Any person desiring to become a floral designer must learn the techniques of funeral piece construction. This chapter will discuss the construction and use of some specialty floral pieces created by flower shop designers. The important items to be discussed are the techniques required for constructing corsages, wedding bouquets, funeral pieces, and special holiday designs.

CORSAGE CONSTRUCTION

Corsages are still very popular florist's items for special occasions. Florists sell corsages mainly for special dances, such as formals and proms. Corsages are also sold extensively during the Christmas,

Easter, and Mother's Day holidays. One of the more important uses for corsages, however, is for weddings.

The construction of florist's corsages resembles the arrangement of flowers in containers. The bow and trim become an integral part of the design and satin leaves may provide the foliage. The principles and elements already discussed for cut flower arrangements still apply. However, a few additional rules need to be followed when constructing corsages.

Scale.   The size and form of the corsage should be compatible with the wearer. A tiny girl would not feel comfortable wearing a large Cattleya orchid corsage.

Emphasis.   The center of interest of a corsage is created at the point where the flower stems meet. This is most often the location where the largest blooms or the bow is placed.

Balance.   The corsage must be constructed so as to be easy to wear. Smaller flowers are placed at the top and edges, with the focal point receiving the most attention. A well-designed corsage should appear as an accessory to the dress or gown on which it is worn. The presence of the corsage should be apparent but not stand out unnecessarily.

Construction.   The corsage must be constructed securely enough to retain its original design. This should be done with a minimum of wires and floral tape to avoid large masses of stems at the base. Flowers should be positioned for best effect and firmness in construction. The designer will learn to position flowers in a corsage so that they will not appear as a tight mass of blooms or as a loose mixture of petals and wired stems. The flowers should be placed in the corsage to be worn as they grow, pointing upward.

The materials used in the construction of a corsage are numerous. Artificial leaves and other accessories are added to enhance the flowers in the corsage. A bow or ribbon is added as a finishing touch at the focal point. Wires and floral tapes replace the flower stems in a well-constructed design. The elimination of the natural stems in corsage designs is to create a sturdy, less bulky framework.

Wiring Corsage Flowers

Corsage flowers are placed on wire stems to make them easier to handle. The wires used for this purpose are green-enameled sections in lengths of 12 or 6 inches. Enameled wire is used to prevent rust formation and staining on clothing. The narrowest gauges of wires (26–32) are generally selected for this purpose to reduce the bulkiness of the corsage. Corsage flowers and foliages are wired by different methods, depending on their size, form, and intended use.

Hook method. The *hook* method (often called *shepherd crook* method) of wiring is used to secure blooms having a flattened head or crown. Daisies, chrysanthemums, and asters may be wired onto stems in this fashion. A medium-weight wire is inserted through the base of the flower near the original stem. The wire is extended beyond the center of the bloom for a short distance (Figure 11–1). A hook or U-shaped bend is placed on the end of the wire before being pulled back snugly into the center of the petals. The longer wire is twisted around the end of the shorter section of wire from the hook and the stem stub that remains to complete the wiring operation. This wired stem may then be wrapped with floral tape.

FIGURE 11–1. Wiring flower stems by the "hook" method; this technique is generally used for blooms with a flattened head or crown.

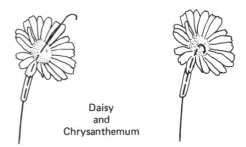

Daisy
and
Chrysanthemum

Insertion method. Flowers having a mass-type head (large, thickened calyx) may be wired by the *insertion* method. The original flower stem is removed at a point immediately below the calyx of the bloom. A heavier wire (22–26 gauge) is inserted through the calyx midway from the base to the petal so that equal lengths of wire protrude from each side (Figure 11–2). This wire is bent to form a stem with both ends of the wire pulled together at the base of the calyx. A thin-diameter corsage wire (28–32 gauge) is twisted around the calyx and original wire before the entire length is wrapped in floral tape.

Wrapped Wire. Delicate flowers that will remain on their original stem and some foliages used in corsage work are wired by the *wrapped wire* method (Figure 11–3). A thin diameter wire is laid along the length of the stem to be supported so that an equal length of wire remains above and below the area to be wrapped. The upper portion of the wire is twisted around the wire that extends below the stem. The stem is completed by wrapping the wire with floral tape.

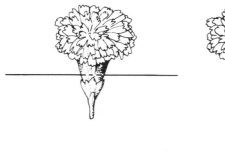

FIGURE 11-2. Wiring flower stems by the "insertion" method; this technique is used for flowers having a mass-type head (large, thickened calyx).

FIGURE 11-3. Wiring flower stems by the "wrapped wire" method.

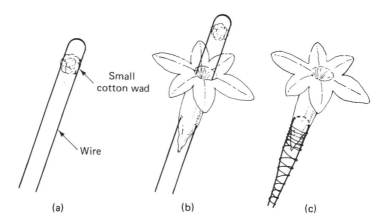

Small cotton wad

Wire

(a)   (b)   (c)

FIGURE 11-4. Wiring flower stems by the "hairpin" method: (a) A small cotton wad is used for support. (b) Insert the wire through the neck of the flower. (c) Wrap the hairpin wire for additional stem strength.

Hairpin. Tubular flowers (such as tuberoses and stephanotis) are supported by folding a wire (30 gauge) into the shape of a narrow hairpin. Include a small cotton wad in the center of the hairpin wire for support. This wire is inserted through the neck of the tube-shaped bloom from the top. A second, lightweight wire is wrapped around the base of the flower and the extended lengths of wire from the hairpin. The stem is completed by wrapping the wires with floral tape (Figure 11-4).

Ivy vines are often used in wedding bouquets to create a cascading effect. The wiry stems of cut ivy tend to remain stiff and in the same position in which they grew on the plant. The designer uses wire to form the ivy strands into lines that conform to the shape of the arrangement. A medium-sized wire (26 gauge) is wrapped around the lower part of the stem. The wire should be extended to a length that covers approximately two-thirds of the ivy stem. The upper end of the wire is folded back on itself to secure it to the stem. Green floral tape is wrapped over the wired portion of the stem to cover the wire. A second wire may be added to the base of the stem to provide added length for arranging.

Corsage tapes.   The wired stems on corsage flowers are not attractive unless covered by floral tape. Floral tapes are available in widths of either ½ or 1 inch and in many different colors. Florists generally use the narrower width of tape in green, white, orchid, and brown. These floral tapes are marketed under such names as Floratape®and Parafilm® White floral tape is used on most wedding work and corsages. Green tape is generally used for rose corsages, boutonnieres, and any design work where green-colored stems are desired. Orchid colored tape is used most often for wrapping stems of lavender Cattleya orchids and brown tape for Cymbidium orchids or dried flowers in fall colors.

Flower stems are wrapped with floral tapes by placing the material near the flower head from behind. The flower stem is held in the left hand and twirled as the tape is being wrapped with the right hand (Figure 11–5). The twirling action feeds the tape onto the stem as the right hand guides the tape placement and stretches it tightly

FIGURE 11–5. Wrapping with floral tape; stretch the tape tightly for best results.

around the stem. After a little practice the stems may be rapidly and smoothly wrapped in floral tape.

Corsage Accessories

The natural foliage of flowers most often used in corsage and wedding pieces does not remain as fresh as do the flowers for extended periods of time out of water. For this reason artificial foliages are created by the florist to add a fuller appearance to the designs. The accessories are added to enhance, and sometimes substitute for, the living flowers and foliages when a corsage or wedding bouquet is being created.

Artificial leaves or foliage may be used by florists to add the appearance of natural greenery to a corsage. These leaves are sold in packages containing up to a gross (144) of the stemmed leaves. Most leaves are taped together in sprays of five stems that may be used together, or split apart into individual stems. Most florists prefer to divide the stems into individual leaves for placement in corsages.

Satin leaves may be constructed by the designer from strips of ribbon. Select ribbon that is 1½ inches or larger in width. The satin ribbon is cut into lengths that are twice its width. The two opposite corners are folded downward to meet at the bottom center of the strip of ribbon. This triangular piece of ribbon is doubled over on itself, leaving the smooth-finished side of the ribbon on the inner fold. The bottom of each side of the triangular pieces of ribbon are gathered (bunched) from the bottom to the tip. The entire ribbon is placed on a wire stem and wrapped with floral tape (Figure 11–6). These satin leaves may be varied in size for use in corsages and wedding work.

Net background material is added to a corsage or wedding bouquet to create a fuller design without the addition of weight. Net materials available for florist design work are sold in different patterns and styles, known as tulle, net, and Lacelon.® These are available in widths ranging from 2⅞ to 8 inches, with the 2⅞-inch (No. 40) and 4½-inch (No. 120) widths used most often in corsage work or wedding bouquets.

These net materials may be cut into lengths equal to their width to create either a net butterfly or fan. A slanted cut across the net is more attractive. The net butterfly is constructed by gathering the net strip into the center and wiring it onto a stem of No. 28 or No. 30 wire. The "fan" is constructed by first cutting a square section of net. This is gathered at the bottom and wired to create a fan-shaped piece of net on an artificial stem. Both the butterfly and fan are finished by wrapping the wired stems with floral tape (Figure 11–7).

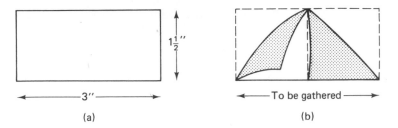

FIGURE 11–6. Construction of satin leaves: (a) Cut stain ribbon twice as long as the width. (b) Fold the top corners into the center and gather at the base. (c) Use a wire to create a stem. (d) Wrap the stem with floral tape to finish.

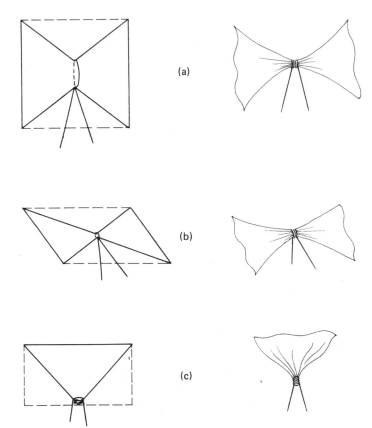

FIGURE 11–7. Net corsage materials: (a) Square net wired in a butterfly shape. (b) Diagonally-cut net wired in a butterfly shape. (c) Rectangular net wired in a fan shape.

The net may be added to a corsage to provide background (fan) or in place of foliage or flowers as a filler (butterfly). The corsage may be constructed with alternate patterns of flowers and net as the design takes shape. Larger wedding bouquets and nosegays may be constructed by first forming a net pillow in the center of the design. Flowers are then added to the bouquet among the folds of the net to create a beautiful, yet simply designed arrangement.

Ribbon is made into bows or streamers for corsages and wedding bouquets as well as for cut flower designs in containers. Ribbon styles vary considerably according to the fashion trends. Satin ribbon remains popular for an elegant addition to any design. Ribbon is available from manufacturers in bolt widths ranging from ¼ inch (No. 1) to 4½ inches (No. 120). Corsage bows are constructed from ribbon that is ⅝ inch in width (No. 3). Satin ribbon and metalline (spray ribbon) are used in widths of 2⅞ inches (No. 40) and 4½ inches (No. 120) for cut flower bouquets, funeral work, and other decorating purposes.

Corsage bows should be constructed in proportion to the size of the design. The streamers and bow loops must not attract attention away from the flowers. The placement of the bow in relation to the flowers creates the focal point of the corsage and is generally at the base or center of the design. Flowers in a corsage should point upward—with the bow at the base—just as they grow.

Regardless of the size of ribbon or intended use of the bow, the construction is the same (Figure 11–8). Most satin ribbons are single-faced or smooth on only one side. The loops appear best when the smooth side is always on top of the bow. The construction of a simple florist bow is quite simple when the following steps are used:

1. First hold the ribbon in the left hand (right-handed persons) so that the smooth side is facing you. The short end of the ribbon should extend about 3–4 inches below your left thumb.

2. A loop is formed above your thumb of equal length as the lower streamer. This is done by bringing the ribbon back to your thumb and joining the two together between the thumb and forefinger. The loop is pressed firmly between the thumb and forefinger to reduce the bulkiness of the bow in the center.

3. The first lower loop is formed in the same manner. However, the ribbon must be reversed to bring the smooth side of the ribbon to the top of the bow before the loop is formed.

4. The second pair of loops (one above and one below the center) is completed in a similar manner. The ribbon is reversed before each new loop is formed. After each loop is completed the ribbon is pressed firmly between the left thumb and forefinger.

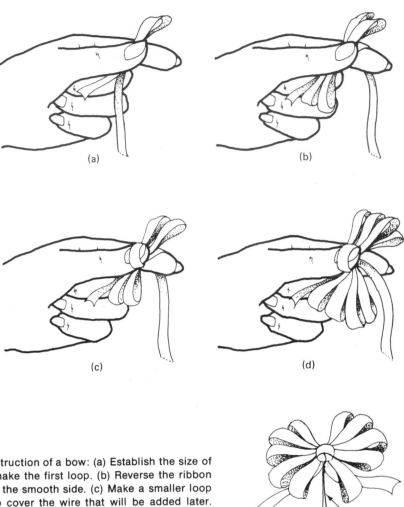

FIGURE 11–8.   Construction of a bow: (a) Establish the size of the bow when you make the first loop. (b) Reverse the ribbon as you work to show the smooth side. (c) Make a smaller loop in the center to help cover the wire that will be added later. (d) Make the necessary number of large loops to complete the bow. (e) Place a wire through the center loop of the bow (see step "c") and twist slightly to secure it to the bow while leaving a tail for securing it to a pick or corsage.

5.   Once two pairs of loops are completed, a small loop is added at the center of the bow to conceal the anchor wire that will be added later. This small loop is formed in exactly the same manner as the others, except that it will be only about 1 inch in length and is formed by simply rolling the ribbon over the thumb.

6.   Two more loops are added and a streamer is left at the end of the bow before the ribbon is cut. Most corsage bows will consist of three pair of loops (three above and three below the center) plus the small loop at the center.

7.   The bow is anchored by running a No. 30 wire through the smallest loop and around to the back of the bow. Equal lengths of the wire should extend backward from the bow. The wire is secured by twisting the two lengths of wire firmly at the back of the bow leaving ample wire to be tied to the corsage.

DESIGNING CORSAGES

Corsage designs are similar to cut flower arrangements in containers, except that they are designed to be worn or carried. Although most corsages will contain a bow, this addition is not always necessary to finish a design. Construction of corsages begins with the wiring of flowers to be used. This topic has been discussed earlier under "Wiring Corsage Flowers." Florists design corsages in many styles and from a large array of floral materials. The basic corsage styles and more typical flowers used will be discussed here. The more common styles for florist's corsages are: shoulder corsages, wrist corsages, and nosegays.

Shoulder Corsages.

A *shoulder* corsage is designed in a crescent shape to be worn on the left side (or sometimes on the right or around the front collar). The basic pattern of these corsages will appear as either a crescent or a triangle that has been reduced in size. The smallest buds are placed at the top (and sometimes also at the bottom) of the corsage. The largest and fullest flowers are used to create the focal point. Several types of shoulder corsage styles may be created easily by the beginning student. These will be described in detail, outlining each step in their construction.

SINGLE CARNATION CORSAGE

1.   Remove the natural stem from a standard carnation. Replace this with a chanille (pipe cleaner) stem by inserting it into the base of the calyx at the point of attachment with the original

stem. Wrap the base of the calyx and pipe cleaner with floral tape.

2.   Place a net fan background behind the carnation bloom. Wrap the stem tightly with floral tape.

3.   Add the bow to the base of the flower by wrapping the two wire ends around the corsage stem. This is done by cranking the wires together firmly. Cut the remaining wire.

4.   Finish the corsage by removing all unwanted wire or stem. Tape all cut ends and cover the bow wire with tape also. Arrange the bow loops for best appearance in the corsage (Figure 11–9).

FIGURE 11–9.   A single carnation corsage; this simple corsage requires one whole carnation. It is quick to construct and is often used when large quantities of corsages are required.

BOUTONNIERE

The *boutonniere* is designed to be worn on the buttonhole of a man's lapel. Since this design is to be worn by a man, it should not be too ornate. There are several types of boutonnieres, but most consist of only a single flower placed on an artificial stem (Figure 11–10).

1.   Stem a rose, carnation, or other single flower as previously described.

2.   Tape the stem with green floral tape.

3.   Bend the stem into a small, flattened circle at the base. Add a boutonniere pin by inserting the point into the base of the calyx.

FIGURE 11–10. A boutonniere and a carnation corsage.

A small leaf from leatherleaf fern or other appropriate foliage, if desired, may be added as background for more formal occasions. No net or ribbon is included in a boutonniere.

DOUBLE CARNATION CORSAGE

The *double carnation* corsage is not a well-designed arrangement. The purpose of this style is to provide a corsage that may be mass-produced and can be worn in either direction without appearing to be upside down (Figure 11–11).

FIGURE 11–11. Making a carnation corsage.

1. Stem two flowers as previously described.
2. Place a net fan behind each bloom.
3. Join the two individual corsages at the center.
4. Add a bow between the blooms and tape all work to be concealed. Add an appropriate corsage pin.

FEATHERED CARNATION CORSAGE

The standard carnation bloom is often too large and bulky for proper corsage construction. The floral designer may wish to split the larger bloom into several smaller portions to be worked into a well-styled corsage (Figure 11–12). Although many florists are now replacing *feathered carnations* with miniature carnations, this technique may be used in various styles of corsage construction. The carnation flower is "feathered" in the following manner:

1. Remove the carnation flower from its natural stem.

2. While holding the flower head by the base of the calyx, press firmly with the thumb and forefinger until the pistil of the flower is removed.

3. Peel the calyx in a manner similar to peeling a banana. This should leave the florets free, yet still attached at the base of the stem.

4. Divide the florets into five or possibly six sections. Allow the sizes of these sections to vary.

FIGURE 11–12. "Feathering" a carnation to create several smaller blooms; wrap the wire tightly without cutting the carnation petals.

5. Place a 6-inch length of No. 30 florist wire an equal distance from the base of the florets.

6. Wrap the wire tightly (but avoid cutting through the florets) from the base of the florets to a point near the top that will form a firm flower head. Wind the tail of the wire back down the stem to the lower wire, which forms the stem.

7. Tape each wired stem firmly to conceal the wire and add support to the florets.

Construction of a simplified feathered carnation corsage will be described. The steps required to arrange this style of corsage may be used with roses, feathered carnations, miniature carnations, pom-pon chrysanthemums, or other small-headed flowers (Figure 11–13).

FIGURE 11–13.   A sweetheart rose corsage; uneven numbers of flowers are used for a more aesthetically pleasing effect.

1. Place a single artificial leaf in the left hand. Behind this leaf add a piece of net fan.

2. The smallest bloom is then placed above the leaf. The leaf should extend about 1 inch above the flower. Tape these stems together. Avoid wrapping the wires around each other, since the sticky tape will hold the stems securely.

3. A second bloom is placed to the right and below the first. A third bloom is then placed below the second and to the left in a manner similar to the arranging of cut flowers in a container. This process continues until the shape of the corsage is established.

4. The stems are taped after the addition of every third wired stem. Artificial leaves and net are added in appropriate locations in the corsage to provide form and to fill spaces.

5. Once the size and shape of the corsage is completed, the bow

may be added. This is attached to the base of the stem. All wires and cut ends of stems are taped to finish the corsage.

6.   The bow may also be added to the side of the corsage, giving the design a crescent-shaped outline. Always complete your design by adding a 2-inch corsage pin into the ribbon and stem.

ORCHID CORSAGES

The Cattleya orchid is the largest of the orchids used by florists for corsages. Because of its size, it is generally used alone to create a corsage. The construction of an orchid corsage is similar to the other designs, except that fewer flowers are used. The wiring of orchids is also rather specific, since these are very fragile flowers (Figure 11–14). The Cattleya orchid corsage is constructed in the following manner:

1.   The Cattleya orchid requires two wires, a heavy one to strengthen and extend the stem and a smaller wire to anchor the flower onto the stem (Figure 11–15).

2.   The heavy wire (No. 18 or 20) is inserted lengthwise through the short stem into the center of the orchid. This wire provides strength and holds the flower in the proper position.

3.   A smaller wire (No. 28 or 30) is inserted through the base of the flower at a right angle to the stem. The ends of this wire are laid along the stem.

4.   The wired stem is then wrapped with floral tape. The stem should taper to the base. This can be accomplished by proper taping of the stem.

FIGURE 11–14.   An orchid corsage; this is a long-lasting corsage if appropriate construction techniques are used.

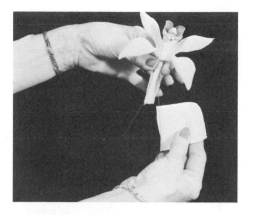

(a) Use the insertion method of wiring.

(b) Wrap a moistened tissue around the natural stem before taping to help keep the orchid fresh.

FIGURE 11–15.   Construction of an orchid corsage.

5.   Three pieces of net "fan" are placed strategically behind the orchid petals to provide a softened background. These stems are taped to form a solid stem for the corsage.

6.   The bow is added below the lip of the orchid. This is taped across the back to prevent the wearer from being stuck by any cut wire ends.

7.   The stem of the corsage is cut to an appropriate length, taped, and gently curled to provide a place to pin the corsage to a dress. (Do not forget to add a corsage pin.)

The Cymbidium orchid is even more popular as a corsage flower than the Cattleya. Cymbidium orchids are wired in the same fashion but are often combined into a corsage by use of two blooms with a bow at the base.

Phalaenopsis and Vanda orchids are quite fragile blooms which require specific support when used in corsages or wedding work. Corsages constructed from either of these orchids are handled in the following manner:

1.   A 6-inch length of No. 26 or 28 wire is tightly and smoothly taped. This is bent to form a hook with a slight crook at the top.

2.   The hooked wire is inserted carefully from above the orchid lip and down the sides of the petals. The wire does not puncture

any part of the flower; rather it provides a base for supporting the bloom.

3. A piece of dampened cotton is placed over the cut end of the flower stem and is held in place by a length of No. 26 wire. The entire stem is taped, including the cotton, to provide a natural appearing stem.

4. The corsage is constructed by combining net, flowers, and a bow as previously described. Be sure to conceal all individual stems with tape as they are added to the corsage.

GLADIOLUS CORSAGE

The gladiolus flower may be used to construct various styles of corsages. Probably the most common use of gladiolus blooms is in the assemblage of a glamellia corsage. The glamellia is fashioned from the various sizes of buds from a gladiolus stem to resemble a camellia flower.

The buds and florets must be placed properly to keep the face of the finished flower as flat as possible. The center bud and each added floret should meet at the same height for proper balance. The number of florets required to complete the glamellia corsage will depend on the sizes of the florets used and the ultimate size of the finished flower desired (Figure 11–16). The following steps are required in the construction of a glamellia corsage:

1. Obtain five to seven graded sizes of fresh gladiolus florets. One of the buds must be tight, without the stamens showing. At least three florets should be fully open. Select five salal leaves. Trim them with a scissors to suit the size of the corsage. Wire them to form stems and then tape the wires.

2. Remove the lower portion of the calyx from the larger bud and florets. This is done easily by inserting the thumb inside the florets. The end of the gladiolus floret is cut at an angle just above the thumbnail with a sharp knife. Each floret is cut in the same manner. The stamens will fall from the bottom of the flower without causing it to split if done correctly.

3. The tightest bud is wired by inserting two 6-inch lengths of No. 30 wire at 90-degree angles to each other and taping them to form a stem.

4. Each successive floret is added in increasing sizes by fastening them securely with two wires. The wires should pierce the folds of the florets to hold them securely without cutting the petals. The wires are bent downward to form the central stem.

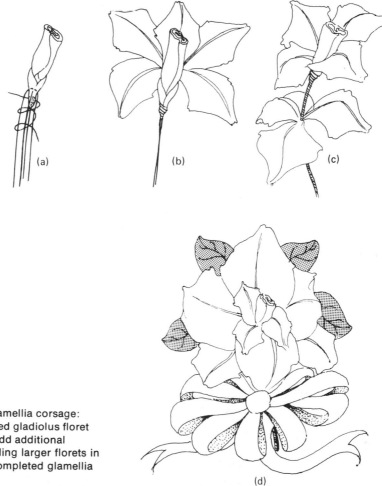

FIGURE 11–16.   The glamellia corsage:
(a) Wire a partially opened gladiolus floret
and clip off its top. (b) Add additional
florets. (c) Continue adding larger florets in
a spiral fashion. (d) A completed glamellia
corsage.

5.   The florets should be added at the proper height to keep the face of the glamellia flattened. The florets are added in a spiral fashion as the flower is constructed. The top ½ to ¾ inch is cut from the center buds to give a more natural appearance to the completed flower.

6.   The glamellia corsage is finished by adding the salal leaves around the back of the flower. The central stem is thoroughly taped and shaped to form a strong support for the flower.

THE FOOTBALL MUM CORSAGE

The football mum corsage has become a tradition with high school and college students, especially during the homecoming football game and school dances. The football mum will generally be finished with a bow and streamers that match the school colors (Figure 11–17). A simplified football mum corsage is constructed in the following manner:

1.   A standard incurve chrysanthemum is selected for freedom from blemishes or bruises. The flower is removed from the stem about 1 inch below the calyx. Most florists protect the flower by a process known as "waxing." A lighted candle is inverted to allow hot wax to flow around the petals from the back of the flower. The cooled wax holds the petals firmly and prevents them from being pulled loose when the corsage is worn.

2.   The chrysanthemum is wired by crossing two No. 26 wires and adding tape, as explained earlier.

3.   Five leaves from salal or lemon leaf foliage are placed on wire stems and taped. These are placed around the mum flower to

(a)

FIGURE 11–17.   The football chrysanthemum corsage: (a) "Waxing" prevents the petals from dropping. (b) A completed corsage.

(b)

add background and support to the petals. These wired stems are then taped to secure them to the central stem. Do not twist the wires any more than is necessary. The floral tape will hold the wires securely without adding unnecessary bulk to the stem.

4.   The bow is constructed from two individual bolts of ribbon having the school colors. The bow is formed in the same manner as explained earlier, except that twice as many loops will result. The streamers on the bow will also be longer to match the size of the flower. The tying of a corsage bow from two strands of ribbon simultaneously may require practice to avoid bulkiness at the center.

5.   The bow is added to the central stem at the base of the flower. The stem is cut to an appropriate length, taped, and a corsage pin is added.

Wrist Corsages

Wrist corsages are very popular with teenage girls for dancing because they are not damaged as easily as those worn on the shoulder. The design of a wrist corsage is the same as one for the shoulder, although most are tapered to a point at each end to conform to the line of the arm. The corsage is worn on the wrist as an accessory, so it must not be too large or it will overpower the dress.

The easiest way to construct a wrist corsage is by arranging two equally sized triangular corsages. These are then joined at the center by hooking their stems. A bow is added to the center before the corsage is attached to a holder. Most florists purchase commercial forms having elastic bands for the arm and metal clips that are used to attach the corsage to the bands. The designer must be especially careful to keep all cut wires away from the wearer and to neatly cover all construction with floral tape. A wrist corsage may be arranged from any flower material, but the larger orchid blooms are generally avoided for this purpose.

Corsages for the Hair

Some women prefer to wear flowers in their *hair*. They may have their hair styled for a specific corsage for a very special occasion. This type of corsage is often used for weddings or elegant parties. It is especially important that the flowers remain flattened and as close to the hair as possible for this design. The corsage is constructed in a tear-drop shape so it will conform to the shape of the head. A bow may be used but is not always necessary. Net artificial leaves replace the foliage used in other corsages.

The hair corsage is attached by use of either bobby pins or it may be wired to a small comb. The corsage flowers will require adjustment once attached to form them into the contours of the coiffure.

Nosegay Arrangements

The *nosegay* is designed to be carried rather than worn. The flowers are arranged in a globular shape on a handle. The flowers are grouped together so they appear to radiate from a single point. Like the wrist corsage, the nosegay is very popular with teenage girls. The nosegay design is also the basis for most wedding bouquets.

Flowers to be used in a nosegay are wired on long stems to facilitate handling and to form a central handle. Many florists prefer to use a net foundation as a center of the nosegay. This is formed by making several loops from net and placing this on a single wire. The individual flowers are then fitted into the folds of the net. The flowers are placed first at the center and gradually spiraled to the outside perimeter (Figure 11–18).

(a) Arrange the flowers around a net foundation center. Wrap taped "handle" with satin ribbon.

(b) A completed nosegay.

FIGURE 11–18.   Construction of a "nosegay."

The nosegay is finished off by placing either net fans as background or satin leaves around the outside edges. Commercial nosegay holders are available for supporting the bouquet. If these are to be used, the wired flowers and net may be pulled through the hole in the center of the form. When all the flowers have been assembled in the nosegay, the holder is secured with pins or wires. The handle is shaped by taping the assemblage of wires together. The handle is then wrapped with satin ribbon, reverse side out, to cover the sticky parafilm tape. A bow and streamers may then be attached at the top of the nosegay behind the holder.

WEDDING DESIGNS

Weddings have traditionally been a very important part of the florist business. Although the size of most weddings has become smaller in recent years, more people than ever are being married. Designing wedding bouquets and planning each detail of a wedding with the bride is exciting and rewarding for most floral designers.

The wedding plans are first discussed between the bride and the head designer at the flower shop. Although the assistant designer may eventually be working with the other designers on the wedding, the consultation requires considerable training and experience. The head designer must be able to sell the bride on the best wedding designs while filling the bride's desires in a ceremony. The common wedding flowers used for a church ceremony include floral arrangements for the altar, bridal bouquets, and corsages and boutonnieres for the wedding party. Other decorations are generally used to enhance the beauty of the ceremony. These may consist of candelabra and various bouquets for the front of the sanctuary, an aisle carpet, bows and aisleabra for the pews, and an assortment of floral pieces to be used for the reception.

The bouquets to be carried by the maid of honor, bridesmaids, and the bride are the most beautiful floral pieces of any wedding. Great skill and training are required for their proper construction. These floral bouquets may be constructed in many styles and from an array of flowers. The bride's desires dictate the type of wedding bouquets to be used.

The designer must be certain that only the freshest flowers are used in the construction of wedding bouquets. Each individual flower or floret added to the design must be wired, taped, and handled carefully to ensure that the finished work will remain as beautiful as possible until after the wedding. The designer's assistants will be responsible for wiring most of the flowers and preparing the accessories to be used in each bouquet. The bouquets used in a wedding are very expensive because the flowers are of the highest quality and require tedious labor in their preparation. The final product of this work is a reward to all who were involved in their planning and construction.

The styles of wedding bouquets may vary from a single long-stemmed rose decorated with a bow to those having elaborate arrays of flowers and cascades of blooms. Since each florist establishes the styles to be used most often in each shop, no specific patterns may be taught to the beginning designer's assistant. Generally, the bridesmaids' bouquets are smaller but similar in design to the bride's bouquet. Each style may be easily constructed once the designer has mastered the techniques of wiring the various types of flowers, forming the different shapes of bouquets, and combining the flowers into a securely attached arrangement.

FUNERAL DESIGNS

The retail florist sees customers on occasions of joy, love, and in sorrow. The circumstances surrounding the death of a loved one brings customers to the flower shop for the purpose of bestowing a floral tribute to the deceased. Flowers portray the traditional emotions of love, faith, and sympathy to the family. Funeral flowers are arranged for the living members of the family and friends that attend the funeral service. These arrangements may be simple or they may be quite lavish. The floral designer will use the same amount of talent and creativity when arranging funeral pieces as with any other specialty designs.

Funeral services and the floral pieces displayed vary according to the tastes and desires of the family. A typical funeral might include displays featuring cut flower arrangements in baskets, a casket piece, sprays, potted plants, and wreaths or fraternal easel pieces. The cut flower baskets are constructed as described in Chapter Ten, in either a radiating or triangular design. Easel pieces are constructed by picking fresh flowers into a moss or styrofoam block of the appropriate design. These are supported on specially designed wire frames for display at the funeral service.

The beginning floral designer will need to learn the basic steps required for constructing flat sprays and casket sprays for funerals. The flowers selected for use in funeral pieces are those that are at the peak of their beauty. Roses, for example, appear best when half-opened or fully opened. Flowers that are in tight bud will not show to their full advantage. Each flower to be used in the construction of the funeral sprays is placed on a supporting wire (18–22 gauge). If these flowers are to be placed in a styrofoam block, a wooden or steel pick is attached to the end of each stem. The quantity of flowers used in each of the funeral pieces will be determined by their price.

Both flat sprays and casket sprays may be constructed on a styrofoam block which provides a base for the design. Flat funeral sprays may be constructed without such a base by simply wiring the stems together in a continuous line to form a spray of flowers and foliage (Figure 11–19). The flat spray may be constructed in the following manner:

1. Select nine standard carnations. Place each stem on a support wire as described in Chapter Ten.

2. Lay three pieces of flat fern in the shape of a fan. These stems are wired with the aid of malin (No. 24 wire) wound onto a stick or spool. The wire is circled around the stems twice and pulled tight.

3. A flower stem is placed on each piece of fern so that the fern tip extends about 6 inches beyond the flower head. The center flower will extend about 4 inches above those placed at the sides. Tie these stems with wire from the spool.

4. Two pieces of filler fern (asparagus fern or plumosis) are laid between the flowers to add depth to the spray. Tie the stems together with the wire, as done previously.

5. Lay a single piece of flat fern in the center of the spray below the first flower so that the back of the foliage is facing upward. Fold the top of the fern frond down to meet the base of the stem. This will create a pillow for raising the next flower to be placed upon it. Tie the fern and flower stem securely with wire.

6. A fern frond is next placed on each side of the center of the spray slightly below the last flower stem on each piece of fern, as was done earlier. Filler fern is then placed between these flowers to fill the spaces. Tie all these stems securely with the wire.

7. A pillow is formed in the center of the spray from a piece of flat fern, as described in step 5. After placement of the next flower stem on this pillow, the stems are again tied with wire.

8. Two additional fronds are placed below the center pillow and on each side to form a background for flowers. The flower stems are placed on top of these fronds and filler fern is then added between them. The stems are again tied with wire.

9. The extraneous stems are cut off at a point about 6 inches below the point where the wires have been tied. Two additional pieces of flat fern are placed at the base of the spray stem facing down, with their stems wired to the main stem of the spray. These appear as tails on the spray.

10. A bow is constructed from No. 120 metaline or satin ribbon. This is attached to the spray with wire above the fern fronds, which were placed as tails on the spray.

The flat spray must be constructed tightly and securely so that it will not bend or flop when handled. The flowers and fern are wired about one-half the distance up the stems to form a strong central stem on the spray. The remaining stems located below the point of tying are broken and folded upward to add support to the center of the spray. Some floral designers construct the tied spray on a bamboo stake to achieve this support. The spray is completed by cutting the wire from the spool and tying the free end securely.

(a) The flowers and fern are wired in a continuous progression using spooled wire.

(b) The stems are cut uniformly at the base of the spray.

FIGURE 11–19.   A tied spray.

(c) A tail formed from the fern is added before placing a bow on the spray.

(d) The completed spray.

A Basic Casket Spray

The casket spray is used to cover the lid of the casket and is often the most elaborate floral piece at a funeral service. The florist takes pride in the design and construction of each casket spray. The flowers may be arranged in floral foam blocks to keep the flowers fresh for an extended period or picked into styrofoam blocks. The support block (either floral foam or styrofoam) is attached to a specially designed casket saddle made for this purpose. All flowers to be used in the casket spray are wired to support the stems and blooms. When styrofoam is used as the supporting medium, each flower stem is placed on a pick to hold the stem in the block. The size of the casket spray is determined by its price, which affects the number of flowers used and the type of accessories that will be added (Figure 11–20). A simplified casket spray may be constructed by following these steps:

1. Attach the support block to be used to a casket saddle (frame) designed for this purpose. Styrofoam blocks must be wired to the frame. Floral foams are placed in watertight frames and secured with waterproof adhesive tape.

2.   Place stems of flat fern or jade leaf greenery around the outer perimeter of the saddle container. These stems of greenery provide a background for the design and conceal the saddle container. The foliage stems are longest along the horizontal axis of the arrangement.

3.   Create a pillow affect across the top center of the spray by inserting loops formed from the foliage stems, which have been curved. The leaves are gently folded without breaking their stems and secured in the holder.

4.   Next, the wired flowers are added to the spray. Flowers are positioned first all around the lower perimeter above the background of foliage. The next level of flowers is placed above the first, with the blooms being on slightly shorter stems. The addition of flowers is continued until the spray is uniformly covered.

5.   Additional greenery or filler foliage may be placed between the foliage if necessary. Filler flowers, such as baby's breath or pom-pon chrysanthemums, may be used to accent the design.

6.   A satin bow is constructed leaving streamers the length or width of the spray. The bow is attached to a wooden pick approximately 6 inches in length. The pick is easily inserted into the support block to secure the bow in the spray.

7.   Once the bow and streamers are neatly positioned, the designer will carefully inspect the spray for faults. Greenery is added wherever a void or space appears. The flowers are checked to make certain that the stems are secured. Before the casket spray is taken to the storage refrigerator, the designer adds additional water to the container on those sprays in foam blocks.

FIGURE 11–20.   A casket funeral piece constructed in floral foam.

(a)   The floral foam block is placed in a casket saddle container.

(b)   Foliage is added to provide a background for the design.

(c) Flowers are wired, picked, and placed in floral foam.

(d) Flowers are placed uniformly throughout the design.

(e)   A bow is constructed leaving long tails.

(f) The casket spray is completed by arranging the flowers and bow. Filler flowers may be added to create a more interesting design.

WREATHS

Wreaths are designed by florists for many different occasions. Fresh flowers may be picked into a circular frame of moss or styrofoam and placed on an easel for a funeral. Wreaths are also used for seasonal decorations, especially during the Christmas season. Circular styrofoam frames are available for use with picked greenery or flowers. Christmas wreaths are constructed more often on wire frames by tying greenery with a continuous strand of wire. This type of wreath is constructed in the following manner:

1.   Bend a wire stake or coat hanger into a circle. A 6-foot length of No. 9 rose stake makes a convenient frame. This wire is firmly held with 26–28 gauge wire from a spool.

2.   Boughs of evergreen foliage are cut into sections approximately 8–10 inches in length. Spruce, pine, and fir boughs make attractive foliage for Christmas wreaths.

3.   The foliage is placed on top of the wire frame in alternate patterns. The first stem is placed at the center of the wire and tied by pulling the wire around the stem twice. A piece of the foliage stem is placed slightly below and to each side of the first stem. These are then tied with the wire.

4.   The pattern of placement of foliage and tying with wire is continued until the entire circle of wire frame is filled. The wire should be tied firmly as close as possible to the center of each twig to provide firm support. If the stems are placed too far apart, the wreath will appear thin and ragged.

5.   The wreath may be made more attractive by the addition of pine cones, holly sprigs, or other ornaments.

6.   Once the circle of foliage has been completed, the wire is cut from the spool and secured to the frame. A bow may then be attached to the frame at the point where the foliage was last added.

The specialty floral designs constructed by floral designers require specific training and much practice before they can be made with ease and originality. Designers' assistants are more involved with the preparation of the flowers and accessories used in the designs than with the actual construction. As designers gain experience and training in the retail flower shop, they will be given more responsibility for creating the more elaborate designs.

The arranging of flowers into corsages is very similar to the construction of beautiful wedding bouquets. When designers' assistants have mastered the art of corsage making, they will be ready to attempt the more difficult designs in the flower shop. Sprays and casket pieces are similar to cut flower arrangements constructed in containers. Designers' assistants will have ample opportunity to observe and aid more experienced designers with the construction of specialty pieces. When the florist feels the assistants have the necessary confidence and creativity, they will be allowed to design the more expensive floral arrangements for the shop.

SELECTED REFERENCES

Benz, M. 1966. *Flowers: Geometric Form*, 3rd ed. San Jacinto Publishing Company, Houston, Tex.

Bode, F. 1968. *New Structures in Flower Arrangement.* Hearthside Press, New York.

Color Supplement—Wedding Presentation Manual. 1965. The John Henry Company, Lansing, Mich.

Cutler, K. N. 1967. *How to Arrange Flowers for All Occasions.* Doubleday & Company, Inc., Garden City, N.Y.

Pfahl, P. B. 1968. *The Retail Florist Business.* The Interstate Printers and Publishers, Inc., Danville, Ill.

Reusch, G., and M. Noble. 1960. *Corsage Craft,* 2nd ed. D. Van Nostrand Company, Inc., Princeton, N.J.

Scanlon, J., and B. Palmer (eds.). 1968. *Album of Designs: Funeral Flowers for Professional Floral Designers,* 22nd ed. Florists' Publishing Company, Chicago.

Tolle, L. J. 1969. *Floral Art for Religious Events.* Hearthside Press, New York.

Vogue in Wedding Flowers Thirteen. 1969. Florists' Publishing Company, Chicago.

TERMS TO KNOW

Boutonniere	"Feathered" Flowers	Glamellia
Corsage	Frond	Nosegay

STUDY QUESTIONS

1. Describe how the principles and elements of design are used in corsage construction.
2. List the various methods used for wiring flower stems in cor-

sage construction and explain how each is adapted to specific types of flowers.

3. Explain why the natural stems of flowers are substituted with wires and florist tape for corsage construction.

4. Describe the process of "waxing" chrysanthemum flowers and explain why this is done.

5. Explain why the funeral business is an important and necessary part of the retail florist business.

SUGGESTED ACTIVITIES

1. Practice the wiring and taping of flower stems to be used in corsages.

2. Construct different types of net accessories for use in corsages.

3. Practice the construction of bows for corsages, bouquets, and sprays.

4. Invite a professional floral designer to your class to demonstrate the construction of various corsages, sprays, and wreaths.

5. Have a professional floral designer visit the class and show various wedding designs and discuss wedding consultations that are conducted with brides.

6. Practice the construction of simple funeral sprays using readily available flowers.

7. Construct Christmas wreaths using available evergreen foliage.

CHAPTER TWELVE

Customer Relations
and Sales

The greatest expense in the operation of any retail store is the cost of maintaining a staff of employees. No matter how skilled designers might be at arranging flowers or planning landscapes, the sales personnel are those employees in contact with the most customers. It is not surprising the customers judge the entire shop by the sales personnel. Whether the person is selling products in a flower shop, garden center, plant store, or department store, they are the store's ambassador in dealing with the public.

The average small business loses 80 percent of its customers during the first 10 years of its existence. One of the primary reasons for this failure to keep customers has been shown to be caused by the indifference and poor service offered by the store's employees. Customers appreciate courtesy, a knowledge of merchandise, a pleasant attitude, and consideration from a sales clerk in a retail store. The sales clerk is in an excellent position for learning the operation of a shop. After a period of training and experience, the sales clerk will be given greater responsibilities within the business. However, no matter what level of management or responsibility the employee holds in the business, the knowledge gained from selling will always remain invaluable.

Although every employee connected with a small business will at some times be directly involved with selling, the sales personnel will have these more specific duties, depending on the size of the shop. In a small store those duties might include (in addition to selling) record keeping, inventory control, plant care, light cleaning, and display area organization. The most important characteristics to be possessed by any salesperson are (1) the desire to learn all about the business and (2) enjoying the work being done.

PERSONAL REQUIREMENTS FOR SELLING

Many salespersons believe their role in the shop is merely that of an "order taker" or cash register operator. A business cannot operate for long when each employee has this attitude. A successful salesperson will create greater profits for the business, and improve themselves, by maintaining the following professional qualities: good personal grooming, a desire to learn, a knowledge of the merchandise, a knowledge of operating procedures, an ability to communicate, and the ability to use business mathematics.

Personal Grooming

Your attitude toward yourself, your job, and the customers' impressions of the store will be affected by the manner in which you present yourself. Since the salesperson is in constant contact with the public entering the store, the first impression they have of you will directly influence your ability to make a sale.

The sales personnel should check their grooming periodically during each working day. Each employee should be appropriately dressed for work. Girls will need to wear appropriate outfits, without appearing too formal or informal in attire. Boys should remain as neat as possible. Change soiled clothes or wear a suitable uniform when working in the shop.

Cleanliness and neatness are necessary for all employees who come in contact with the public. No matter what length or style of hair, it should be kept clean, cut, and styled appropriately for work. The workers will need to keep hands cleaned as often as possible, since the hands are readily noticed while demonstrating products to the customers.

Desire to Learn

Ambitious employees who enjoy their work will be most successful. A salesperson who aspires to be promoted to a position of greater responsibility will be willing to put forth more effort than is expected by the manager. Sales clerks can improve their sales potential by learning all they can about the merchandise. The sales personnel in a garden center, for example, can sell more plants if they are familiar with the names and care of the plant products being sold. This personal attitude will be recognized by the employer and will be rewarded.

Knowledge of Merchandise

A knowledge of the merchandise being sold in the store is required of all salespersons. A sale depends upon the ability of the sales clerks to anticipate the needs of customers through the demonstration of appropriate products. A thorough knowledge of pricing structures, availability of stock, and related items of merchandise will aid the sales clerks in deriving the greatest potential for sales. This knowledge allows each salesperson to "up-sell" the store's products. An example of up selling might be shown by the following:

> A customer desires a potted plant for her home. After deciding which plant might be most desirable for her purposes, the sales clerk suggests other related items to accompany the sale of the plant. He might suggest a pretty pot or macrame hanger for the plant to be displayed in the customer's home. He might also suggest how the plant will grow best for the customer when given the proper care. This discussion may lead to the sale of fertilizer and insect sprays for the plant. Such selling brings more profit to the store and is a direct result of the employee's knowledge of the merchandise.

The sales staff of a flower shop must be familiar with the designs available from the florist. They should be able to suggest appropriate color schemes and combinations of flowers for suitable arrangements. Often these designs will be available in a range of prices. It is the responsibility of the sales clerk to obtain the highest value from the sale, while giving the customer complete satisfaction. Most florists provide brochures with pictures of various designs and suggested price ranges. The sales personnel must then be familiar with the types of flowers available at that time and their retail prices. Most sales, however, may be made by suggesting to the customer that a beautiful design will be arranged for the agreed-upon price. The designer may then create an appropriate arrangement from flowers currently on hand.

Knowledge of Operating Procedures

To be able to do the best possible job at selling, the sales personnel must be familiar with the business practices of the firm. The sales staff must be able to make intelligent decisions concerning refunds, credit, price reductions, and services available to customers. Each member of the sales staff should be informed of the methods for setting prices on each item of merchandise. An informed employee may then be able to handle specific selling situations that may involve pricing of merchandise.

The sales staff will be most directly involved with finalizing purchases. They must be familiar with delivery costs and procedures, charges required for gift wrapping, and any guarantees or replacement policies offered at the store. Many small stores extend some type of credit to preferred customers. It is helpful for each of the sales clerks to be familiar with the credit plans made available. Most stores are now using bank credit plans, such as Master Charge or Visa, to alleviate the need for in-store credit services. The procedures required for handling credit purchases is a highly important part of selling today.

Communication

An employee sells products in a store through communication. Without the ability to use language correctly, a salesperson's selling potential is lessened. Often the correct selection of words paints a mental picture in the mind of the customer. This is particularly true when the employee is making a sale over the telephone. The sales clerks should build up a speaking vocabulary so they might express ideas exactly and colorfully. For example, when describing carnations to customers over the telephone, they should say: "They have the fragrance of clove and are a beautiful deep red in color." This is in contrast to what might be said, such as: "They are red and have a strong odor."

The language and use of words by each employee is as important as their personal appearance in formulating an impression on customers. The customer will most easily be sold on merchandise if the sales clerk's language is free of grammatical errors and incorrect pronunciation. The employee of an ornamental horticulture business has the additional requirement for learning the names and pronunciation of all plant materials sold by the firm. Each employee should be able to recognize these plants and use their correct names repeatedly during conversations with customers.

Using Business Mathematics

One of the major functions of a sales clerk is to total sales on the cash register. Large department stores may have sophisticated computerized cash registers, but these are rarely found at small retail stores. The sales personnel will be required to use adding machines and cash registers properly. Even when making sales, the salesperson must be able to make simple calculations mentally. The retail salesperson should be able to make these simple calculations:

ADDITION

1 dozen roses	$20.00
Packaging charge	1.00
Delivery charge	1.50
Total	$22.50

Some sales may require division and multiplication, as well as addition:

3 carnations at $10.00/dozen .... $10.00/12 $\times$ 3 = $ 2.55

$+$

7 roses at $20.00/dozen .... $20.00/12 $\times$ 7 = 11.90

Total $14.45

The figuring of sales taxes to be added to each purchase also requires the use of mathematics. Most states require that sales tax be collected from each sale and sent to the state treasurer. These taxes are added to the purchase price after all price adjustments or credits have been made. The taxes vary from one locality to another. Sales taxes may be derived from standard tax charts for the state, but these are not always readily accessible. The sales staff should be able to make the necessary tax computations, such as in the following example:

State sales tax is 4.5%

Then:

Amount of sale	$12.75
Tax rate	$\times$ 0.045
Total tax	$ 0.57375 (or $0.57)

The sales slip would show:

Amount of sale $12.75
Tax 0.57
Total price $13.32

These computations will become easier as the sales clerk becomes familiar with the sales routine. Some employers require that each purchase be recorded on a cash register tape showing such separate informations as employee number, amount of purchases, taxes collected, special charges added (such as delivery), and total price of each purchase.

In addition to the more customary cash sales, the sales staff will encounter charge sales and payments requiring money to be removed from the cash register. Credit sales are generally recorded either on a special form used for this purpose or by use of a bank card. Credit forms are filled out in triplicate and one copy is filed with the credit accounts. One copy of the form is used for filling the order at the store and the third is given to the customer as a receipt of purchase or as a bill for their records. Credit card sales are handled in a similar manner, except that a national credit card issued to the customer is stamped onto the purchase form in a machine designed for this purpose. One copy of this slip is given to the customer, with the others filed with the bank card accounts. The manager then periodically assembles all the credit charges and bills the customers or the bank credit firms for collection of these accounts.

Money will occasionally be required for withdrawal from the cash register during the course of a normal working day. Funds are provided by the management to take care of these needs. This is called a *petty cash* fund and is used for paying out money for such items as postage stamps, C.O.D. (cash-on-delivery) shipments, and other incidental items related to the operation of the store. When money is removed from this fund, a petty cash slip is filled in completely and filed either in the cash register or in the petty cash accounts file.

PERSONAL SELLING TECHNIQUES

The sales staff in a retail store is constantly in contact with customers on a one-to-one basis. This personal contact allows the sales clerk the opportunity to fit the product or service to the needs of the customer by determining what products they desire. This can be done by observing customers' actions and listening carefully to what they have to say. The salesperson is likely to encounter the following types of customers: the just looking type, the undecided customer, and the decided customer.

The "just looking" customer. Many customers who enter a flower shop or garden center are interested in plants, but do not have a specific need to buy at the time. They may have entered the store for the purpose of seeing what is offered. They may be looking for a bargain but do not wish to be pressured into a purchase. This type of customer is hardest to sell, since his intentions are to merely browse through the shelves of merchandise. However, this person is a potential customer. Since he was interested enough to come into the store to see the merchandise, he may eventually be pursuaded to purchase. This shopper may not have a *need* at present for the merchandise offered or he may wish to shop further before deciding upon a purchase. This type of customer should be treated courteously and patiently but should be allowed to look around for himself. Often the customer will find items of interest and may return later to make a purchase.

The indecisive customer. The *undecided customer* has already determined that he has a need. He has not decided yet exactly what he wishes to purchase to fulfill that need. This customer may be looking for a suitable gift for his wife or a friend but wishes to see all the merchandise before making a purchase (Figure 12–1).

The indecisive customer is the type most in need of personal selling from the sales staff. By carefully listening to what the customer is saying or feeling, the salesperson may then select appro-

FIGURE 12–1.   The undecided customer may require the special assistance of the sales employee before he can make a decision for a purchase.

priate products to show. The customer will be easily confused and will change his mind about purchasing if shown too many items. Often the salesperson can close a sale easily by first determining the customer's need and then demonstrating a narrow line of merchandise. The husband who desires a sentimental gift for his wife, for example, may be shown a line of crystal glassware. The sale may be further enhanced by suggesting the addition of a cut flower arrangement to be placed in the container. The flowers will add a special meaning to the occasion, while the crystal bowl may be used for years afterward.

The decided customer. The *decided customer* has already made up his mind concerning the type of purchase he wishes to make. He may have already determined what price he is willing to pay. This customer is by far the easiest to sell. He does not require the amount of persuasiveness normally required of the undecided customer.

The decided customer wants to be shown the exact item he desires. For example, a man may wish one dozen long-stemmed roses to be delivered to his wife. The salesperson merely has to show him the selection of long-stemmed roses on hand and arrive at a suitable price. This customer will then pay for the purchase and leave completely satisfied. In some instances the customer does not specify a price he is willing to pay. The sales clerk may begin the sales approach by showing him the higher-priced line of merchandise. If the customer shows no objections to the higher prices, a sale can be made that otherwise might not have occurred. This technique is another form of up-selling, since the customer is persuaded to purchase a better line or quality of product when he might have selected something a little less expensive.

On occasion the merchandise requested by the customer will be out of stock or out of season (in the case of flowers or plant material). In this case the sales clerk will find it necessary to do *substitute selling*. This means the customer should be shown plants or stock that comes closest to the customer's specifications. A sale will be lost if the sales clerk merely reports to the customer that the item desired is out of stock.

Determine Personality Traits

Each customer entering the store will exhibit a different personality characteristic. When greeting a customer, be specially receptive to their mannerisms and speech—in other words, recognize their personality traits. After a period of training and experience, the sales clerk will recognize customer temperaments and will adopt the proper selling techniques to serve each one. Some of the customer personality types that might be encountered in a small retail store are: the hurried customer, the quiet customer, and the conversationalist.

Hurried customer. The *hurried customer* has determined his needs and wishes to make a purchase as quickly as possible. The salesperson should be available to serve the customer as soon as he enters the shop. When it is not possible to wait on the hurried customer immediately, he should be told that he will be helped in a moment. At times the customer will appear to be in a hurry merely to command respect or to impress the sales staff. This type of customer should be treated with respect and efficiency so that his ego will not be hurt.

Quiet customer. The *quiet customer* may be silent in personality or merely wish to be left alone to browse unpressured by a salesperson. The sales clerk may find it necessary to approach this type of customer from a distance and ask questions that will help to "draw out" the customer. The quiet customer may desire to be shown a variety of plants or goods, so that he may make comparisons. This customer does not ask questions readily, so the salesperson will have to be able to read his facial expressions or other mannerisms. As soon as the customer begins to respond favorably to the sales demonstration, he will readily make a purchase.

The quiet customer may act in this manner because he feels ill at ease under the circumstances. A man may feel uncomfortable purchasing flowers for a girlfriend for the first time, for instance. He should be made more comfortable by the sales staff. This can be done easily by explaining to him the various types of flowers and arrangements available. Allow him to smell their fragrance and touch their petals to give him confidence in his decision to buy. Selling to this customer is a matter of teaching him about the product or services he is considering.

Conversationalist. The *Conversationalist* is the sociable customer who enjoys talking. The sales staff should listen intentively to him and respond to his conversation. However, the sales personnel are busy so should bring his attention back to the subject of the sale. This should be done tactfully to avoid appearing too rushed to be bothered with him. The goal of every sales clerk is to keep the customer's attention centered on the merchandise.

RETAIL SALES TECHNIQUES

Sales are created in retail stores in various ways. Effective advertising promotions aid in bringing in new customers to the shop.

Once the customers arrive at the store, they may be further convinced of a purchase through the use of effective merchandise display methods. However, the sales personnel formulate the most important part of the first impression a customer has of any retail store. The salesperson can do much to guarantee a sale to each customer who enters the store by following the basic rules of selling. Each sale may be broken down into four basic steps: approach, interest, desire, and the closing.

Approach. The *approach* to a customer should be warm, genuine, and personal. If the sales clerk knows the name of the customer, this should be included in the greeting, such as: "Good morning, Mrs. Anderson." This approach to selling will surely put the customer at ease and make him more willing to make a purchase. Often the sales clerk will not know the customer's name. It is not common practice for the sales personnel and the customers to introduce themselves. However, an attentive sales clerk may be able to learn the customer's name during the sales conversation. Whenever this occurs, his name should be used often during the remaining discussion.

The sales approach most often used is: "May I help you?" This greeting does more to alienate the customer than it does to put them at ease. The customer obviously desires to be helped, so the use of this greeting is superfluous. The greeting "Do you want something?" is also a poor introduction by the salesperson, for it can be assumed from their presence in the store. When the customer's name is not known, the addition of "sir" or "madame" may be the best way to address them.

Interest. Once the customer has been placed at ease through a proper greeting, the sales clerk should get him *interested* in making a purchase. The customer's attention is immediately directed toward the merchandise that interests him most. The sales clerk should establish the customer's confidence early in the course of making a sale. The customer should be shown that the sales clerk is familiar with all the merchandise and can aid in making a selected purchase. Once this confidence is established, the customer may be more willing to accept the sales clerk's judgment in product selection. In the case of plants or flowers of interest, the sales clerk's knowledge of them may convince the customer of the wisest selection. The aim of the salesperson during this stage of the sale is to determine the needs of the customer and match these with the correct product.

Desire. Once the customer's needs have been determined, he must be convinced to make a purchase. The salesperson can now create a *desire* to purchase by providing a sales presentation. The merchandise can be demonstrated or shown to the customer. The merchandise may be handled carefully by the clerk and customer, or it may be necessary to demonstrate how the product is used.

The salesperson can arouse interest and create desire in the customer by having a list of "selling points" in mind for the different types of merchandise carried by the store. A good sales presentation introduces the most important selling points early in the sales discussion. The salesperson should provide the customer with this information in a manner that shows how the product will fill his needs. Often the most difficult part of the sales presentation is overcoming the objection to prices. The purpose of the sales presentation should be to acquaint the customer with the product and how its value is worth the price.

Occasionally, the salesperson may have misinterpreted the needs of the customer by demonstrating the wrong product. When the customer says, "That is not quite what I was looking for," the salesperson must overcome this resistance in selling. By showing the customer either other merchandise or different quality of the same product, the customer may be convinced of his need to purchase. It is important that not too many unrelated products be demonstrated to the customer. Presentation of a wide variety of merchandise only causes the customer to become confused and may destroy his desire to make a purchase.

The customer may have entered the store with no particular needs in mind or a desire to purchase at that time. These "just looking" customers may object when pressured into a sale under these circumstances. The sales personnel should use careful discretion in discerning whether a customer wants to really make a purchase, merely browse through the shelves of merchandise, or look over the plants in the store. These "just looking" prospects may be turned into buying customers by the skilled handling of the salesperson, however. The sales clerk may interest a customer enough to create a desire to purchase through the careful demonstration and explanation of the product or plant. When the customer expresses that he wishes to "think about it" when shown a product, the sales clerk should not press him further for a sale. The salesperson might suggest that the customer browse through the various items of interest and that the clerk will be available for additional assistance, if needed.

Closing the sale.   A sale is not completed until the customer has actually made a purchase. Often the salesperson may easily proceed through the initial steps in selling, only to lose the customer during the closing. The customer must be made to feel that he has decided to purchase a product on his own. The most common failure in closing is the attempt by the sales clerk to do so before the customer has made up his mind. Often the sale's closing is reached simply when the customer asks "How much is it?" The right moment for the salesperson to close a sale is when he feels the customer is convinced that the product will fill his needs.

The salesperson can skillfully bring about the closing by suggesting to the customer, "I can have this on the early afternoon delivery for you." If the customer has made up his mind, the sale is completed. An experienced salesperson will find the closing of the sale is nearly automatic when a clear and convincing sales representation is tailored to the customer's needs.

Once the customer has made a selection for purchase, the salesperson should conclude the selling process quickly and professionally (Figure 12–2). Either the item may be wrapped, or the order written up for the designers (in the case of flower arrangements). The money exchanged should be accurately recorded on the cash register and the change counted carefully to the customer. When making change, the salesperson should make certain the customer sees each coin or bill and is satisfied that the change received is the correct amount. The proper method for making change is shown by this example:

Amount of purchase .... $ 4.38
Amount given by customer 10.00
Difference .... $ 5.62

FIGURE 12–2. Once a purchase has been decided upon, the product is either decorated or wrapped for ease of transportation by the customer.

The sales clerk should place the $10.00 bill above the cash drawer for later reference if a question about the currency handed him should arise. He then removes from the drawer 2 pennies, 1 dime, 2 quarters, and a $5 dollar bill, in that order.

This change is counted back to the customer in the following manner: "The purchase price was $4.38. You gave me a $10 bill, so your change is $5.62." This is counted into the customer's hand in the following steps:

1. $4.38 + 0.02 = $ 4.40   (2 pennies)
2. $4.40 + 0.10 =   4.50   (1 dime)
3. $4.50 + 0.50 =   5.00   (2 quarters)
4. $5.00 + 5.00 = $10.00   ($5 bill)

Promptly tell the customer that you appreciate his business by saying, "Thank you, come in again." The customer may have paid for the purchase by either a check or credit account. These will bear the customer's name. Make a mental picture of the name so that you might remember the customer on his next visit to the store. This simple courtesy may make future sales much easier.

MAKING TELEPHONE SALES

Some retail stores rely heavily on telephone orders as a major portion of their business. This is more true of the retail flower business than many other horticulture enterprises. The retail florist may generate nearly one-third of his/her business in this manner. Each member of the sales staff must be able to receive these calls in a highly professional manner. An inexperienced employee may require additional training before being capable of handling telephone orders. The sales personnel should be familiar with the selection of flowers and plants available, their prices, and what services may be provided by the store.

Answering the telephone. The telephone should be answered as promptly as possible by a member of the sales staff. It should not be allowed to ring more than three times before being answered. When the sales clerks are occupied with other customers in the store, they should politely excuse themselves to answer the telephone. It is very discourteous to customers, both those in the store and the caller, to allow the telephone to continue ringing during a sale. Customers will accept this interruption obligingly, for they would expect the same courtesy if they were placing the call.

When answering, the salesperson should state clearly and distinctly the name of the shop and the clerk with whom they are speaking. The opening of the sale, for example, would begin with, "McDaniel's House of Flowers, Mary speaking. How may I help you?" This approach to selling aids in answering some of the customer's initial questions and places him in a buying mood. If the call was the result of a wrong number being dialed, the salesperson should courteously acknowledge this fact and end the conversation positively.

Obtain all the information. The majority of customers making telephone orders are already at the point in the sales presentation where they are prepared to make a purchase. These customers merely wish to place an order, and little selling is required. The salesperson should have an order form or pad of paper and a pencil available by the telephone at all times for taking orders. The order is recorded in the following manner on the form:

1. Name of the caller, address for billing, and the telephone number for future reference.

2. Name and address of the party who is to receive the order. (Have the caller spell the name for accuracy.)

3. List the item(s) being ordered and any other pertinent information relative to the sale, such as the delivery date.

4. List the names and any message to be placed on the delivery card. (Have each name spelled to you when necessary.)

The experienced salesperson will repeat the order back to the caller for verification, including the total charges for the sale. Giving the delivery date or time of day when delivery will be made gives the customer confidence that the order will be handled conscientiously. In some instances the caller will not have determined his exact needs yet. The salesperson may make several suggestions once the customer's desires have been expressed. It is absolutely necessary for the salesperson answering the call to have adequate knowledge of available merchandise when selling on the telephone. A list of flowers, plants, and merchandise items may be posted near the telephone sales desk for easy reference. This procedure will aid the inexperienced sales clerk in assisting with telephone orders.

The better flower shops are affiliated with one or more of the national flower wire services. Some of the more familiar of these are: Florists' Transworld Delivery Association (FTDA), Teleflora Delivery Service (TDS), and Florafax. Through its members, each of these wire services is able to provide telephone orders from any location in the United States and several countries. The flower shop employee may be required to receive or place these orders daily. The procedures for receiving a wire service order are similar to those of any telephone order, with some exceptions. The caller will identify the order as being a wire service order. He will then state the name of his shop and his own name (which may be used for future referrals concerning the order). In addition to the basic information provided by the caller, he must also provide the shop's wire service number for billing purposes. The remainder of the order is recorded in the same fashion as other telephone sales. An example of an order placed from one shop to another would be handled in this manner:

1. "Hello, this is an FTD order."
2. "This is Mary speaking from the 'Pot Shop' in San Antonio, Texas."
3. "Our FTD number is 0034567."
4. The order is continued in a normal manner.

Closing the telephone order.   Once the order has been recorded and repeated back to the caller, the sale is generally completed. The salesperson may be able to generate further sales by suggesting related items that might accompany the order. Use the "up-selling" technique whenever applicable. It may be necessary for the salesperson to make arrangements for the payment of the order during the final conversation with the customer. Normally, the order will be charged automatically to the caller unless other arrangements are made. The sales clerk should not assume that the caller is a steady customer having a good credit standing with the store when writing up the order. This should be checked with the store manager before the order is filled by the designers.

When closing a sale over the telephone, the salesperson will use the same positive approach as with face-to-face selling. The customer should be thanked for the order and invited to make future purchases from the store. An example of a closing for a telephone order might be: "Thank you, Mrs. Jones. Your order will be handled immediately. Please call again." When the caller is ordering flowers for a funeral or severely ill friend, the closing must be tactful and in good taste. A closing for this type of order might be: "Your order will be given our special consideration."

The sale is not completed until the caller has hung up the telephone receiver. The salesperson who assumes the sale is over when they thank the caller for the order may be missing out on additional sales. Occasionally, the caller may remember additional items, or important information, at the last moment. If the sales clerk hurriedly hangs up the telephone, this additional part of the sale is lost. A wise sales clerk always allows the caller to end the sale by hanging up the receiver first.

SELLING THROUGH VISUAL MERCHANDISING

The use of visual merchandising techniques allows a shop to create a favorable first impression on customers and perhaps generate sales without requiring much personal selling from the sales staff. Window displays are used to invite customer curiosity in the products offered by the retailer. They are used to "pull customers in from the street." Once in the store, the customer may find the merchandise displayed in a pleasant and orderly fashion for appealing to the customer's desire for making a purchase. These displays are designed to create "impulse" purchases through the use of elegant backgrounds, massed groupings, or tastefully decorated merchandise. The sales staff, designers, and the manager will all be involved in creating displays that project a desire to make a purchase in the customers entering the store. The use of window and interior displays by retailers will be discussed.

Window Displays

The purpose of the window display is to create a desire in potential customers to enter the store. The window display area is used to feature seasonal items of merchandise and create a setting for the image of the store. The window displays create the important "first impression" that a customer gets of the store. These window settings contribute to the store's volume and in-store traffic, when the merchandise and ideas are effectively and attractively displayed.

The window displays are changed often, reflecting the mood of the season, the manager, and the merchandise to be featured. Stores in large cities may find it advantageous to create new window settings every week. The merchandise, in any case, should be rotated (replaced or changed) from a window display often to prevent a shop-worn appearance. Only the most desirable or freshest products should be kept on display for the inspection of passers-by. The frequent changing of displays also exposes potential customers to a greater variety of merchandise. In high-traffic areas, people come in contact with the store-front displays more quickly.

A window display is like a salesperson, as it is expected to (1) *attract attention* (greetings), (2) *arouse interest*, (3) *build desire*, and (4) *lead to action* (in this case to bring the customers into the store). To accomplish this, the display area should be planned with the following elements in mind: *background*, *display materials*, *props*, and the *merchandise* itself. Every window setting will need to carry a central theme or idea which is easily conveyed to the public. An Easter bunny and decoratively colored eggs molded from styrofoam instantly convey an Easter season message. This naturally will attract potential customers with Easter gift buying plans, when appropriate merchandise is also on display.

Window displays should be designed using the basic principles of art. *Balance* involves the arranging of merchandise in either symmetrical or asymmetrical patterns. *Proportion* in window display relates the shape and size of the display to the size of the window area. Above all, *unity* is required to pull the entire display, with its merchandise and props, into a concise theme.

Successful displays must be easily seen in all their parts quickly from a distance. Store windows on a busy, fast-moving street must remain simple, with large printed signs or designs. The average motorist may be able to view these displays for 5–7 seconds only. The entire concept of advertising for the shop must be comprehended by these potential customers in this short period of time. Store fronts on pedestrian walkways may contain more information and merchandise. Here, the potential customer may be allowed to browse over the selected merchandise before deciding to make a purchase in the store.

Color may be used to create a desire to purchase. Attractive color schemes may create an immediate response through color psychology. Since certain colors affect the emotional responses of the public, they become a vital element in merchandising. Some psychological emotions stimulated by various colors are:

Neutral Colors: White, black, gray, and brown create a serenity best used in the background of the display area. When a single hue is used with the neutral background, an accent is created to attract attention to the display.

Green: This color emphasizes the natural green shades of plants. It may be used in large masses with flowering plants to set off the blooms. Green is a good color to use for any retail horticulture business display.

Yellow: Yellow is a stimulating color that suggests sun and cheerfulness. It may be seen from great distances, so will attract attention easily. Yellow is used to create an emotion of springtime. When used during late winter in displays, yellow creates a buying motive in winter-weary shoppers.

Red: Although red is most often used for seasonal displays (such as Christmas and Valentine's Day), it may be used wherever a dramatic effect is desired. Being a warm color, red can create interest in otherwise drab mechandise.

Orange: When used with black, orange naturally offers the theme of the Halloween season. Orange is used in floral displays during the autumn months. Colors most often used with orange are brown, black, red, and yellow.

Interior Displays

The purpose of interior display cases and shelves is, of course, to display the store's merchandise. This is done in a manner that will help to sell the customer on the various products as they browse through the shop. These displays are in part designed to aid in making sales to the "just looking" customer. Use of self-demonstrating displays and attractively arranged floral designs in fancy containers helps to promote a buying desire in the customer.

The display cases and shelves are used to aid the salespeople in selling merchandise by providing additional sales appeal to customers. Related merchandise is grouped for easy inspection. Even within the small retail store, related items may be placed together on shelves, tables, or glass cases. For example, in a flower shop, the flowering plants will be arranged in a display near the front window for light and to attract the attention of customers as they enter the store. Candles may be located in a darker location to avoid the fading effects of sunlight on their bright colors. The expensive lines of pottery and glassware will be displayed in a central location to catch the eyes of customers as they look over other merchandise. This type of merchandising aids the customer in selecting the items desired without the direct assistance of a salesperson.

Grouping of merchandise into related products is of great assistance to the sales staff in up-selling. When a major store product is displayed near its accessories, these might easily be tied into the sale, thus bringing more money into the store. An example of this type of merchandise display technique might be the placement of pots or containers, potting soils, leaf polish, fertilizers, and insecticides near the foliage plant displays. Any or all of these types of products may easily be suggested to complement the sale of the foliage plants selected by a customer. The placement of these related products in the same area serves as a reminder to the customer of possible additional needs that otherwise might be overlooked.

The sales staff will find it beneficial to look over the merchandise in the store's sales rooms as often as possible. Whenever the sales area is free of customers, the employees may rearrange the shelves, restocking items that have been sold or soiled, and listing all items that have been reduced in inventory to the point where reordering is required (Figure 12–3). The employees should dust the shelves daily as they inspect the merchandise. Some items may be in need of washing to remove soil from dust and handling. Every piece of merchandise to be displayed must be free of blemishes and soil. The store's reputation is determined to a large degree by the aesthetic appeal of the merchandise and display techniques used.

FIGURE 12–3.   The sales staff must be familiar with all the merchandise in the store. Whenever they are not occupied with other duties, the employees should take time to clean, inspect, and familiarize themselves with the products on display.

BECOMING PROFICIENT IN SELLING

Every employee of a retail store, whether a clerk, a designer, or a manager is at one time or another a salesperson for the business. All must do their part to provide service to keep satisfied customers. All members of the staff should be enthusiastic about their job and dedicated to creating a pleasant atmosphere in which to work. The manner in which the employees conduct themselves in the store is reflected in their ability to generate sales. Employees who enjoy working together in a shop make the task of selling merchandise much easier. The customers can easily sense the tenseness created by ill-will among the staff and retreat quickly out the front door.

Most inexperienced personnel who are newly employed in a retail store are given the duties of salesperson. Since this job requires the most contact with the buying public, it is imperative that the salesperson become proficient at selling as quickly as possible. Some points to consider in review of retail selling techniques are:

1.   Greet the customer in a friendly manner. Use his name often in the selling conversation, whenever you know it. Avoid such trite statements as "May I help you?" Be positive in your first impression and place the customer's mind at ease.

2.   Take control of the sale early in the presentation. Determine the needs of the customer, then make suggestions as to how his needs may best be met.

3.   Create a desire to purchase. Use your hands to demonstrate how a product may serve his needs. Appeal to his sense of smell when demonstrating flowers or plants. Let the customer handle merchandise that is not extremely fragile. Make them feel they must possess the product.

4.   Close the sale on a positive note. Show the customer he has made a wise selection. Suggest added services or related products to accompany the purchase. Package the purchase and offer to carry it to the customer's car, if requested and convenient.

5.   Record and ring up the sale in a professional manner. Make change quickly by counting each coin as you place it in the customer's hand.

6.   Leave the customer satisfied with the purchase. Send him away courteously with a personal "Thank you" and an invitation to "Come in again."

The inexperienced salesperson should adapt the policy of reviewing every sales presentation immediately after it is made. This is particularly important whenever the sale was lost. Each salesperson must be able to analyze his failures and eliminate his mistakes before he can become a successful sales clerk. Gaining a thorough knowledge of the merchandise and services offered by the store may be the best way to begin training in sales.

The experience gained from working as a salesperson in a retail store will do much to provide the employee with a good future in the business world. The ability to deal with other people is an important phase of obtaining maturity in character and in providing sound business sense. These tools may prove invaluable later when a person chooses to begin his or her own business.

SELECTED REFERENCES

Ernest, J. W., and R. D. Ahmun. 1973. *Salesmanship Fundamentals*, 4th ed. Gregg Publishing Division, McGraw-Hill Book Company, New York.

King, B. B. 1974. *Retail Sales Techniques Self-Taught*. National Retail Merchants Association, New York.

Pederson, C. A. 1971. *Salesmanship: Principles and Methods*, 5th ed. Richard D. Irwin, Inc., Homewood, Ill.

Pfahl, R. B. 1968. *The Retail Florist Business*. The Interstate Printers and Publishers, Inc., Danville, Ill.

Richer, G. H., W. G. Meyer, and P. G. Haines. 1962. *Retailing Principles and Practices*. Gregg Publishing Division, McGraw-Hill Book Company, New York.

Russell, F. A., F. H. Beach, and R. H. Buskirk. 1963. *Textbook of Salesmanship*, 7th ed. McGraw-Hill Book Company, New York.

Walsh, L. A., R. D. Joy, and N. K. Hoover. 1971. *Selling Farm and Garden Supplies*. Gregg Publishing Division, McGraw-Hill Book Company, New York.

TERMS TO KNOW

Inventory	Petty Cash	Substitute Selling

STUDY QUESTIONS

1. List the personal requirements needed to be a sales clerk in a retail shop and discuss their importance in selling.
2. Give examples of up-selling products in a store.
3. Discuss how the sales clerk's selling techniques would be different when dealing with the just looking, undecided, and decided customer.
4. List the important information required when receiving telephone orders, and give examples of each.
5. List the four basic steps in making a sale and explain their importance.

SUGGESTED ACTIVITIES

1. Set up a simulated shop and practice sales techniques through role playing. Each student should have the opportu-

nity to be both sales clerk and customer. Students should attempt to provide an accurate description of the products and use positive-speaking vocabulary when selling.

2. Practice making change from a cash register to become skilled in handling money transactions.

3. Make a sales tax table for use in class sales projects.

4. Learn to make the following types of calculations:
   (a) If roses are sold at a cost of $20 per dozen, how much would each single rose cost?
   (b) If candles are sold in a box of one dozen for $6, how much should you charge for four candles?
   (c) If carnations cost $12.50 per dozen, how much should you charge for 15 flowers?

5. Practice receiving simulated telephone orders in class. The local telephone office may be contacted for assistance in learning telephone communication skills.

PART FOUR

The Nursery Industry

CHAPTER THIRTEEN

Production of Nursery Stock

The commercial production and marketing of nursery products is the function of the nursery industry. The crops grown by commercial nurseries include trees, shrubs, vines, and other plants that remain in a landscape for more than 1 year. The nursery industry includes many different types of businesses. Some nurseries grow many different crops, while others specialize in only a few. A nursery may produce plants for sales to retail stores, while others may grow their own plants for sales at the nursery or to be used in landscape contracting jobs by the nursery. The various types of nursery businesses and related fields of interest include wholesale nurseries, garden stores, mail-order nurseries, and landscape contracting firms.

The rapid population movement from the large cities to the suburbs and smaller communities is causing a surge in the construction of new homes in America. The building contractors and homeowners wish to provide trees, shrubs, and lawns for these new homes in order to make their living areas more pleasant. This increased need for plant material in this country has caused a similar growth in the nursery industry. New and expanding commercial nurseries and garden stores require competent employees who are capable of meeting the needs of this industry. A person who seeks employment in a commercial nursery should have a good understanding of production methods, plant growth requirements, and a genuine interest in the business. Nursery production offers the employee an opportunity to work outdoors and to be involved with many phases of the production of nursery products.

THE WHOLESALE NURSERY

Wholesale nurseries do not sell their products directly to the general public, but sell to retail stores or other plant material producers. Some of these wholesale nurseries specialize in the propagation and production of only certain types of plants. For example, some nurseries in the southeastern states specialize in the propagation of azaleas and hollies. These plants are often sold before they are ready for use in a landscape. These plants are sold to other plant producers, who grow the plants to a size suitable for landscaping.

Wholesale nurseries grow most of the woody plant material that is used in landscapes in the United States. They produce plants to be sold in large quantities to businesses that ultimately sell them to the consumers. Some of these nurseries may grow several hundred different types of plants, while others grow only a few specific crops. The operation of a wholesale nursery requires a great amount of labor and skilled employees to carry out the tasks required to handle the large numbers of plants being grown.

A typical wholesale nursery will consist of several related divisions arranged according to the various phases of the production of the plant materials. A *propagation area* is included for the purpose of starting new plants from seed, cuttings, or by grafting. The facilities required for this area might include greenhouses, sashhouses, cold frames, seed beds, and areas for growing the stock plants used for propagation. The nursery will have areas that are used for growing the plants to a larger size once they have left the propagation area. These areas consist of open land that is used for planting the crops directly in the ground (*field stock*) or for holding plants that are being grown in containers (*canned stock*). These areas constitute most of the space occupied by the nursery.

Additional buildings and facilities are required to handle the large quantities of nursery stock being grown. Buildings are required for the storage of the nursery stock once it is ready for sale. These facilities usually consist of a *grading room,* a *packing shed, cold storage rooms,* and *storage sheds.* Adjacent to this area is found the loading area, where trucks and semitrailers may be loaded. A nursery requires considerable amounts of large and small equipment in its operation. This ranges from tractors or crawlers to the gasoline-operated tillers and hand tools that are used. These are stored in an *equipment and storage building.* An *office building* is generally centrally located in the nursery so that all operations can be coordinated. Employees must learn to operate this equipment skillfully and efficiently.

NURSERY STOCK PROPAGATION

The production of nursery stock begins with the propagation of the plants. Many wholesale nurseries specialize in only the starting of seeds or vegetative propagation of specific types of plants. These small plants are then sold to other nurseries for further growth. Most large nursery businesses include a propagation area for starting seeds, rooting cuttings, and grafting plants for further growth in containers or in the fields. The propagation of nursery plants is directed by the head propagator at a nursery. He has the responsibility for the propagation of a large number of different types of plants for the nursery.

FIGURE 13-1.   Lath-covered shade houses are widely used in nursery stock production for seed germination, liner production, and plant protection.

The propagation area of a nursery includes greenhouses, shadehouses, or hotbeds, which are used for germinating seeds, making grafts, or rooting cuttings (Figure 13-1). Areas used for the rooting of vegetative cuttings are equipped with mist beds to maintain a high moisture content of the air. The same type of mist system is used for nursery stock as it is for greenhouse plants. The standard rooting medium consists of an equal volume of peat and perlite, although other rooting media are sometimes used. The rooting medium is kept warm by use of electric heating cables located under the surface at a depth of 4–5 inches (10–12.5 centimeters). The rooting plants are kept alive by intermittent application of a fine water mist over the foliage to prevent drying until roots are formed. This mist is applied by fine mist nozzles that break the water spray into small droplets which remain suspended in the air for a short time. This causes the air to be filled with moisture, so the leaves remain moist at all times. The mist nozzles are connected to a water supply that is regulated by mechanical or electrical valves. Electric time clocks or specialized misting timers are used to regulate the frequency of the misting cycle.

PROPAGATION BY VEGETATIVE PLANT PARTS

When new plants are started from their vegetative parts rather than from seeds, the technique is called *asexual propagation*. The various methods for obtaining new plants by asexual methods include *cuttings, layerage*, and *various types of grafting*.

Cutting Production

New plants may be started by the rooting of individual plant parts, such as portions of the stems, roots, or leaves. The use of cuttings for propagation ensures that the new plants will be exactly like the parent, since all the cuttings will be exactly alike. This method of propagation is the best way by which a nurseryman can guarantee that the plants will be maintained true to a patented cultivar. When seeds are used to start new plants, the young seedlings may not have the same characteristics of the parents. Some plants are also poor seed producers. Nurserymen require large quantities of plant material, so they prefer to use cuttings rather than rely on seeds from poor seed-producing plants. Certain plants are propagated asexually for other reasons also: (1) the seeds germinate slowly, so cuttings are used to speed up production of plants; and (2) with some plants specific sexes are desired, such as males of Ginkgo, since the female produces fruit having a foul odor.

Cuttings are generally made from the terminal stem segments of actively growing shoots. This region of the plant includes a portion of the stem, leaves, lateral buds, and an actively growing shoot tip (Figure 13–2). Cuttings made from this portion of the plant are easiest to root because they are still actively growing. The age or maturity of the plant part determines to some extent the ease in rooting cuttings.

The soft succulent growth at the shoot tip of most plants will root rapidly without special treatment of the cutting. These cuttings are generally taken from stock plants in late spring while the plants

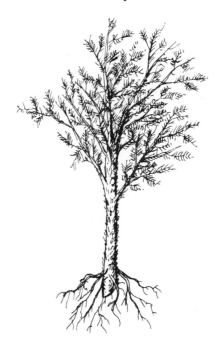

FIGURE 13–2.   A terminal
cutting.

are growing most rapidly. These *softwood* cuttings are taken in early
morning when the stems are filled with water and the air is cool.
Cuttings are removed from the stock plants by breaking them from
the plant at a length of 3–5 inches (7.6–12.5 centimeters). When these
cuttings are ready to be stuck in the mist beds, the lower third of the
leaves are removed to prevent rotting of the foliage that would be
immersed in the propagation medium. Cuttings taken after the first
flush of active growth has ceased in midsummer are called *greenwood*
cuttings. These require that the stems are cut, rather than broken off,
and they usually root more slowly than do softwood cuttings.

Hardwood cuttings are used for many evergreen and deciduous
woody plant species. These cuttings are taken from stock plants in
the late autumn or early winter after the plants have been subjected
to freezing weather. The plants are in a dormant condition at this
time, so new shoot growth is inhibited. The cuttings are generally
placed in moist sand or peat at 33°F (0.6°C) during the winter. This
treatment allows the cut ends to heal, or form callus tissue, which
will later root rapidly. These cuttings are then placed directly in nursery beds or in the field to root and form new shoots.

Cuttings that do not normally form roots easily are often treated
with chemicals to speed the rooting process. These chemicals are
used to ensure better root formation from tissues that would nor-

mally root under proper conditions. The chemicals normally used are naphthalene acetic acid (NAA) or indolebutyric acid (IBA). These chemicals are applied as either a dry powder dip to the cut ends or by soaking the lower portion of the cut stems in the chemical solution. Difficult-to-root woody stems are often soaked overnight in a solution to ensure adequate penetration of the rooting chemical. When the cuttings are dipped in the powdered chemical, it is important that the excess material is shaken from the base of the cutting. Overtreatment with rooting hormone powders may kill the tissues and allow stem-rotting disease organisms to enter the plant. Most commercially prepared rooting hormone powders include one or both of the root-stimulating chemicals, talc powder as a carrier, and a fungicide to reduce the incidence of diseases during rooting.

Layerage

Certain types of plants are best rooted while still attached to the parent stock plant. The stock plant continues to provide nutrients and moisture to the new plant while roots are forming. This procedure often speeds the rooting of some rather difficult-to-root species. Two types of layerage are commonly employed in the nursery. These are: simple layerage and mound layerage.

Simple layerage is accomplished by bending the branches from a stock plant so that a portion of the middle of the stem may be covered with soil. Rooting is enhanced by making a cut through the bark or outer tissue on the lower surface of the stem, which will be covered with soil. A stake or wire hook is placed in the ground around each stem to ensure proper soil contact during rooting. When it is determined that proper rooting has occurred, the branch connecting the parent plant to the newly rooted plant is severed and the layered plants are dug. These are then transplanted to either containers or nursery rows in the field to continue further growth on their own roots (Figure 13–3).

Mound layerage is similar to the simple layerage method of propagation. In this case the stock plant is pruned severely to reduce the branch height to just a few inches from the soil line. A mound of soil is placed over the crown of the plant so that the newly formed branches must penetrate several inches of soil before reaching light. This treatment encourages root formation in the zone of the stem below the soil line. When ample roots are formed, the new plants are dug and placed in containers or lined out in the field to resume growth (Figure 13–4).

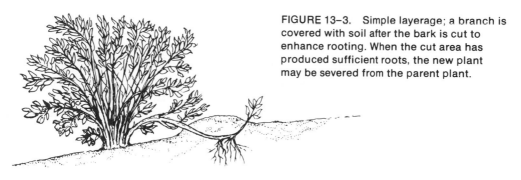

FIGURE 13–3.   Simple layerage; a branch is covered with soil after the bark is cut to enhance rooting. When the cut area has produced sufficient roots, the new plant may be severed from the parent plant.

FIGURE 13–4.   Mound layerage; the shrub is severely cut back and mounded with soil. As the new shoot growth penetrates through the soil, roots are also formed. These newly rooted shoots may then be removed from the parent plant and grown in the field or in containers.

Propagation by Grafting or Budding

Grafting and budding is used to create plants which will produce one type or cultivar of plant from the roots or stems of another. This technique of propagation is most often used in the production of a plant that does not come true to type from seed or does not root readily from cuttings. Other reasons for using grafting methods for propagation include (1) the use of winter-hardy roots on desired plants to extend their range of growth, (2) the use of disease- or insect-resistant roots on desired cultivars to reduce losses in the landscape, (3) to create interesting shapes or types of plants (for example, the creation of a tree rose by budding hybrid tea roses onto the canes of a tall-growing rose), and (4) to create dwarf plants (usually fruit trees) by placing the desired cultivar on a root system of another type which creates a slow-growing dwarfed tree.

Grafting of woody plants is an exacting science and is generally done only by skilled propagators at a nursery. The skilled grafter must be able to select the correct sizes of wood to match for the top (*scion*) and bottom (*understock*) of each grafted unit. The most important skills to be learned are to make the proper cuts on both the stock and scion pieces and then to fit these components together in such a way to make the maximum contact with the *cambium* layer of the plant parts. The cambium of a plant is the actively growing layer of cells, which lays down new xylem (water-conducting cells) and phloem (food-conducting cells). It is from the cambium region that connecting tissues are formed to permanently join the understock with the scion piece that has been grafted onto it.

Various types of grafting techniques are employed in the production of specific types of plants at a nursery. Much of the grafting is done during the winter with understocks or rootstocks and scion wood which has been stored in cooled rooms until grafting is begun. This type of grafting is often referred to as *bench grafting*. The most common method for making a bench graft is by the *whip-and-tongue* graft. This grafting method is often used to join a hardy or dwarfing tree rootstock with a desired scion cultivar. This graft is begun by selecting two plant parts (understock and scion) which are of nearly equal size, usually under 1 inch in diameter. A diagonal cut is made across the top of the understock and bottom edge of the scion for a distance of about five times the diameter of the pieces to be joined. The tongue is made by positioning the grafting knife about one-third the distance from the outer bark. The pith of the understock is cut at right angles to the original diagonal cut for a distance of about ½ inch (1.3 centimeters). A similar tongue is then made on the scion piece. The two cut pieces are then interlocked to make cambium layer contact and are secured in place with grafting tape (Figure 13–5).

To provide a faster method of grafting fruit trees and other plants, the *splice graft* is often used. The understock and scion are prepared as with the whip-and-tongue graft, except that the tongue cut is not made. The understock and scion are not held firmly with this method, so more care must be taken to ensure proper cambium contact of both grafted parts (Figure 13–6).

Most conifer species and broadleaved evergreens are propagated by the *veneer* or *side grafting* method. Potted understock plants are grown in greenhouses during the autumn and winter in preparation for grafting in late winter. To prepare the understock for a side graft, the region immediately above the pot rim is rubbed free of buds and side shoots. A long slice of the bark, which includes a small part of the wood, is removed from the side of the understock. The scion piece is prepared by removing all buds and lateral shoots for a distance of 1–2 inches (2.5–5 centimeters) from the base. The base of the scion is cut so that a slice of equal length as that made on the understock is removed. A shorter slice is then removed at a 45-degree angle on the opposite side of the base of the scion. The two pieces

FIGURE 13–5. The whip-and-tongue graft: (a) A diagonal cut is made on the stock and scion pieces. The tongue is formed by making a slice in both the stock and scion. (b) The stock and scion are joined securely. (c) The graft is completed by wrapping the union securely with grafting tape.

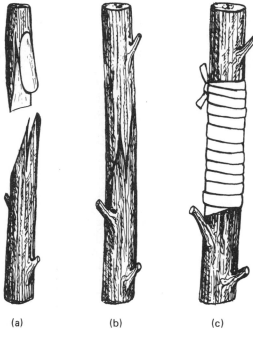

(a) (b) (c)

FIGURE 13–6. The splice graft: (a) A diagonal cut is made on both the stock and scion pieces. (b) The two grafting pieces are placed together. (c) The graft is completed by securing the stock and scion with grafting tape.

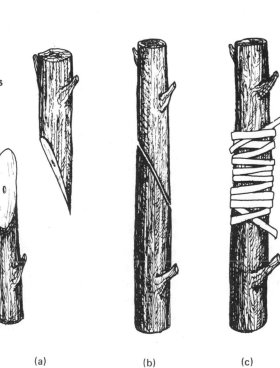

(a) (b) (c)

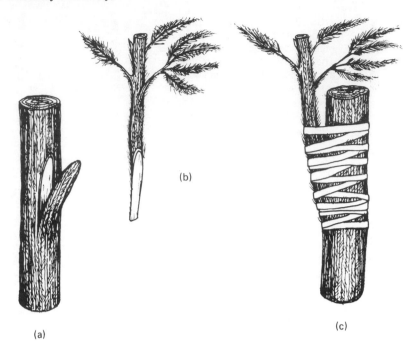

FIGURE 13-7.   The veneer or side graft: (a) A long slice of bark is made in the side of the understock. (b) The base of the scion is sliced to match the cut made on the understock. (c) The two pieces are joined and tied securely with grafting tape or rubber strips.

are then positioned together to match the cambium layers and secured with grafting tape or a rubber budding strip (Figure 13–7).

Bud grafting is a method of grafting that is normally done in the field or on potted understocks. The scion that is used in this method of grafting consists of only a single bud rather than a twig or stem. Bud grafting is used for the rapid propagation of roses and many fruit tree species in the nursery field. This method is more economical of scion wood, since only single buds are required for each graft. Certain types of tree species can best be grafted using the bud graft method, because they produce a wound gum that interferes with the healing of grafts that cut through the xylem tissues of stems (such as peach and cherry).

A bud graft is generally made during the spring or early summer when the understock is growing actively. At this time the bark may be easily separated from the wood. This condition is often referred to as the period when the bark *slips* easily. A single bud is inserted in the bark of the understock between the nodes. When the bud graft has joined sufficiently, the understock is cut immediately above the bud graft and the original top is removed. This stimulates the growth of the desired bud and a new "top" results (Figure 13–8).

The bud graft is prepared by first selecting an appropriate bud to be inserted in the understock. A quick slicing cut is made from above the bud, under the bud region, and extended below the bud. A T-shaped cut is similarly made on the understock 2–3 inches (5–7.6 centimeters) above the soil line or pot rim. This cut extends to a depth to include only the bark and not into the wood of the stem. The bud shield scion piece is fitted into this T-shaped cut area by lifting the bark carefully with a knife handle designed for this use. Once the bud and understock are appropriately joined, the grafted area is securely wrapped with a rubber budding strip. Rapidly growing species, such as roses, are trimmed of the understock top to force the growth of the scion bud after 2–3 weeks. Slower-growing trees are allowed to grow with the original top growth for the remainder of the season. The top growth is then removed above the bud graft the following spring. During the time that the bud graft is healing, the scion bud should swell to indicate that good cambial contact has been made. If the bud does not swell, but shrinks or turns black, the bud has died and must be discarded. The budded plants normally will remain in the nursery field for one or more seasons following grafting before they are sold.

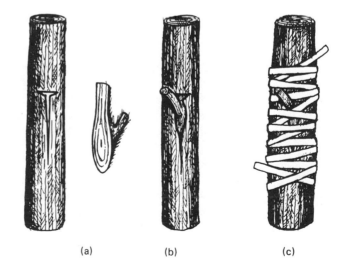

(a) A "T" incision is made in the understock and the scion is prepared by slicing off a single bud from a dormant shoot (bud stick).

(b) The scion bud is slipped into the understock incisions.

(c) The two pieces are securely joined by wrapping them tightly with a rubber strip.

(a)   (b)   (c)

FIGURE 13–8.   The bud graft

PROPAGATION OF NURSERY PLANTS FROM SEEDS

Propagation of nursery plants by seeds is preferred by nursery-men whenever this may be done. Seed propagation is more rapid and economical for those plant species which produce ample seed quantities and when the seedlings are all similar to (but not exactly like) the parent plants. Most nurserymen obtain their nursery stock seeds from various sources. Some seeds are collected from mature plants grown at the nursery. This is especially true for those species of plants considered to be winter-hardy or to hold a special appeal in a particular climatic region. Some seeds are obtained from distant or foreign sources. Local nurserymen purchase these seeds from commercial seed collectors. Many tree and shrub seed sources may be obtained by collecting the seeds from the wild or from plants in landscapes.

When seeds are obtained from plants at the nursery or else-where, it is necessary that they are collected at the correct stage of maturity and stored under the proper conditions to retain the seed viability. The seeds must be allowed to mature fully if they are to germinate. The fruits should be observed daily as they ripen and must be collected at the fully ripe stage, but before the seeds fall from the plant. Some seeds (such as certain plants in the rose family) will germinate immediately when harvested before they are fully ripened. If these seeds are allowed to ripen completely, they must be subjected to several months of cold treatment before they will germinate. Most seeds, however, must be fully ripened before germination is possible.

Seeds are separated and cleaned as soon as possible after collection. Pulpy fruits (such as barberry, juniper, viburnum, yew, and others) must be depulped in a hammer mill to remove the seeds in a stream of water. Separated seeds are screened and washed briefly to remove chaf and debris. The cleaned seeds are then dried to the desired seed moisture content, stored in watertight glass containers, and labeled by seed source, date of collection, and variety of plant seed.

Many seeds cannot be germinated immediately upon collection. For many seed types, a period of exposure to cold temperatures in a moist environment is required to break the *rest period* imposed on the seed. This treatment is called cold *stratification* of seeds. Most seed types are stratified in alternating layers of seed and damp peat moss in containers held at 35–41°F (1.5–5°C). The length of the stratification period varies from 1 to 4 months depending on the plant species. These conditions are normally met naturally by the over-wintering of seeds outdoors.

Some seeds germinate slowly or not at all because the seed coat provides a physical barrier to the entrance of water. Hard seed coats may be made more penetrable to water by a process called *scarification*. The seed coats are scratched with abrasive materials to wear down the impermeable seed coat. This is often done with a revolving drum lined with sandpaper or a concrete mixer filled with sharp sand which may easily be separated from the seeds. Some seeds may be made to germinate more readily by soaking them in very hot water. Boiling water will kill most seeds, so care must be taken to prevent overheating. Seeds soaked in this manner must be planted as soon as possible, for when they dry the treatment is nullified. In some cases extremely hard seed coats are softened by placing the seeds in concentrated sulfuric acid for 1–12 hours. This method is very effective in getting seeds of this nature to germinate, but the propagator must be very careful to avoid injury when handling dangerous acid solutions.

Field Germination of Seed-Propagated Plants

Direct seeding of many nursery stock species is more economical, since less handling and greenhouse space is required. This method also allows the plants to grow unchecked, with no loss of growth from transplanting. The type of equipment used to prepare a suitable seedbed for seed germination will depend upon the size of nursery. Large nurseries will prepare the field in early autumn by plowing, disking, and harrowing after a suitable cover crop has been grown. Smaller nursery firms will employ rototillers to prepare small seed beds.

Seeds are planted in the autumn or the spring, depending on the germination requirements of the seed species and the practices used by the nursery. Seeds are planted with automated seed planters whenever this is feasible; however, many large-seeded plants (such as walnut) require hand sowing. Large-seeded tree species are sown 12 or more inches apart in rows, while medium-sized seeds are spaced much closer and thinned following germination. Prior to sowing of seeds, the rows are often treated with chemical sterilants and allowed to remain unplanted until all the effects of the chemical have dissipated. This treatment aids in reducing insect, disease, and weed pests in the seedling nursery.

Following the sowing of the seeds, a thin layer of sand or light soil is applied to cover the seeds. The seed beds are given a light application of fertilizer, if this was not incorporated before planting. A 6–12–12 or 5–10–10 fertilizer is most often used. Automatic irrigation is generally used to supplement natural rainfall to ensure proper germination and seedling growth. Specially constructed shade coverings of lath or Saran cloth are used to protect the newly emerged seedlings. Some species, such as broadleaved evergreens and others, may require special protection for the first 1–2 years of their growth in nursery beds.

Many field-germinated plants (such as roses and fruit trees) are grown for 1 year in the field and then used as understocks for bud grafting. Other plant species that have been grown at close spacings in a seedling nursery bed may be transplanted after 1 or 2 years to containers or to other field blocks to resume growth until ready for sale.

Greenhouse Germination of Seeds

Greenhouses are used for the germination of fine-seeded plants that would not ordinarily survive when directly planted in the field. The same basic germination methods are used for nursery plants as for greenhouse species. A fine-textured soil mixture is selected for seeding. This may consist of one part each of soil, peat, and vermiculite or perlite; or it may be completely without soil. Species requiring an acid seedling mixture, such as azalea, rhododendron, or mountain laurel, are sown in a sphagnum peat and vermiculite medium. Other sowing media may be selected by various nurserymen.

Whether soil or an artificial soil mix is used for a sowing medium, best results are obtained when the medium is sterilized first. When steam is available, this may be the most satisfactory method for eliminating soil diseases, insects, and weed pests. Steam is admitted into a closed container or covered pile of soil mix until the temperature reaches 180°F (82°C) for 30 minutes at the deepest point. The mix is then cooled before the seed is sown. All equipment and tools to be used in the sowing operation should be cleaned before use to avoid contamination of the soil. Where steam is not available, methyl bromide gas or chemical sterilant liquids are used according to labeled instructions.

Seeds are sown in rows made by pressing wood slats into the flats filled with seeding medium. The seeds are sown thinly to prevent overcrowding and then lightly pressed into the medium to ensure good soil contact. The surface of the seed flat is then covered with sphagnum moss to prevent rapid drying of the medium and to reduce the incidence of root-rotting diseases. The flats are watered thoroughly following sowing by immersing them in water to a depth that is slightly less than the soil level. This first watering usually contains a fungicide to aid in the prevention of "damping-off" diseases, which kill newly germinated seedlings.

The newly sown flats are then placed in the greenhouse in specially constructed humid chambers. These chambers are often made of polyethylene sheets formed like tents to conserve moisture. Heat will build up under these germination chambers on a bright day, so adequate ventilation must be available when necessary. The temperature should be checked routinely to be sure that the air does not surpass 90°F (32°C) at any time. The side curtains or the top covering should be raised slightly to allow air movement into and out of the chamber on warm days. Water should be applied to the flats with a fine mist nozzle whenever it is necessary. Once the seedlings begin to emerge from the soil, water may have to be applied more frequently.

When the seedlings have germinated uniformly and have established their first true leaves (not cotyledonary leaves), they are transferred to a cool, humid, shaded location in the greenhouse. When these seedlings become crowded in their flats, they are either transplanted to other flats at wider spacings or potted individually. If the newly established plants are not moved to outdoor beds in the fall, they are overwintered in cold frames and then transplanted to nursery rows in the spring. These plants are then placed in their final spacings in the field to reach a salable size. Occasionally, these small-seeded plants must be grown under lath or Saran-covered structures for added protection until the plants are of sufficient size to survive in the open fields.

FIELD PRODUCTION OF NURSERY STOCK

Field production of trees and shrubs is the oldest and still the most common method used to obtain the larger specimen plants. These woody plants are placed in the fields as *liners* from the propagation area and may remain in the field several years before being sold. The stock fields generally cover a large portion of the nursery. Fields are separated by roads or dirt drives into individual *stock blocks*. These blocks may vary in size but are used to separate the plantings into units. Those plants which are planted at the same time and of the same species or variety are planted together in these blocks so that the entire space may be harvested at the same time (Figure 13–9).

The various plant species are planted on yearly rotations in these stock fields to ensure a constant supply of finished stock several years later. Some of the slower-growing shade trees may require 8–10

FIGURE 13–9. A block of junipers grown in a nursery field.

years of growth in the fields, while fast-growing shrub plants may remain in the field for 2 years or less. Once a field block is cleared of plants, the area is planted to a cover crop which will later be plowed under to provide organic matter for the next crop of nursery plants. This is an important phase of the production because the digging of the nursery stock removes some of the topsoil from the field. Also, the nursery stock depletes the soil of some of the minerals that are required for plant growth. Soils that contain relatively high amounts of organic matter are required where plants are to be dug with the roots in a ball, or else the soil will fall easily from the roots.

Soil preparation of the stock blocks in the field is one of the most important phases of the field production of nursery plants. The fields are plowed in the fall of the year to prepare for spring planting of the blocks. Fall is the best time for this because the freezing and thawing of the soil aids in improving the structure of the soil. Insects and disease organisms are exposed to the chilling temperatures and are often destroyed. Fall plowing also allows the land to be planted earlier in the spring and relieves the amount of work which is required at that busy season in the nursery.

Plowing must be done at the proper time to prevent the destruction of the soil structure. If the soil is worked while it is too wet, the soil particles will compact into tight clods which will not break up easily. When the soil is too dry for plowing, the tractor will not be able to pull the machinery through the field. Plowing should be done only when the soil may be lifted by the hands and broken easily with the fingers. Following the plowing, the soil is given a final preparation in the spring before the field is planted. This is done by disking or harrowing the field twice. Most large nurseries use crawler-type tractors for this because the wide tracks cause less compaction (packing) of the soil.

Fertilizers are added to the fields prior to planting in the spring. Representative soil tests are taken of the prepared area and are sent to the state soil testing laboratory. The report that is returned from the soil testing laboratory will give recommendations for the addition of fertilizers and the adjustment of the acidity (pH) of the soil. Nurseries that have animal manures available will apply this material to the soil before the area is plowed. Animal manure supplies the necessary organic matter, as well as some important minor nutrients, to the soil.

Agricultural lime is added to the soil when the report of the soil acidity shows the need. Some plants, such as rhododendrons or azaleas, require an acid soil (pH 5.0–6.0) for best growth. Most nursery crops, however, grow best when the pH is between 6.0 and 7.0. The soil pH is raised (made more alkaline) by the addition of up to 30 pounds of ground limestone per 1000 square feet. When the pH must be lowered for acid-soil-requiring plants, 10–15 pounds of ground sulfur is worked into the soil.

When a cover crop or animal manure is turned under by plowing, the soil nitrogen is used by the soilborne bacteria to break these materials into minerals. This action ties up the fertilizer so that the crops are unable to obtain sufficient nitrogen. Agricultural fertilizers containing nitrogen, potassium, and phosphorus plus other minerals are applied to the soil in the spring as recommended by the soil test report. These are applied by broadcasting inorganic fertilizers onto the soil prior to the final preparation of the field. Fertilizers are applied again each fall after the leaves have fallen from the deciduous plant material, to provide adequate nutrition of the plants over the winter months. The fall fertilization is very important in southern states, since the plants are growing for a longer period each year. Each spring, fertilizer is again added to the nursery blocks to provide nutrients to the plants as the new shoots begin to grow.

Lining-Out Nursery Plants

The plants removed from the propagation areas are transferred either directly to the field blocks or to cold frames, transplant beds, or healing-in beds for intermediate growth before being moved to the fields. Some plants may be stuck directly into the fields as callused cuttings and will root in the field during the early spring. Grafted plants that have healed are also handled in this manner.

Most plants are moved to the field blocks after they have been sufficiently rooted. These plants may have been rooted under mist or have been growing in pots or cold frames for a period before they are planted in rows in the field. Some evergreen species cannot be planted directly to open fields after they leave the propagation area. These plants require protection from the sun and wind, so are placed in transplant beds or under plastic-covered frames for 1–2 years. Once these plants have reached sufficient size, they may be grown satisfactorily in the open fields.

Seedling plants are transplanted to the field when they have reached a suitable size and root structure to survive in the open. Some plants may be grown in greenhouses or in transplant beds for 1–2 years before they are planted in the nursery blocks in the fields. Planting of the liners and transplants is done mostly by hand. This is done by cutting furrows in rows along the blocks and workers placing the plants at the proper spacings. In some cases, transplanting machines may be used to place the plants in the ground. The transplanting machine requires two operators for placing the plants on the planter wheel as the machine cuts a furrow, places the plants at the proper spacing, firms the soil around the plant roots, and applies a small quantity of water to each plant.

Once the nursery stock is planted in the field, it must be watered periodically during the growing season. This is done by providing water from trench irrigation, overhead sprinklers, or by soaker hoses. The soaker hoses allow water to leak from them very slowly, soaking the soil around the roots of the plants. The most common method for irrigating nursery stock is from overhead sprinkler devices. For most plants that are grown, about 1 inch of water is required each week, either from irrigation or from rainfall. Species that grow continuously during the summer months require more water than do those which produce only one flush of growth in a season, such as slow-growing trees.

Training Field-Grown Stock

The training and pruning of nursery stock in the field is an important part of the production of landscaping plants. The plants must be spaced properly in rows to prevent overcrowding and allow for the best development of each plant. Specimen trees and shrubs are given more space and attention than those plants being grown for mass-market sales. These specimen plants will be allowed to grow in the field an extra year or two to develop a more mature shape or size.

The training practices used on the field stock are different for shade trees, fruit trees, hedge shrubs, and flowering shrubs.

When deciduous shrubs are first planted in the field, and each year thereafter, approximately one-third of the previous season's growth is pruned. This practice causes the plants to form new side branches lower to the ground, making the plant more compact in growth habit. The removal of the top one-third of the growth may be done with hedge shears, hand clippers, or in some cases with the sickle-bar attachment on a tractor. Although this type of pruning is not the best for the plants, the large quantities of shrub plants requiring pruning in the nursery fields prevents any individual attention in pruning (Figure 13–10).

FIGURE 13–10. Nursery shrubs are often pruned with machine-driven cutting bars to form a more dense habit of growth. When done in this manner, the plants are pruned into a somewhat globular shape.

Shrubs normally grown for their upright growth habit are not pruned by removing the top of the plant. These plants are shaped by removing the long lateral branches (called "heading back") to strong side shoots. Narrow-leaf evergreens (yew, juniper, hemlock, etc.) are pruned before new growth begins in the spring and again in midsummer, if needed. Pines, such as the shrub-form mugho pine and tree-form types, are pruned while the new growth is soft and succulent. This new growth is called the *candle* stage. The candles are pinched or clipped to shape the plants and create more compact growth. This period of new growth occurs in the spring. Flowering shrubs are generally pruned in the spring or early summer following flowering to provide the best vegetative growth and formation of new flower buds for the shrubs to be sold the following spring.

Shade trees are trained to a single main leader or trunk. A strong shoot is selected when the tree is planted and all others are removed. These young shoots are called *whips* because they do not support much weight and "whip" in the wind. Some tree species become curved and crooked from the action of this whipping in the field. These trees require staking for the first year or two to provide support to the developing trunk. The stakes are carefully tied to the tree whip by slipping a loose-fitting plastic band or soft twine anchor to the stake and tree trunk. The stakes should be removed after 2 years so that the trees develop strong wood fibers. The stake ties should not be allowed to fit snugly around the tree trunk, for as it grows in diameter it will be girdled by the rope or plastic band. The girdled area will become weak and will eventually break in the wind. Staking is especially necessary when grafted trees are placed in the field. The stake supports the shoots at the graft union until the graft has healed sufficiently to support the tree.

Fruit trees and small ornamental trees are grown as single-stem whips during the first year or more. When the whip reaches a height of 6–8 feet, the tip of the whip is removed. This practice causes side branches to form on the trunk. The pruner selects 3–5 main branches which are spaced uniformly around the tree and removes all others. Some ornamental and fruit trees are grown in the nurseries only as whips to be sold to mass-market firms or for mail-order sales. The pruned and trained trees are grown in the field for 2–5 years and will be sold as balled-and-burlapped plants.

Root pruning of trees and shrubs is also practiced to develop compact root zones and stimulate active root growth in a confined area. This is necessary for plants that will be dug in balls (balled-and-burlapped) or removed as bare-root stock. If these roots were not pruned, they would spread for several feet around the plant. When a non-root-pruned tree or shrub is dug, the major portion of the root system will be left in the ground. These plants will not survive as well as those which have most of their roots concentrated in a smaller space in the field.

The roots are pruned on the field grown nursery stock in the fall of the year after the leaves have dropped from the deciduous plants. This is done during the dormant season so that the new growth in the spring can compensate for the reduced root system and the plants will not be severely stunted in growth.

Digging Balled-and-Burlapped and Bare-Root Plants

Most of the digging of plants in a commercial nursery is done in the fall of the year after the plants have become dormant. This is done at this time for several reasons. Plants dug while they are still actively growing do not establish easily in a landscape and may not survive. The field is generally cleared at one time, so the soil must be prepared for planting a cover crop before wet weather occurs in the winter. Also, most of the grading and packing of nursery stock is done in the winter from the storage sheds, since this is the time of year when little outdoor work can be done.

Many small shrub and tree stock are sold without soil *(bare-root)*. These are bundled or packaged and stored in sheds over the winter months. Digging of these plants is done by tractor-pulled U-shaped blades which cut the soil and roots around the plants for a distance of 12–18 inches (30–45 centimeters). The same blade is used for the root-pruning operation, but for digging it is fitted with forks that lift the plants from the ground. Field workers lift the plants from the furrows, remove excess soil from the roots, and bundle the nursery stock. The bundled plants are then quickly moved to storage sheds to prevent excessive drying of the roots.

Plants that are dug along with soil around their roots and wrapped in burlap cloth are called *balled-and-burlapped* nursery stock. Some plants (such as magnolia, flowering dogwood, and evergreen plants) may not survive transplanting if the soil is removed from their roots. Larger trees and specimen plants also must have the original soil around their roots to establish new growth in a landscape.

Balling-and-burlapping may be done by specially designed machinery when the nursery is large enough to afford its use (Figure 13–11). Most often the digging is done by two field workers using

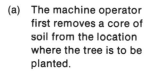

(a) The machine operator first removes a core of soil from the location where the tree is to be planted.

FIGURE 13–11. A tree digging machine.

FIGURE 13–11. (*continued*)

(b)   The tree roots are cut
      with the slicing blades
      of the machine.

(c)   The tree is then lifted
      from the hole and
      transported to its new
      location.

spades to dig a circular trench around the plant. The pitch of the trench is directed toward the plant until the soil ball is completely undercut. Burlap is placed under the soil ball by raising first one side of the ball and then the other. The ball is wrapped tightly with the square of burlap cloth and is fastened securely with box nails and twine. The balled-and-burlapped plant may then be lifted from the hole by handling the soil ball. The plant should not be carried or handled by the stem or trunk, since this loosens the soil from the roots and may cause the plant to die. Larger balled-and-burlapped plants and trees are carried by placing them on large squares of burlap that may have to be handled by several men (Figure 13–12).

(a) A circular, sloping trench is dug around the plant.

(c) A burlap sheet is placed under the soil ball and pulled tightly around the root ball.

(b) The soil ball is shaped and the plant roots are undercut.

(d) After the burlap is firmly secured with nails, the plant may be lifted from the hole.

FIGURE 13–12.   Balling-and-burlapping nursery stock.

Grading and Storage of Field-Grown Stock

Field stock must be handled very carefully and moved quickly from the field to the storage areas. The root systems have been disturbed, so care must be taken to provide a moist and shaded area for the plants as soon as possible. For short-term storage of balled-and-burlapped nursery stock, the plants may be placed under lath or other shading and the root balls covered with moist peat, sawdust, or shingletow (wood shavings). Bare-root stock may be *heeled-in* in shallow furrows of soil in a shaded location. The plants are set at an angle pointing toward the south to prevent excessive sun damage, and soil is placed over the roots. These furrows must be kept wet to prevent the drying of the plant roots. Mist watering systems are often set up above the bare-root and balled-and-burlapped holding areas to provide moisture to the tops of the plants.

Long-term storage of plant material is done in refrigerated rooms at 35°F (2°C) or in cool storage sheds which provide protection and moisture for the nursery stock. Bare-root plants must be stored under high humidity and shipped with moist shingletow around their roots to prevent them from drying. Certain plants (such as roses) dry out easily in cold storage areas so are dipped briefly in hot paraffin to protect the stems from water loss. The bare-root plants are treated with fungicides to control diseases and then placed in separate bins in storage according to species and variety to facilitate easy retrieval and identification. During the winter months, graders remove these plants and grade them according to size and health. At this time the plants are pruned to remove dead branches and roots or to trim out crossed branches. The plants are rebundled according to grades and plant labels applied. These labels identify the nursery, the grade of the plants, and their names.

CONTAINER PRODUCTION OF NURSERY STOCK

The nursery industry is moving rapidly to widespread use of container-grown plants. Plants are grown in containers rather than being grown directly in the field. Many advantages can be listed for the use of container-grown plants, both to the nurseryman and to the customer. Many woody shrubs and trees do not transplant well when dug from the field unless they retain a ball of soil around their roots. When grown in suitable containers the roots are not disturbed, even when planted in the landscape. The heavy soil ball is reduced to a compact soil and root unit, so handling and shipping costs are reduced. Plants grown in containers require less space in the nursery, so more plants can be grown. Since most nurseries are now growing some or most of their plants in containers, it is necessary for all employees to become familiar with this phase of nursery production.

Types of Nursery Containers

The containers used for nursery production are generally larger than those used for greenhouse plants. They range in size from 1 to 12 gallons in capacity and are constructed of plastic, metal, or sometimes of wood. The most familiar nursery containers are dark green in color, light in weight, tapered from top to bottom, and are durable enough to last during the period the plants are in the nursery. The cans provide drainage through holes cut in the bottom or sides of the container.

Metal nursery cans are more durable than are other types of containers. These containers do not collapse when plants are stacked in trucks for shipping. The metal protects the plant roots, since there is little opportunity for the soil ball to be disturbed in a metal can. Plastic cans have become popular since they are light in weight and are durable enough to support the plants while in the nursery. Although the plastic cannot withstand as much handling abuse as metal can, the plastic offers the roots more protection from heat and cold extremes.

Soil Mixes for Container Crops

The soil mixes used for containerized nursery stock are similar to those used for greenhouse production. The soil medium used should retain nutrients and water well, yet provide adequate drainage and aeration to the root zone. Since most nursery stock is grown to a larger size than most florist crops, the soil mixes must be heavier to support the plants. This weight is often provided by including field soil, sand, or gravel in the basic soil mixture. The standard container mix includes one part soil, one part peat or bark, and one part sand or vermiculite.

Field soil is included in most nursery container mixes for several reasons. Besides the addition of weight for container stability, the soil particles provide a reservoir for the storage of moisture and nutrients. During freezing weather the soil pore spaces provide insulation to protect the fragile roots from cold damage. Under periods of water stress, such as hot and windy days, the soil medium provides the roots with moisture for a longer period. Artificially synthesized soil media containing peat, perlite, vermiculite, fired clays, bark, and other materials may be used satisfactorily when handled properly. Since soil offers some nutrients not available in most artificial media, close attention must be given to the needs of the plants when they are used. It is very important that the personnel in charge of watering and fertilizing the container stock be familiar with each soil mixture being used. All watering and fertilizing practices should be adjusted according to the type of soil mix, climate, kinds of plants being grown, and the special requirements of each plant.

When synthetic soil mixes are used in containers, it is necessary that the ingredients be mixed thoroughly to provide a homogeneous medium. Mixing may be done in small engine-powered plaster mixers or in flail-type spreaders. Many large nurseries currently use large truck-mounted cement mixers for this purpose. Once the growing medium is properly mixed, it is sterilized to eliminate most weed seeds, nematodes, and other soilborne pests. This may be done by injecting a sterilant gas or live steam into a pile of soil mix under a protective cover. When steam is used, the steam is allowed to penetrate the soil until the temperature of the soil mixture has reached 180°F (82°C) for 30 minutes. Many nurseries use dump trucks specially fitted with porous pipes in the bed for sterilizing soil mixes with steam.

Planting Container Stock

Some large nurseries utilize semiautomatic potting machines for the purpose of planting nursery stock in containers. For most potting operations, hand labor is required. The container is first filled to a level one-third to one-half full with soil mix. The plant to be potted is then placed in the center of the container so that it will be at the same level as it was previously growing. If the plant is bare-root (without soil from a propagation bed), the roots are spread evenly over the soil mix in the container. The soil mix is then firmed and additional medium added to fill the container to the top. After the soil medium is again firmed, the container is watered liberally. When properly potted in the container, the plant should be located in the center of the pot, with the soil level approximately 1 inch below the rim of the pot. This will allow a basin for holding water until it can penetrate the soil mixture around the roots.

Most container stock is grown in open fields. Since the containers are placed on top of the ground, some problems may require attention. The roots of the plants in containers will eventually root through the drainage holes of the cans and into the ground when placed on bare soil. To prevent this from occurring and to make working conditions more suitable, containerized plants are placed on sheets of roofing felt or black polyethylene plastic. A suitable non-specific herbicide is used to kill all vegetation in the area before the ground is covered. The sheet of plastic or felt prevents the rooting through of the plants and growth of weeds in the nursery block. In some cases where permanent beds are required, the entire area may be covered with crushed rock or asphalt paving.

Fertilizing and Watering Container Stock

Fertilizer nutrients are added to the soil mixes in nursery containers in one of several ways. Some nurserymen fertilize their container nursery stock with slow-release fertilizer granules. These slow-release fertilizer products disintegrate slowly to provide nutrients to the plants for several months. The breakdown of these slow-release fertilizers is controlled by the frequency and amount of water added to the soil. Therefore, whenever the plants are watered, they are also being fertilized. This system of fertilizing nursery stock allows use of overhead sprinkler systems for irrigation, while providing adequate fertilization of the nursery stock.

Many nurserymen prefer to apply fertilizer nutrients to their container nursery stock from soluble liquid feeding systems. These automatic fertilizer injector systems are similar to those used in commercial greenhouse ranges. An automatic fertilizer injector is used to combine a concentrated fertilizer stock solution with water. This diluted fertilizer solution is then pumped through irrigation pipes to each container plant in the nursery block. The irrigation pipe is usually constructed of black polypropylene pipe with small leader tubes extending to each container to be fertilized. This method of fertilization ensures that the plants will receive adequate fertilizer each time the plant is watered. When this system is connected to an automatic irrigation control panel, the routine watering and irrigation of container nursery stock can be nearly labor-free.

Fertilizers may also be applied as a dry powder top dressing to the nursery containers. This is done by adding a complete-analysis fertilizer to the soil surface after the plants have been allowed to grow for a short time. These are applied by broadcasting the fertilizer with a cyclone-type fertilizer applicator while the containers are still pot-tight (pot-to-pot) in the nursery bed. The soil is moistened slightly before the fertilizer is applied and the plants are watered immediately following the fertilization.

Fertilizers applied in this manner are usually those having a low nutrient analysis, such as 10–10–10 or 6–12–12. These fertilizers are often mixed with the soil at the time the soil mixture is prepared. More often, slow-release fertilizer is mixed with the soil at one-half the recommended rate. Following planting, the containers are then provided additional fertilizer from either the top dressing of dry fertilizer or from soluble liquid fertilizers applied with each watering. This method of applying fertilizer is most popular because nursery stock growing in containers requires a steady supply of fertilizer.

No matter how the watering and fertilization of the nursery stock in the containers is accomplished, it is still necessary for nursery workers to observe the growth of the plants routinely. Automatic irrigation systems sometimes become clogged with dirt, so some of the plants may not receive adequate water. The small leader tubes should be checked often to make certain that they are working properly. Occasionally containers placed farthest from a sprinkler will not receive proper water as a result of reduced water pressure or wind. These should be handwatered occasionally or relocated before severe damage to the plants results. Timing of the application and amount of fertilizer should also be coupled to the amount of growth occurring on plants.

Plants are fertilized heavily in the spring of the year to enhance new root and shoot growth. Fertilizer applications are reduced during summer on those species which show only a single flush of growth in the spring. Fertilization is again increased in late summer for proper food storage as a preparation of overwintering. The fertilization is gradually reduced and finally stopped by late August or early September to prevent the formation of soft vegetative growth. The gradual hardening of the plants prevents excessive winter damage during the dormant season. Growth of the root systems resumes during warm winter days, however, so the plants will require occasional irrigation during winter.

Pruning and Training Canned Nursery Stock

Plants growing in containers are trained in a similar manner to those grown in the field. The purpose for the pruning of the nursery stock is to stimulate the lateral branching of shrub species or to correct the shape of trees and other plants. Those plants which normally grow with a single main leader or trunk, such as upright evergreens or shade trees, are pruned by cutting back any long side branches which may appear. Pines are pruned in early spring when the new growth ("candles") appears. By cutting such new growth to one-half their original lengths, new side shoots will emerge below the cut. This allows the plant to develop a fuller branching habit. Deciduous shrubs are also trained by cutting the larger stems (or canes) to a height of one-half or one-third that of the original length. The resulting new growth remains lower on the plant and develops a much fuller-appearing plant. Fruit trees and shade trees are generally grown as a single-leader whip during the first 2–3 years. Once the whip reaches the proper branching height, the terminal is removed to promote branching. The following year three to four desirable side branches are selected for further growth and the remaining branches are removed at the trunk.

Some plants are normally sheared to create special effects or to promote a fuller growth. Some plants are sheared to create topiary ball effects. These plants are trained on a single main trunk and then allowed to branch in several tiers, which are kept clipped into round balls of foliage.

Azaleas are most often grown in containers because they do not withstand transplanting from the field to the home landscape. These plants do not grow naturally into compact well-branched specimens. For this reason, they require periodic shaping or pruning to maintain a well-branched appearance. Many growers use grass shears or hedge shears to remove unwanted shoot growth from the plants. Chemical pruning materials may also be utilized to stimulate more liberal side branching of the azalea plants. These chemicals are applied to the foliage to kill the growing point at the shoot terminal. Such techniques as this reduce the amount of hand labor required to operate a modern nursery efficiently.

Providing Winter Protection for Canned Nursery Stock

Nursery stock grown in containers is more subject to winter injury than are the same plants grown in the field. The soil volume around the roots of plants in containers is small and offers little protection from freezing temperatures. Most container stock may be protected from fluctuating temperature extremes by placing the pots as close together as feasible. Bales of clean straw may be stacked around the perimeters of the nursery block to protect the outside rows of plants. Additional straw may then be placed over the plants when exceptionally cold weather threatens. The straw covering provides ideal nesting areas for field mice and rabbits. These rodents often feed upon the tender bark and twigs of nursery stock during the winter, so poison baits should be set out to deter plant losses.

Most nurseries have found it necessary to erect unheated or partially heated plastic covered greenhouses to protect certain species of nursery stock during winter. These structures are covered with polyethylene plastic film in late autumn to house the more tender plants, such as certain broadleaved and coniferous evergreen species. The plastic surface collects and traps the heat from the sun during the day and holds the heat inside the structure during the night. Most sensitive plant species can be grown successfully under these structures during winter. When warm days occur in late winter it is necessary to ventilate these houses to prevent rapid warming. Warm days may stimulate early plant growth which will be subject to damage should the temperature again drop below freezing.

Harvesting and Handling Container Stock

Most canned nursery stock is sold when it has been grown for 2 years in the nursery. Some slow-growing plants, such as broadleaved evergreens or conifers, may require an additional year to reach an appropriate size for sale. Whenever plants are not sold during the scheduled season, they must either be transplanted to larger containers or destroyed. The root system becomes bound in the container, causing the roots to girdle themselves. This condition prevents the proper rooting of plants held additional years in the same pots once they are transplanted to the landscape.

Pest Control in the Nursery

Pest control in the nursery begins with sanitation. The nursery beds and fields should be kept clean of weeds wherever practical. The weeds not only reduce the salability of the nursery stock but harbor harmful disease and insect pests. Weeds may be reduced in the field and around container beds by the use of preemergence and postemergence herbicides. Preemergence herbicides should be applied before the growth of weeds in the spring. Winter annuals, perennials, and other noxious weeds are best treated with postemergence systemic herbicides once growth of the weeds has begun. These sprays are generally nonspecific for plants that are affected, so great care should be taken to prevent damage to the nursery stock when applying postemergence herbicides.

Weed seeds, insect grubs and pupae, and many soilborne disease-causing organisms may be eliminated through proper sterilization of soil mixes used with container grown plants. The effects of this treatment are short-lived in the nursery, but the pest populations are at least reduced initially. Since containers are kept in the nursery beds for up to 3 years, additional weed control methods are necessary. Grassy weeds may be controlled by selective herbicides around and in the containers. Broadleaved weeds pose a greater problem, but carefully applied systemic herbicides may control them effectively.

Disease and insect pests are a constant problem to nurserymen. The problem is compounded by the large numbers of similar plants being grown in one location. When a susceptible host plant is predisposed to a particular insect or disease pest, the entire crop is in danger of widespread damage. This situation may be prevented only when all personnel associated with the nursery keep a close watch on any potential problems. Early detection and recognition of the pest followed by the proper control measure is the best way to avoid these losses of plant material. Recommended chemical control measures are changed periodically, so all nursery workers should be aware of current information provided by the state extension service in their area.

STANDARD GRADES FOR NURSERY STOCK

The size, quality, and condition of all plants sold in the United States are standardized under the grading standards established by the American Association of Nurserymen, Inc. These standard grades establish the minimum standards used to describe the various grades of plants sold by nurseries. A basic knowledge of the standard grades for the various classes of nursery stock is required of all nursery personnel involved in this phase of the production.

Most field-grown shrubs and other bare-root stock are graded in the packing shed as they are prepared for cold storage or shipping. Container stock is graded in the nursery bed, since these plants can be sorted and moved more easily. Large specimen plants and trees are generally graded in the field prior to digging. The size of the tree trunk or height and spread often determines the size of the ball that will remain on these plants. The specifications outlined by the AAN for standard grades are extremely detailed and should be consulted for further study.

SELECTED REFERENCES

American Association of Nurserymen. 1973. *American Standard for Nursery Stock*. AAN, 230 Southern Building, Washington, D.C.

American Association of Nurserymen. 1971. *Career Opportunities in the Nursery Industry*. AAN, 230 Southern Building, Washington, D.C.

American Association of Nurserymen. 1976. *Sources of Plants and Related Supplies*, 1976–1977 ed. AAN, 230 Southern Building, Washington, D.C.

Baker, K. F. (ed.). 1957. *The U.C. System for Producing Healthy Container-grown Plants*. Division of Agricultural Sciences, University of California, Berkeley, Calif.

Furuta, T. 1968–1970. *Nursery Management Handbook*. Agricultural Extension Service, University of California, Riverside, Calif.

Garner, R. J. 1958. *The Grafter's Handbook*. Oxford University Press, New York.

Hartmann, H. T., and D. E. Kester. 1968. *Plant Propagation Principles and Practices*, 2nd ed. Prentice-Hall, Inc., Englewood Cliffs, N.J.

Horticultural Research Institute, Inc. 1968. *Scope of the Nursery Industry*. HRI, 833 Southern Building, Washington, D.C.

Mahlstede, J. P., and E. S. Haber. 1957. *Plant Propagation*. John Wiley & Sons, Inc., New York.

Patterson, J. M. 1969. *Container Growing*. American Nurseryman Publishing Company, Chicago.

Pennsylvania State University. 1971. *Nursery Production—A Teacher's Manual*. Teacher Education Series, Vol. 12, No. 4T. Department of Agricultural Education, University Park, Pa.

Pinney, J. J. 1967. *Your Future in the Nursery Industry*. Richards Rosen Press, Inc., New York.

TERMS TO KNOW

Balled-and-Burlapped	Cuttings	Scarification
Bare-root	Field Stock	Scion
Callus	Greenwood Cutting	Softwood Cutting
Cambium	Hardwood Cutting	Stock Block
Candle	Heading Back	Stratification
Canes	Heeled-in	Undercut
Canned Stock	Hotbed	Understock
Cold Frame	Layerage	Whip
Cultivar	Liner	

STUDY QUESTIONS

1. Explain why cuttings are rooted in beds that are heated at the bottom and have mist nozzles above the foliage of the plants.

2. List the various reasons for propagating plants by vegetative cuttings rather than from seeds.

3. Compare the rooting and time that propagation is done with softwood, greenwood, and hardwood cuttings.

4. Explain how rooting chemicals are used in plant propagation.

5. Describe the techniques of simple layerage and mound layerage in plant propagation.

6. List the reasons for using grafting in the propagation of nursery stock.

7. Explain why most nursery stock is currently grown in cans rather than as field stock.

SUGGESTED ACTIVITIES

1. Start a school nursery for growing seedlings and grafted plants into larger landscaping sizes.

2. Make a display board that illustrates the different types of grafting techniques using understock and scion pieces.

3. Design a poster illustrating the proper methods for pruning and shaping nursery stock.

4. Practice the techniques of balling-and-burlapping nursery plants from the school nursery.

5. Compare the ability of cuttings to form roots both with and without the use of rooting chemicals.

CHAPTER FOURTEEN

Landscape Design

A major function of the landscape nursery business is the design and installation of residential and commercial landscapes. The landscape nurseryman plans the landscapes and installs the plants for homes, subdivisions, office buildings, and larger commercial complexes. Although each of these types of landscapes may appear to be entirely different, each is designed using the same basic design principles. Every employee of a landscape nursery should become familiar with these principles of landscaping before attempting to do any design work alone.

To become familiar with the goals of landscaping, observe the residential and commercial properties in your community. Also compare the newly constructed homes without completed landscapes with those having more mature plantings. You will find that most residential landscapes are either overplanted or underplanted. New homes have harsh lines in their design which often require modification by ornamental plants (Figure 14–1). Landscaping assists in making a building a part of the environment. Proper landscaping can make a dwelling appear to have been constructed in a more natural setting.

PLAN BEFORE YOU PLANT

A properly landscaped property begins with a plan. The designer must consider the landscape to be an extension of the home or business. Landscaping involves the creation of an environment that is functional as well as attractive. Consider the view from each room to

379

FIGURE 14–1. Newly constructed homes are generally devoid of trees and shrubs. The harsh architectural lines will require softening from a well-designed landscape planting.

determine the assets and liabilities of the landscape. Perhaps there are unsightly views which require screening by solid fences, groups of trees, or shrubbery. Other views may be pleasant or even enhance the value of the property. These views should be preserved when landscaping.

The first requirement for designing a landscape is to create a plan of the existing property. A surveyor's plot plan of most building sites is available from the local city or county engineering office. A copy of this plot plan will assist you in landscape planning. It will show the correct boundaries of the property, the actual location of the existing structures, and the locations of any utility easements connected with the property. Locate these on your plan first, using a suitable scale (for example, ⅛ inch = 1 foot of actual property dimension).

You will need to compare the surveyor's plot plan with the exact boundaries of the property. If the boundary pins cannot be located, a registered surveyor could be hired. Often the front property line actually is located inside the boundaries of the front lawn. The front property line may coincide with the line of a sidewalk or may be 10–15 feet inside the lawn area. The easements for utilities are found inside the property lines. These are generally located across the back of the property, with 5 feet (1.5 meters) belonging to one lot and the other 5 feet on the neighboring lot to the rear. The land is owned by the homeowner but has been set aside for utility construction and maintenance. In some communities it is required that no permanent features be erected on the easement. Whether it is a restricted area or not, it is wise not to plant shrub borders or erect fences in the easement zone, since this area may be disturbed later by utility crews. However, since it is a part of the landscape, it is important that it be designed to increase the aesthetic value of the property.

Next locate all existing utilities on the plan. The electricial and telephone wires that cross the property above the ground are easily found. Underground utilities, such as buried cables, sewer lines, septic fields, water, and gas lines may be more difficult to locate. Contact the various utility companies responsible for these. This information will be helpful when determining where large trees or excavations will be located and will prevent serious mistakes in the landscape plan.

Sidewalks, driveways, and parking areas require special planning. Traffic surfaces that already exist may not be easily changed. However, where these features may still be included, it is important that they be located properly on the landscape plan. Sidewalks appear best in a landscape when they are placed parallel to the house or other structures. Those walks that run directly from the front entrance of a building to the street appear to divide the landscape into equal parts. Place sidewalks where they will be functional and plan for them to handle the expected traffic. A residential sidewalk requires a minimum width of 4 feet (1.2 meters). This will allow two persons to walk together comfortably. A business sidewalk may require a width of 8 feet or more to handle the expected traffic. To provide adequate planting space between the sidewalk and the home or building, place the walk at least 5 feet, preferably 6 feet (1.8 meters), from the building. This will allow adequate space for full development of shrubbery used for the foundation planting and will prevent unnecessary maintenance of the landscape.

Driveway and parking areas require careful planning. Large American automobiles require plenty of space for maneuvering. A single-car drive should be a minimum width of 8 feet (2.4 meters). Most modern residential driveways are constructed with a 10-foot minimum width, since the driveway provides for both automobile and pedestrian traffic. Space must be allowed for car doors to be opened without striking obstructions. When more than a single car is to be parked in a driveway, each parking space will require 12 feet (3.7 meters) of unobstructed width and 21 feet (6.4 meters) of length.

ADDING DESIGN TO THE LANDSCAPE

Shrubs, trees, flowers, and groundcovers are used in landscaping to provide foliage and flowers which will tie the house or other structures to the total environment. You should strive to include good design in every landscape. Simplicity in design is aesthetically more pleasing than an overplanted landscape. Attempt to be original with each design, but restrain from overusing ornaments, such as statues or gazing globes, and from spotting shrubs with brightly colored foliage indiscriminately throughout the landscape.

Consider a few simple principles when creating a pleasing design: unity, proportion, emphasis, balance, and variety. *Unity* pulls all plantings together to create a desired feeling. *Emphasis* will feature a particular area of the landscape. *Balance* establishes a pleasing visual weight distribution. *Proportion* adds aesthetic restfulness because of desirable size relationships. *Variety* prevents monotony and adds dimension by providing continuous interest as the seasons change.

The house or building is the most important feature of any landscape. Proper landscaping will enhance strong points and direct attention away from weak points while creating unity. A house should appear as if it were placed in a naturalistic environment. The landscape includes the following elements: the house, lawn, trees, shrubs, and flowers that reach to the very edge of the property. A portion of the landscape also includes the larger picture, extending beyond the property. This might include such views as the surrounding fields, a view of the city lights, or a background of hills.

Symmetry is created in a design when the same trees and shrubs are placed on each side of the house. An asymmetrical design will create a more informal appearance to the landscape. A medium-sized shrub might be used to soften one corner of the house, while a small ornamental tree could be planted near the other corner. This plan will allow each landscape design to be unique and attractive (Figure 14–2).

FIGURE 14–2.   Use of a medium-sized ornamental tree placed away from the corner of a house allows some privacy and shade on the windows, while providing a softening effect to the lines of the corner.

Variety is achieved by the use of different types of plant material in the design. Select shrubs having different heights, flowers, fruit, and foliage colors. A properly designed landscape plan will include enough species of trees and shrubs for nearly continuous blooming or ornamental fruit production from early spring to autumn frost. Added interest may be designed into the landscape by including plants that will provide fruit for winter color and attract wild songbirds to the yard. All these features will add interest to the planting, yet they must be combined harmoniously to create a successful design for the total landscape.

DESIGN FOR FAMILY PERSONALITY

A home landscape should reflect the personality of the family that will live there. Before planning any permanent developments, it is wise to first consider the family's needs. Every family has a different idea of how they would like to have their home grounds designed. A family with small children will require outdoor recreation space in the yard. Another family may prefer to have a large area of the yard devoted to outdoor entertainment. Still others may wish to keep the landscape as simple as possible to prevent time-consuming upkeep on the yard.

A typical residential landscape consists of three parts: the public area (or front yard), the recreation area (such as patios, back yard, or pools), and the utility area. The *public area* is designed to call attention to the house. This area is planted for the public view and so should remain uncluttered and simple. The public area includes the front lawn, approach to the front entrance, and off-street parking. The *recreation area* is designed to accommodate the outdoor hobbies and recreational needs of the family. This may include a patio area, a swimming pool, children's outdoor recreational equipment, or flower gardens. This area should be screened to provide privacy and protection, while maintaining pleasant views. The family recreation area should constitute the largest portion of the landscape. The utility area is reserved for those unattractive, yet necessary elements of the landscape. The trash storage area, vegetable gardens, clothesline, or woodpile should be located where they are convenient, yet screened from view.

LANDSCAPING PRINCIPLES

Trees, shrubs, and groundcovers are used in a landscape to enhance a building's architecture. Specimen plants are selected for *accenting* an entryway or border, others for *softening* corners and harsh lines, or to provide a low *transition* from one point to another. Plants may be useful for dividing spaces by *separating* use areas in the yard. Shrubs are used with trees to *screen* unpleasant views or to protect the yard from strong winds.

Shrubs in the Landscape

Foundation plantings.   The selection and placement of shrubs along the foundation of a building requires more planning than any other area of a landscape. Indiscriminate planting of the foundation may result in a dense mass of overgrown specimens that detracts from the lines of the house and provides a real maintenance problem. A few suggestions will assist you in planning a foundation planting.

Softening harsh lines.   The strong vertical lines at the corners of the building need to be *softened*. Shrubs that will soften these lines have a *round, oval,* or *irregular* shape. Such shrubs will break the strong vertical lines at the corners of the house, rather than calling attention to them.

If the house or building does not have windows within 3–4 feet of the corners, you might select a shrub to soften the lines. Select shrubs that will grow no taller than 6–8 feet (1.8–2.4 meters) at maturity, unless the corner is taller than a single story. Avoid the use of plants having strikingly colored foliage or those having a coarse textural appearance. Plants that create an accent will attract unwanted attention to these architectural lines. Do not use the same variety of shrub on both corners of the house; otherwise, the loss of one plant may destroy the planting design.

Use groupings of low-growing shrubs on the lower corner of a split-level home to help balance the weight of the building. This plant grouping should be one-and-one-half times as broad as the plants are tall and should form a broadly curving line around the corner. You might use a small ornamental tree to break the lines of the higher wall in this situation. Whenever trees or shrub grouping are used, it looks more appealing if you face the plantings with a mass of groundcover. Use a widely curving line around the plantings and the corners of the building (Figure 14–3).

Taller shrubs may be used to soften the corners of two-story buildings. Many of these tall shrubs will require the use of low-growing facer plants, since they often do not have sufficient foliage near the ground to hide their stems and branches. Select tall shrubs for corners of buildings only where they will not overgrow the eaves of the house. A good mature height for shrubs or trees planted at building corners is two-thirds of the height from the ground to the eave.

Avoid the selection of upright-growing pyramidal evergreens, such as junipers or hemlocks, for planting at the corners of your home. The shape of these plants creates an *accent*, which will detract from the foundation planting. Instead of softening the harsh vertical lines at the corners of the house, upright pyramidal plants will em-

FIGURE 14–3. Shrub planting areas along the foundation do not have to be arranged in straight lines. The use of widely sweeping curved lines around the corners of the foundations with groundcover facings provide an informal appearance of the landscape.

FIGURE 14–4. Shearing plants into unusual shapes creates an accent. The landscape planting shown does not have a central accent to create attention, since the plants have been sheared throughout the lawn.

phasize them. Another reason for not selecting these plants is that many varieties will grow to a height of 15–20 feet or more unless clipped continually. This not only increases the amount of time required to maintain the landscape, but creates an even stronger accent at a location that should not be emphasized. These plants are very useful in other situations and should not be discounted for their landscaping values.

Entryways. The entrance to a home or other building is the ideal location to use shrubs that attract attention. The entryway is the primary area of the landscape for creating a strong *accent*. Accents may be created by:

1. *Height*—taller plants will catch the eye first.
2. *Color*—brightly colored foliage or stems always attract attention.
3. *Texture*—coarse leaf or stem textures provide strong contrasts.
4. *Form*—pyramidal plants or shrubs sheared into unnatural shapes will create accents (Figure 14–4).

The most common method used by professionals to create accents is with plants having brightly colored foliage or stems. If a plant is to be used successfully as an accent specimen, it must provide color patterns for at least two seasons of the year. The blue-foliaged junipers, for example, retain their blue color all year and are a strong contrast when used with typically green junipers. For this reason, brightly colored shrubs should be planted only in areas where a definite accent is desired. Irregular placement of brightly colored shrubs causes a spotty appearance, generally referred to as the "measles" effect.

Coarse-textured foliage or bark on plants can also create a desirable accent when used with plants having a fine-textured appearance. The winged-euonymus, for example, provides bright red foliage during the autumn and then displays the coarse-textured winged stems during winter to provide a pleasant visual accent.

The form or shape of a plant can create an accent if it is unusual. The pyramidal forms of many upright junipers, yews, and hollies create an accent when used as a single specimen (Figure 14–5). The narrow profiles or sharply pointed tops of some cultivars draw attention from other plantings. A pyramidal holly not only provides accent by providing colored foliage and a coarse-textured appearance all year but also displays a pyramidal form. All these characteristics provide a desirable accent to the landscape.

FIGURE 14–5. The formal appearance of the upright evergreen plant creates a desired accent at the entrance to this home.

The placement of plants at the entryway will depend on the building architecture and upon the direction in which the sidewalk approaches the front door. When the sidewalk leads directly from the curb to the door, accent plants may be located on either one or both sides of the entryway (Figure 14–6). Use shrubs that will not overgrow their surroundings and place them far enough from the steps or walks that they will not grow over them when mature. When planting on both sides of the steps, create a strong accent to one side and a less dominant accent on the other side of the entryway. Attempt to keep the planting harmonious without making it completely symmetrical.

The most pleasing entryway planting can be created when the front walk approaches the door along the front wall from a driveway (Figure 14–7). Placement of the accent area is then at the end of the walk at the outside edge of the front-door area. This will create a visual, as well as a physical, barrier to traffic approaching the entryway. Use plants that will create strong accents in this area and repeat a smaller number of dwarf accent plants on the other side between the walk and foundation. This will help tie the planting areas together.

FIGURE 14–6.   The creation of an accent ed entryway when the sidewalk approaches directly from the front curb may best be accomplished by use of taller more striking plants to one side and smaller accent plants to the other side of the front steps.

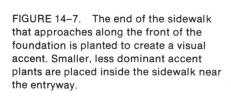

FIGURE 14–7.   The end of the sidewalk that approaches along the front of the foundation is planted to create a visual accent. Smaller, less dominant accent plants are placed inside the sidewalk near the entryway.

Transition plantings. Plants used as a transition are simply those shrubs or groundcovers that pull together the plantings at the corners and accented entryways. These plants are also intended to *soften* the lines of the building along the foundation. Most modern homes have the facing material extended to the ground with little foundation exposed. Low-growing plantings will make a *visual transition* from the vertical line of the wall to the horizontal line of the ground. Where there is 6 inches (15 centimeters) or more of foundation showing, taller shrubs may be used to conceal the concrete. Use shrubs that will not call attention to themselves by having bright foliage or coarse leaf texture. Transition shrubs should be low-growing or dwarfed in size. If the building wall facing extends to the ground without leaving the foundation exposed, a groundcover may be adequate to add depth to the foundation area.

Select shrubs for the foundation planting that will make the house or building appear appropriate for its site. Choose plants that will not get large enough to require heavy pruning to restrict their sizes. Pruning to restrict height destroys the natural shape of plants.

Narrow-leaved evergreens. Narrow-leaved evergreen shrubs add a soft contrast to the foundation of a building. Their green color during the winter will provide an attractive cover when most other plantings are dormant. Evergreens provide excellent foundation plantings when the brick or wood facing of the building is brown, white, or red. Green-stained or painted walls have a tendency to blend with the greens of most evergreen shrubs, therefore diminishing the contrast in colors. This can be used to advantage wherever you wish to soften the appearance of the structure.

Juniper and yew are most often used to soften foundations in the northern states. Junipers are adapted over a wide climatic range in the United States and are available in a wide variety of shapes and colors. The most commonly used juniper for landscaping is the Pfitzer juniper. The Pfitzer juniper generally becomes much too large for foundation plantings. It has a wide-spreading habit of growth, often reaching 8–10 feet in height and 15–18 feet in spread.

It is possible to restrict the growth of these junipers by constant attention to pruning, but this practice requires much physical labor. A more practical solution is to select slow-growing and dwarf varieties of spreading junipers. Compact Pfitzer juniper, for instance, may eventually grow to the size of the common Pfitzer juniper. However, its growth rate is so much slower that a single annual pruning is all that is necessary to maintain a height of 4 feet and similar spread. Such a juniper lends itself well to nearly any foundation planting.

Modern homes are constructed so that little or no foundation is visible above the ground level. The creeping or prostrate evergreens may be used to provide a groundcover foundation planting for these homes. Many of the prostrate junipers remain quite low and fit nicely under very low windows. These plants also make excellent facer plants for other foundation shrubs or in borders. Their evergreen foliage provides a smooth transition from the border or wall to the lawn.

Yews also may be used in a similar manner as juniper, where they may be grown. Yews grow best where they are protected from hot, drying winds and winter desiccation. Generally, they grow well along foundations on the east and north sides of the building.

Whichever type of shrub planting is used, it is important that they are planted far enough from the foundation to provide them with adequate space when they are mature. If the mature diameter of the plant is 6 feet (1.7 meters), it should be planted at least 3 feet (0.9 meter) from the foundation. However, a space of at least 1–1½ feet (30–45 centimeters) should remain between the foliage and the wall to prevent heat reflection damage to the plants. In most instances, shrubs will be placed 4–6 feet (1.2–1.8 meters) from the foundation. Many landscape designers make the mistake of placing the shrubs too close to the foundation simply because the plants are very small when planted. All plantings should be arranged to accommodate the *mature* sizes of the plants.

Spacing shrubs in the landscape.   The proper spacing of shrubs is nearly as important as the selection of shrub varieties. The most serious mistake that a designer can make when developing a landscape is not allowing adequate room for the plants to grow properly. The distance that shrubs should be placed apart depends upon the desired effect and the habit of growth of the plants. Generally, you will not want to place shrubs closer together than the normal spread of the plants when fully grown. The shrubs you select will appear better when set far enough apart that each plant can display its characteristic shape.

Closer shrub spacings may be desired in areas where more dense plantings would be useful. A "living wall" in the shrub border or along the front foundation of a house might be a location for close shrub placement. Space these shrubs so that the branches from individual shrubs will cross slightly once they have reached their mature sizes. An example of this method of spacing would be a foundation planting using a shrub having a mature spread of 4 feet (1.2 meters) when planted alone. For a close spacing, set each plant at a distance of 3–3½ feet (0.9–1.1 meters) from the one placed next to it. Ordinarily, these shrubs would be set a minimum of 4 feet (1.2 meters) apart (Figure 14–8).

Flowering shrubs planted close to the house or building should be set far enough away from the foundation to prevent the foliage from touching the walls when mature. Space between the shrubs and walls permits air circulation, which lessens the occurrence of disease infections. In a sunny location this space also prevents the plant from being dried out by heat reflected from the walls. Avoid planting any shrub under an overhanging roof where rainfall never reaches the plant. The best location for shrubs is to place the planting hole outside the drip line of the roof. This will prevent the plants from being damaged by falling icicles or piled with snow (Figure 14–9).

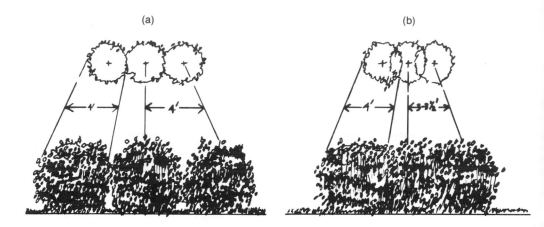

(a)   (b)

FIGURE 14–8.   Spacing shrubs in the landscape: (a) Normal shrub spacing places plants at distances equal to their mature spread. (b) Close shrub spacing places plants at distances less than their mature spread. These plants form a more continuous planting as they mature.

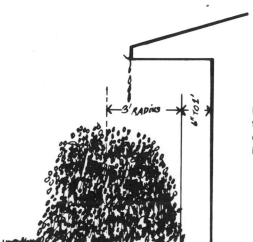

FIGURE 14–9.   Place plants outside the dripline of the roof to avoid dry conditions and winter damage from ice and snow.

Wherever groups of the same species of shrubs are to be planted along a foundation or border, you should strive to give the planting an informal appearance. This may be achieved rather easily if the rows are not planted in an exact straight line. Use wide-sweeping curves with the shrub planting to give the appearance of less formality. You might also use a staggered planting system.

In the staggered planting scheme, each successive shrub is placed either ahead or behind and to the side of the shrub next to it. If the first row of shrubs is placed 3 feet from the foundation, the front row of shrubs would be placed an additional 1–2 feet ahead of this row and between these plants (Figure 14–10).

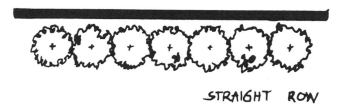

STRAIGHT ROW

(a)   The straight row planting scheme is more structured and formal.

(b)   The staggered planting arrangement provides a more pleasing appearance and adds depth to a shrub border or foundation planting.

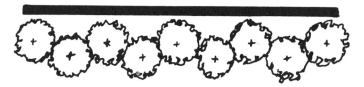

STAGGERED ROW

FIGURE 14–10.   Planting arrangements for shrubs.

Designing a Shrub Border

When planning an outdoor living area, think of the yard as a cube of space. It has a floor (lawn) and a ceiling (the sky and the canopy of shade trees). You may wish to create an outdoor living area by enclosing the sides (walls). Fences or shrub borders may be used to enclose the yard completely, or you might strategically place shrubs and trees to provide a more open feeling.

The common concept of a shrub border is the old-fashioned hedge. These are generally kept clipped at an even height. Clipped hedges not only require high maintenance, but offer few flowers or

FIGURE 14–11.   The clipped hedge requires constant maintenance but provides little flower and fruit production or privacy.

fruit to brighten the landscape. The only advantage these hedges have is to provide some screen and space separation (Figure 14–11).

Another concept of the shrub border is the use of a single species along all sides of the yard. This is a better method for creating a border than the clipped hedge but has obvious drawbacks. Let us assume that a border consists entirely of forsythia. This shrub will bloom for 2–3 weeks during the spring and then will have no other color for the remainder of the season. A diversity of shrub varieties which bloom throughout the season will give the shrub border much more appeal.

The initial step in designing a shrub border requires an analysis of the view into and away from the yard. Plan to conceal objects that are offensive to the view, such as telephone poles, neighbors' yards, and busy streets. Corner lots require special attention to the provision for privacy screening.

Use tall shrubs or small ornamental trees to screen out objectional views. These may also be used to assist in enframing desirable views, much the same as a frame sets off a picture. Plant the taller shrubs to the rear of the border and in odd-numbered groupings. Group several of each selected variety together (perhaps three in one group and five in another). Repeat each variety at several locations throughout the shrub border to obtain a unified effect (Figure 14–12).

Determine next which portion of the view should be retained. Select low-growing shrubs for this area. These will conceal the immediate area behind the border without blocking the desirable view beyond.

Vary the depth of the shrub border to coincide with its height. Design the border so that it will be widest where the tall shrubs are located. Employ wide sweeping curves in the design, rather than sharp angles or straight rows. A staggered placement of the shrubs will give the border more depth (Figure 14–13).

FIGURE 14–12.   The shrub planting in a landscape should provide beauty as well as privacy and protection. The shrub border may be combined with a section of fence to obtain maximum privacy for a yard.

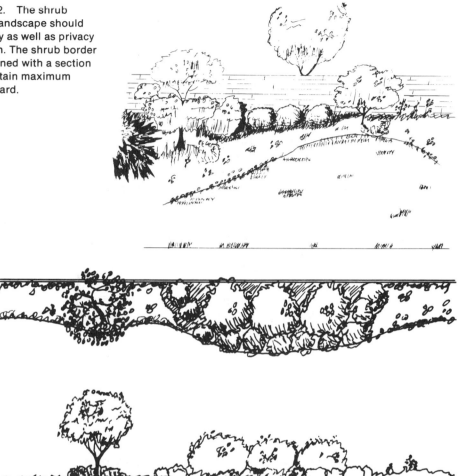

(a)

(b)

FIGURE 14–13.   The depth of the shrub border is varied according to the heights of the shrubs used. Sweeping curves will eliminate the monotony of a straight hedge row.

Fences may be used instead of a shrub border to provide screen and privacy for the smaller yard. Fences have the advantage that they will provide screen immediately upon construction. Shrub borders may take 5 years or longer to achieve the desired height and fullness.

Fences require very little space in the landscape. The average wooden fence will take up less than 8 inches of space. Shrub borders, however, must be planted so that they will not encroach on neighboring property as they mature. A tall shrub that will eventually acquire an 8-foot spread must be placed inside the property line at least 4 feet. The total space occupied by the border may require 10–15 feet of space along a deep border area. This is not a disadvantage where ample lawn area is available, but the loss of this space from a small yard would not be desirable. A combination of shrubs and smaller lengths of fencing across areas requiring immediate screening might be a better compromise for the small yard. There is little that can compare with the beauty and changing interest of a well-designed shrub border.

Provide a center of interest for the outdoor living area. Select a site along the border or fence which can be viewed from both the home and outdoor patio areas. Create an accent at this location using a grouping of accent plants, a single piece of statuary, a garden pool, or a single specimen plant. Avoid the use of plants having strongly accenting characteristics in the border at any other location. The result you want to achieve is to lead the eye to this single point.

Whether you are selecting shrubs for foundation plantings around a house, office building, or for a shrub border, choose shrubs that will bloom or display other interesting features at different seasons. Many of the flowering shrubs bloom for only a short period during the spring, summer, or fall of the year. Once some shrubs have finished blooming, they have no other ornamental features. Combine landscape plants to provide a continuous blooming period throughout the warm seasons.

Trees in the Landscape

Trees will often remain as a part of a landscape long after the home or building is gone, if they are planted properly and placed where they will have adequate space to mature. Before selecting specific trees for a design, consider the functions that trees provide in the landscape. Trees provide shade, help to give enframement to the landscape, produce a background for the house, and may offer wind and snow protection to the landscape.

Shade.   Most trees are planted for no other purpose than to provide *shade* for a home or yard (Figure 14–14). Although this is one of the most important functions of a lawn tree, it is not the only purpose for planting them. Trees used for shade should be placed in locations where the shade is most needed. The south and west exposures of a building benefit most from some type of shade.

Even a modern house with insulation and air conditioning will suffer from the glare of the hot summer sun through unshaded windows. Locate trees where they will protect the house during the hot-

FIGURE 14–14.   The shade provided by trees in a lawn makes them an irreplaceable asset to any landscape.

test part of the day, so that some savings in cooling costs will be realized.

Enframement.   Trees are used to enframe buildings and views in the landscape. *Enframement* trees are those which are planted off the front corners and to the sides of a building or home. The placement of trees at the sides of the house assists in breaking the horizontal line of the roof and leads the eye back to the ground. Consider these trees as the frame for the landscape. They are used to focus attention on the structure or a desirable view. Medium and small trees are best used to provide enframement for single-story homes.

Trees used for enframement may also serve as shade trees. It is seldom good practice to locate a tree directly in front of the house, even though it may face to the south or west where shade is important. Placing a tree in a direct line in front of a house will divide it into segments and will destroy its architectural attractiveness. Shade will eventually be provided by trees placed on a diagonal line extended from the front corners of the building.

Background.   Trees can provide *background* for a home. Trees placed behind a structure should grow to a mature height that will provide foliage above the roof line. This foliage will visually soften the appearance of the roof and will provide a backdrop for the house when viewed from the front. These trees may also effectively shade the roof of the building, therefore cutting air-conditioning costs during the summer (Figure 14–15).

Screen objectionable views.   Trees are useful for *screening* views from a yard (Figure 14–16). When used in groupings, they become objects of beauty in color or form and draw attention away from un-

FIGURE 14–15.   Trees provide shade, enframement, and background to the home and yard.

FIGURE 14–16.   Small ornamental trees create shade, soften lines of corners, and add beauty when placed off the front corner of a foundation. They must be spaced far enough from the building to avoid crowding when they are mature.

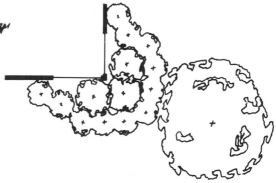

desirable views. Trees may also be planted on either side of a desirable view to help draw attention to it. This is another example of making use of trees for more than just shade.

Wind protection.   Evergreen and deciduous trees are efficient *wind barriers* when planted on the southwest exposure of the yard. Dense plantings of tall evergreen trees will reduce wind velocity, filter dust from the air, and reduce snow drifting along traffic areas around the property.

Windbreaks for rural landscapes are typically planted in straight rows along the north and west sides of farmsteads. These windbreaks consist of four or more rows containing shrubs, deciduous trees, and evergreen trees. This arrangement allows maximum winter wind and snow protection for the rural environment. However, the design is not limited to straight lines.

You may add attractiveness to a windbreak by curving groups of plant materials. These may be arranged to provide protection as well as background for plantings. Cluster a small grouping of flowering trees or plants with brightly colored foliage in one area to provide a focal point. This will break the monotony of a solid planting.

Evergreens or deciduous trees and shrubs may be planted along the south and west sides of a farmstead to shelter the landscape from hot summer winds. Plantings of this type should not be as dense as those along the north side of the farmstead. It is desirable to reduce the wind velocity from the southwest, yet allow cooling breezes to enter the yard. Deciduous trees and shrubs or pines will adequately reduce southerly winds. Doing this might make it possible to plant certain shrubs and flowers in the landscape that otherwise would not be recommended for the climatic region. Windbreaks of this size require considerable space. They must be kept far enough from the circulation system of drives and walks to prevent snow from drifting. A distance of 75–150 feet will usually be sufficient to prevent serious snow deposits from forming.

Urban residential lots are not large enough to accommodate such extensive screen or windbreak plantings. However, groupings of three to five pines spaced 8–12 feet (2.4–3.7 meters) apart will adequately screen views and provide some wind protection. Be careful to place these plantings far enough from streets or drives to prevent snow accumulation around the house or from impairing street vision.

Achieving Balance and Scale with Trees

The proper selection of trees for any home grounds should take into consideration the proper *balance* and *scale* between the house, the size of the lot, and the mature height of the trees to be planted. Scale is an important consideration when planning for trees because their size may either dwarf a dwelling or cause it to appear massive. The house or building is the most important feature in the landscape, but it must appear to belong in the overall design. This is accomplished by selecting trees of the proper height that will be in proportion to the size of the house and lot.

The typical residential lot is between 10,000 and 20,000 square feet (900–1800 square meters) in size. By the time the house, driveway, front yard, and recreational areas are subtracted from this space, little area remains for the placement of mature trees. Generally, only one or two good shade trees and several smaller trees may be spaced properly in lots under 1 acre (0.4 hectare) in size. Medium-sized trees (30–70 feet in height) lend themselves best to this situation. A taller tree is properly placed in landscapes that provide adequate space for development, such as parks, cemeteries, school grounds, or farmsteads.

The trend in building today is toward the construction of homes having only a single story. These homes are long and low in profile. Split-level homes are also very popular. Rarely are three-story homes constructed in new housing developments. A tall tree placed close to a single-story ranch-style home would make the house appear even smaller. A small tree would likewise make a three-story house appear much taller than it actually is.

For a tree to provide the correct balance and scale for the house, it should branch at or slightly above the roof line of the house. Ideally, trees planted around a single-story home should be no taller than 50 feet (15 meters) in height at maturity. Split-level homes are usually treated the same as single-story homes when selecting trees for the landscape. Consider how the overall height and spread of the trees will affect the landscape. Single-story homes require small and medium-sized trees. Multistory homes or buildings require tall and medium-sized trees to achieve the proper scale.

Shade trees. Deciduous trees are considered the best shade trees. These trees provide a canopy that blocks out the sun and cools the areas shaded during the summer months. These trees lose their leaves and are open during the winter to allow the sun to heat the house and landscape in cold weather. Evergreen trees are not as effective for providing good shade. These trees do not produce as dense shade and do not drop their foliage to let in the winter sun.

The form or shape of a tree determines to some extent its ability to provide shade (Figure 14–17). Select trees having a broad, round, or spreading habit of growth to obtain adequate shade. Trees having an oval shape may be grouped together to provide the best shading in a yard. Pyramidal trees make poor shade trees. These trees have very narrow crowns and are widest at their bases. These trees often appear best when their branches and foliage are allowed to sweep the ground. Pin oak and Redmond linden make very poor street trees because the lower branches continually droop, no matter how high the branches are trimmed.

Some shade trees are very shallow-rooted and compete with grass for soil moisture and nutrients. The hackberry is typical of the shallow-rooted trees. To maintain grass under these trees, more water is required than when deeper-rooted species are used. Other trees

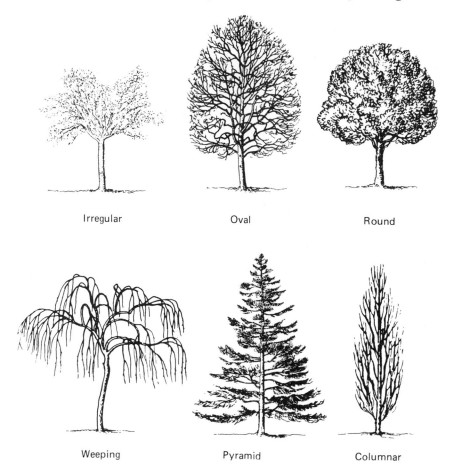

Irregular   Oval   Round

Weeping   Pyramid   Columnar

FIGURE 14–17.   Trees have different forms. Shown in this figure are some typical shapes of trees.

provide such heavy shade that only shade-tolerant plants can be grown under their canopy.

Most homeowners desire a fast-growing shade tree for their landscapes. Trees that grow rapidly are generally very brittle trees and will break up badly in high winds or under heavy ice. Such trees are also prone to many more insect and disease attacks. The better shade trees require more time to reach their mature size but are superior for landscaping purposes.

Small trees for the landscape.   Many small ornamental trees are available for landscaping. Some produce interesting flowers, fruit, colorful foliage, or a combination of these features. Others provide low shade for patios and garden borders. Because of their height and form, they can screen objectional views. They also provide interest and variety to shrub plantings.

When small ornamental trees are used to add height to a shrub border, the width of the border at that location should be one-and-one-half times as wide as the tree is tall at maturity. Create a smooth curving line around the tree or clump of trees in this area. Small shrubs or groundcover plantings used as a facer for the border will create depth and absorb the litter of fallen fruits and leaves from the trees.

Small ornamental trees can be used to *soften* the corners of houses. They are particularly useful for softening corners on split-level homes, where the corner may be taller than the usual single story. Select small trees that will grow to a mature height of two-thirds the distance between the ground and eaves. These trees should be placed on a diagonal from the corner of the building at a distance of no less than the radius of the mature tree spread. For most plantings this will require a distance of 10–12 feet (3–3.7 meters) from the house wall.

Low shrubs placed around the planting area in a wide sweeping curve will add greater depth and aid in visually softening the corner of the structure. The front corners of the house should not be planted exactly the same. If a small tree is planted at one corner, a shrub grouping might be used to soften the opposite corner. Small trees having brightly colored foliage should not be used for the purpose of softening corners. The bright foliage on such trees draws too much attention away from the accent at the entryway of the house.

Some small ornamental trees may be used to *accent* a shrub border or the front entryway of a split-level or multistory home or building. When small trees are used for this purpose, more planning is required than when accent shrubs are used. Trees having a coarse texture in leaves or bark produce accent by contrasting with the finer-textured plants used around them. Some trees have bright red or purple summer foliage, which creates a very strong accent. Purple-leaf plumb and the red-leaf maples create strong accents at entryways. These trees ordinarily do not grow so tall that they overpower the landscape. When used near an entryway, these trees appear best when grown with multiple trunks. This practice will prevent the trees from becoming too tall and will give them a more attractive character.

If there is a possibility that the small tree selected for accenting the front entryway will block windows and views, a shrub planting should be used instead. The selected tree could alternately be planted 10–12 feet (3–3.7 meters) in front of the entryway to allow vision around and through the tree. The space between should be filled with groundcover plants or low-growing accent shrubs. Keep the accent planting simple and uncluttered.

Guidelines for spacing trees.   When preparing a landscape plan on paper, first locate the trees to be planted. Determine the sizes and shapes of the trees to be used in the landscape. Classify the trees according to their intended purposes on the plan. Examples of this might be (1) shade and enframement, (2) shade and background, (3) screen and accent, and (4) accent only.

Plan tree placement carefully. When using a scale drawing to plan a design, it is easy to place trees properly in the landscape. Show a tree on the plan as a circle equal to the mature spread of the tree. By placing tracing paper over the basic plan, the possible locations of trees can be plotted in relation to the house and its room arrangements.

Spacing of trees in the landscape is determined by the placement of the circles on the plan. Strongly branched shade trees may be placed close to the house to provide shade over the roof and walls. A tree having a mature diameter of 50 feet (15 meters) could be placed approximately 20 feet (6.1 meters) from the house, if the tree has strong branches and does not create excessive litter.

The location of trees in a landscape is dependent upon the placement of underground and overhead utilities. Never place a tree close to overhead wires unless the tree will reach a mature height that will be lower than the wires. It is not advisable to place trees over a septic tank, its lateral field, or sewer lines. Any tree that is lacking in water will seek out these lines and may cause expensive clogging problems. Many of the fast-growing species (elms, poplars, birches, cottonwood, and willows) are best planted well away from these utilities. Space all trees at least one-and-one-half times the diameter of the mature tree from these sewage systems.

Weak-wooded or trashy trees (silver maple, cottonwood, willows, poplars, etc.) should be placed far enough from a building to prevent damage to the structure. Place these trees at least two times their mature spread away from the house. This will eliminate the problem of clogged gutters and perhaps damage from fallen limbs on the roof. When several trees are to be used together in the yard, their spacing is determined by the circles plotted on the plan. Broadly rounded and spreading trees produce a more dramatic effect when allowed to grow without competition from other trees. Plan to use only a few trees to suit specific needs, rather than planting too many trees in the landscape.

Oval-shaped trees often appear best when planted in groups. Ashes, for example, lend themselves nicely to planting in groups of three. Place the circles for these on the plan so that their perimeters meet but do not cross. Pyramidal trees, such as the blue spruce and pin oak, may be placed alone or in groupings. When planted alone in an area large enough to accommodate their full development, these trees are accented by their pointed shapes and provide excellent specimen trees. These trees appear best when their branches sweep the ground, so they should not be placed too close to walks, drives, or streets. When planted in groups, the pyramidal shapes are softened and offer excellent screen.

Groupings of small ornamental trees make excellent enframement trees when planted on a diagonal line in front and to the sides of the house. Use groups having three trees of the same species. Locate them far enough apart that they will not become crowded. Place these trees on the plan by using wide triangular or smooth curving designs with the circles so that the trees will not be in straight rows. Allow the circles to cross slightly to provide a massed effect. An alternative method of creating a unique design with small ornamental trees would be the use of three enframement trees of the same species planted together, but using a variety in the center that produces a different flower color than those of the other trees in the grouping.

GROUNDCOVER PLANTS FOR LANDSCAPING

Groundcover plants enhance a landscape while minimizing many maintenance problems. Groundcover plantings are particularly valuable for areas where turf grass is difficult to maintain. Shaded areas under large trees or around those trees having heavy surface feeder roots are common problems in established landscapes. Groundcovers may be planted on steep banks to hold the soil, rather than constructing retaining walls.

The term *groundcover* applies to any vegetation that will blanket the soil. For landscaping purposes, a groundcover means any group of plants used as a substitute for lawn grasses or to provide a "living mulch." For an average-sized home, a low-growing groundcover might be any plant that attains a height up to 12 inches (30 centimeters).

Groundcovers can be used to advantage in areas where mowing is difficult. Some groundcovers will grow in locations that are wetter or drier than those most grass species will tolerate. Some groundcovers tolerate (or prefer) dense shade to bright sunlight and are also able to compete successfully with shallow tree roots for the available nutrients and moisture.

Groundcover plants can be used to stabilize areas where soil erosion is a problem. Once established, their dense root systems will hold the soil while their thick foliage breaks the force of heavy rains. They also provide an excellent mulch for shrub borders. These plants should be used around plantings which grow best in cool soil conditions. Groundcover plants may also be used in beds of spring-flowering bulbs. Here they will provide a background for the blossoms and will disguise the fading foliage after blooming has ceased.

A groundcover mulch may be used wherever hand weeding or clipping is to be reduced or eliminated. Examples are around trees or next to the foundation of the house. Planting strips around trees are particularly helpful because they allow uniform clipping of grass, while protecting the tree from injury by the mower.

Groundcovers, including the prostrate junipers, make an excellent facer for shrub borders. These are used effectively to provide a smooth transition from shrubs to the grass. An effective design will include curving or sweeping lines rather than straight borders. These groundcover areas will also allow mowing around the shrub borders without hand clipping.

The architecture of many homes being constructed today provides an opportunity for use of groundcover plants to their best advantage. When a home is constructed with no obvious foundation visible, a minimum of taller shrubs may be used to soften harsh vertical lines. These may then be tied together with an attractive groundcover facer planting. Together these plants will provide the necessary accent or transition to give the house the required finishing touches in landscaping.

No single groundcover will be adapted to all locations. Selection of a plant to be used as a groundcover is dependent upon the landscape requirements and the conditions under which it must grow. Select several different types of groundcovers for a landscape to provide a diversity of bloom or foliage color and various textural contrasts.

Qualities of a Good Groundcover Plant

A good groundcover plant should have an attractive appearance throughout the year while covering the soil. The narrow-leaved and broad-leaved evergreen plants suit this purpose very nicely. However, because of site limitations or because one wishes to have flower or fruit production, herbaceous perennials are also desirable.

For rapid coverage of an area, the groundcover should have the ability to spread by itself. Species that reproduce by rhizomes or stolons (runners) are usually best for covering the ground. To achieve good weed control, the plants should branch heavily or produce dense foliage close to the ground.

No single species of groundcover can be used to solve all landscape problems. Also, no groundcover species is completely free of disease or insect pests. Before a groundcover is selected for a particular situation, several factors must be considered.

Adaptability. Not all groundcover plants sold by nurseries are capable of growing in all locations or climatic zones. Select species that are well adapted to the environmental conditions where they are to be used.

Exposure. Plant species vary in their ability to withstand sun or shade. If the groundcover plants normally form attractive blooms, the amount of flower production may be dependent upon the quantity of light available. Furthermore, protection from winter sun and southwest winds in late summer may determine whether a species is suited to a landscape.

Traffic. Very few groundcover species will tolerate being walked upon.

Slope. Steeper slopes are difficult to cover adequately with groundcover plants. Select species that will spread rather rapidly and possess dense, fibrous roots to hold the soil.

Area to be covered. Most areas to be covered by groundcover plants are small and restricted. To maintain balance in landscaping small areas, low-growing groundcovers are best used. Usually plants growing no taller than 1 foot (30 centimeters) are best. In larger areas or on extensive, steep slopes, taller plants may be used.

Maintenance of Groundcover Plantings

Although groundcover plantings can reduce the amount of maintenance around landscape plantings, they will require some care. These plants will occasionally have to be dug and separated when they have grown too dense. Groundcover plants should be watered frequently, and occasionally they may require spraying for insect or disease pests.

Groundcover plants will respond favorably to the periodic application of nitrogen fertilizers during the growing season. This is best done in early spring and again in early autumn. If the plants retain their foliage during the winter (such as junipers or euonymus), they should be irrigated at this time also. This should be done when the weather is dry and the temperature is above freezing. The larger groundcover plants also have a tendency to collect trash and paper which is blown in from the surrounding area. Some time will be required to keep the area free from litter.

Planting Groundcover Plants

Select groundcover plants to suit particular landscaping requirements. If a groundcover is to be used instead of grass because the area is too shaded, plants that will tolerate heavy shade must be chosen. Steep banks will accumulate very little soil moisture, so plants that can withstand dry soil conditions must be used for this purpose.

Spacing of the plants will depend on how quickly the area is to be covered by the dense foliage of the groundcover, the ultimate spread of the plant, and the amount of money budgeted for this purpose. Groundcover plants are generally much more expensive than grass, but less expensive than covering the area with a bark chip mulch. Naturally, the closer the plants are spaced, the quicker the ground will be covered. The number of plants required to cover an area of ground may be determined by the following formula:

$$\frac{\text{area to be covered}}{(\text{distance between plants})^2} = \text{number of plants required}$$

Example

If you are planting *Vinca minor* (periwinkle) at 18-inch (1.5 feet) spacing, and you wish to cover an area of 100 square feet, you will need

$$\frac{100 \text{ square feet}}{(1.5 \text{ feet})^2} = \frac{100}{2.25} = 44.4 \text{ or } 45 \text{ plants}$$

or

$$\frac{9 \text{ square meters}}{(0.45 \text{ meter})^2} = \frac{9}{0.20} = 45 \text{ plants}$$

*Note: 100 square feet = 9 square meters.

Normally, the smaller groundcovers are placed in a grid spacing of 1 foot (30 centimeters). On slopes or banks these plants should be spaced even closer; otherwise, the soil will become eroded before the groundcover is established. Many of the spreading groundcovers will cover bare soil more rapidly because they root at the nodes on the stems as they trail along the ground. Some examples are the prostrate junipers, rock cotoneaster, and periwinkle (*Vinca minor*). It is often the custom to use rock or bark mulch around the groundcover plant.

This makes the edging or border more attractive than bare soil. If this is done, a layer of black polyethylene plastic may first be placed over the ground to prevent weed emergence through the mulch. The plastic mulch is slit at the points where plants will be located. The slits in the plastic are opened to allow holes to be dug. Once the plants are inserted in the holes, the soil is replaced and the plastic replaced around the base of each plant. A shallow basin is left to catch and channel water to the plant roots. The slits in the plastic will allow water penetration.

A metal or wood edging around borders and tree plantings will make a clean, clear definition between the groundcover area and the lawn. The edging material should be no taller than 1 inch (2.5 centimeters) above the soil line to allow easy mowing without ruining the edging. The edging material should also extend into the ground 3–5 inches (8–13 centimeters) to prevent cool season grasses from invading the border areas. These edgings are not effective against the invasion of Zoysia and Bermuda grasses.

PERENNIAL VINES IN THE LANDSCAPE

Perennial vines find many uses in the landscape, but they should be carefully selected and placed only where they will enhance the overall design. Vines are useful for covering walls and fences, for use as screens, for covering arbors and trellises, and some may be permitted to trail along the ground to provide a groundcover. Some of the climbing vines, such as clematis and wisteria, can become screens of living color. Wisteria possesses a unique daintiness that is useful for softening harsh garden walls. Some vines, such as euonymus and bittersweet, have colorful winter berries which are attractive and attract wild songbirds to the landscape.

Vines are primarily used for screen and for creating decorative effects. As a screen, vines create shade and privacy in the small areas of a landscape. They can be grown on fences or other suitable support to screen out an objectionable view. They are particularly valuable where space does not permit the use of shrubs.

Vines for Walls

Vines can relieve the monotony of large wall surfaces. If the wall surface is of stone or masonry, vines may be selected that climb without additional support by clinging to the wall surface. It is not advisable to cover a wall completely with vines. A balance should exist between the vine and exposed wall surface.

Most perennial vines are too rank and vigorous for the average house wall. Those vines which will easily attain a height and spread of over 30 feet (9 meters) are not recommended for planting next to small buildings or houses. These vines will soon cover windows and make doorways a jungle trail. These climbing vines will also deteriorate wood siding, if allowed to ramble unchecked. They are not as much of a problem on masonry walls, but their presence will hide the architectural beauty of the brick or stone.

The smaller, shrubby vines are best supported on a trellis that is placed next to the wall of the house. This is much better than allowing them to ramble along the wall wherever the new shoots choose to creep. Climbing roses and clematis are better used for these landscape purposes.

Vines as Groundcovers

Several of the trailing or climbing vines found in the nursery trade provide excellent groundcover. When provided with the proper environment, these plants will give excellent rapid cover. A few of these vines which also provide a groundcover are:

Euonymus. The evergreen wintercreeper or creeping euonymus is an excellent selection for rapid groundcover, where it is adapted. When allowed to grow flat, it makes a quick-growing groundcover for sun or shade. This plant will tend to climb trees and walls when placed in close proximity. This plant is also highly susceptible to euonymus scale infestation. The purple-leaved variety "coloratus" is more often used than other euonymus. It has the same leathery leaves, but they turn purple in autumn and wine-red in winter in cooler climates.

Ivy. English ivy and its selections are very popular for use as a groundcover. These are evergreen and can withstand rather severe winters. Baltic ivy and Engelmann ivy should be planted in areas where they will be shaded from the winter sun in northern climates.

Honeysuckle. One of the fastest growers of the permanent groundcovers is Hall's honeysuckle. However, it may become too rank for small areas and is recommended only for larger spaces. The flowers are extremely fragrant and change from a white to yellow color with age. It is adaptable to banks or level areas, in severe heat or dry soils. The branches root as they make contact with the soil and will climb trees or walls if placed too close to them.

Planting and Care of Vines

Early spring is the best time to plant perennial vines in most areas of the United States. These may be planted from started cuttings or potted plants and placed within 6–12 inches (15–30 centimeters) from the object they are to climb. The planting hole is dug twice as wide and 6 inches deeper than the root system of each plant. Most vine plants grow best when the roots are kept cool, so a mixture of topsoil and peat (3 parts soil to 1 part peat) should be used for filling the planting hole. Soil from a compost pile is excellent for this purpose. This mixture is backfilled into the hole to place the cuttings at the same level as they were growing originally. Settle the plant with water and then apply a 5-centimeter (2-inch) layer of mulch over the planting hole.

Many vines have stems that are too weak to support their foliage. Clematis, wisteria, climbing roses, and some other vines require support, such as a trellis or fence. Others will climb on stone or brick without the requirement for tying. If a trellis is used next to a building, place the trellis several inches from the wall to provide adequate air circulation on all sides of the foliage. New shoots may have to be supported by tying until they catch hold by themselves.

Perennial vines should be fertilized at the same time as shrubs in early spring and again in autumn. Apply the fertilizer sparingly so that new roots will not be burned. The newly planted vines will not require an application of nitrogen fertilizer.

ROCK GARDENS IN THE LANDSCAPE

Rock gardens may take on various forms in our landscapes. One of the most common forms for rock gardens is the rock wall or terrace. This is generally referred to as a retaining wall and makes an excellent location for the appearance of a more natural landscape.

The more formal arrangement of rocks in the traditional Japanese garden is also an excellent setting for rock gardens. In locations where the ground is monotonous, a rock garden may be created to give the landscape a new dimension. However, the simple piling of rocks in the yard does not create an effective rock garden. It requires much planning and creative arranging to produce a rock garden. The final result of these efforts should give the impression that the rocks were placed in the yard by nature many centuries ago.

Designing a Rock Garden

There are many more examples of poorly designed rock walls and gardens than there are of good ones. Builders often fail to follow a few simple rules when designing the rock garden. It is very simple to construct a monstrosity and call it a rock garden. We are able to identify these attempts at one glance. The first example of a poor rock garden might be called the "rock pile" type, because it appears that the rocks were merely dumped in the lawn. With another type, the rocks resemble a set of large false teeth placed in rows. The last type is not as neatly arranged, with large and small stones placed in erratic patterns, resembling markers.

The rock garden should not be used to store or display collections. This type of garden would consist of a heterogeneous assemblage of rocks which may have been collected from many different locations of the country. Sometimes these gardens are decorated with a multitude of ceramic figures, pottery, birds or other animals. Another type is an attempt to recreate towns, farms, or mountains. These often include miniature buildings and cars. The rock garden is not the place for any of these items.

The most important rule to consider when building a rock garden is that it should not appear to be man-made. Select the rocks carefully from similar rock formations. For example, if the common rock formations found in the area consist of sedimentary rocks of sandstone, shale, or limestone, one of these types should be used. If you can find an outcropping of these rocks, use them in sizes ranging from large to small rocks. Attempt to recreate this outcropping in the landscape. You will have to scale the rock outcropping down to a size that will be in proportion to the landscape design. The important thing is to use only one type of rock.

Whatever form the rock garden takes, it must be a true representation of natural rock scenery. Keep rock gardens away from walls and buildings and avoid constructing them on a level site. A natural slope is the ideal area for a rock garden. It is easier to work with a slope or a bank that naturally lends itself to rock outcroppings than with a flat area. The rocks should be used like a retaining wall, where each rock is placed with a slight tilt backward. This will give stability and allow rainwater to run into the bed behind. Most rocks should be buried one-third of their height in the ground and placed in a natural design of outcroppings and ledges.

When arranging these rocks in a garden, remember that nature covers the rocks with soil and fills in the low areas. Expose only the weather-worn surfaces, rather than the freshly cut sides. Rocks already covered with some lichens or moss appear more natural. When fitting the rocks into the setting, use larger rocks as the backbone and then place the smaller rocks into the design. These rocks should not be fitted together too tightly. A good rock garden will have many soil pockets left between the rocks for plants.

Wall Gardens

Wall gardens are among the most practical and useful forms of rock gardens. These are not constructed with mortar, so are called "dry" walls. The results of planting a dry-laid rock wall will often transform a simple retaining wall into a naturalistic focal point for the landscape.

Rock walls may consist of a single row of horizontally laid rock to an extensive wall of 6 feet or higher. Low walls up to 3 feet in height can be constructed directly upon the ground (Figure 14–18). Because these walls are constructed with soil, they do not require the frostproof foundation required for masonry walls. For taller walls it is best to place a layer of larger rocks just beneath the surface of the soil.

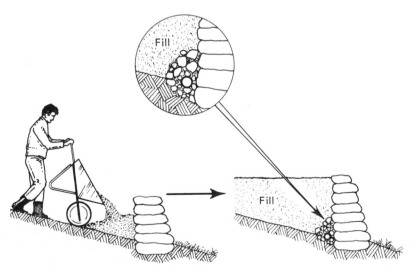

FIGURE 14–18.   Construction of a dry rock well. The rocks are placed on top of each other with a fine layer of soil between them. The pitch of the wall is sloped backward slightly. The area behind the wall is filled with loose drainage material and soil.

The individual layers of rock must slope slightly to the back of the wall. This will prevent the rocks from slipping out of the wall from frost heaving. This also directs rainwater to the plant roots behind the wall. The wall should have a height and slope ratio of 12:1 (12 inches to 1 inch). The top of a 4-foot (1.3-meter) wall should be set back about 4 inches (10 centimeters) from the base layer of rock.

Each layer of the wall should be layed nearly horizontal. Sifted soil is placed over each layer of rock before the next layer is constructed. The fill soil behind the wall is added and compacted simultaneously with the construction of the wall. The successive layers of rock should be placed so that their joints will not coincide with those of the layer below. This process is repeated until the desired height of the wall is reached. The spaces between the rocks should be large enough to insert small plants but not so large as to allow the loss of soil during rains.

Planting the Rock Garden or Wall

A rock garden provides an opportunity to use plants that may not be suitable for another location in the landscape. The more vigorous perennial plants used in shrub borders would likely crowd out these smaller plants having dainty flowers. Groundcovers and vines that are well suited for banks and rocky slopes are too vigorous for the rock garden.

Rock garden plants should be those small hardy perennial flowers, dwarf evergreens, and low shrubs. Vines should be used cautiously or not at all. Annual flowering plants are generally too vigorous for the rock garden. An exception to this would be sweet allyssum, which has a small and compact form. Use plants having small, dainty flowers and fine-textured foliage. The sedums and dwarf phlox are attractive in both the rock garden and planted between the rocks of a retaining wall. Avoid the use of plants that will grow so large as to cover other less vigorously growing plants.

Adequate drainage is probably the most important factor to consider in the planting of any rock garden. Generally, a well-designed rock garden or wall will provide adequate drainage for any plants to be used. Most plants suitable for rock gardens will require sun at least half of the day. However, some protection should be provided from the hot, dry southwest winds. Some shade may be provided by trees, but do not locate trees within the design of the rock garden itself. Tree root competition can be more of a problem than bright sun. Avoid the use of plants that ultimately become too tall for the rocks in the garden. The best design gives the appearance of the plants being dwarfed by the rocks. Some of the plants may be concealed under rocks and deep into crevices. This treatment will then give the appearance of plants in their natural habitat. Carefully select the rock garden plants according to the season of bloom to provide continuous color throughout the seasons.

Rock walls requiring a height of 4 feet (1.3 meters) are better constructed as two walls. The areas between the walls are ideally suited for simple rock gardens. This area may be planted to groundcover, a small rock garden designed, or planted to small bush fruits (such as strawberries or raspberries). Unless these areas are large enough for use in entertaining, they should not be planted to grass. This will aid in the reduction of maintenance of the landscape.

SELECTED REFERENCES

American Association of Nurserymen, 1976–1977 (or current edition).*Sources of Plants and Related Supplies (Nursery Stock and Supplies)*. Allied Landscape Industry Service, AAN, 230 Southern Building, Washington, D.C.

Hoover, N. K. 1973. *Approved Practices in Beautifying the Home Grounds*, Elwood M. Juergenson (ed.), 4th ed. The Interstate Printers and Publishers, Inc., Danville, Ill.

Kramer, J. 1970. *The Complete Book of Patio Gardening*. G. P. Putnam's Sons, New York.

Kramer, J. 1971. *Gardening and Home Landscaping*. Harper & Row, Publishers, New York.

National Landscape Association. 1971. *Landscape Designer and Estimator's Guide*. Washington, D.C.

Nelson, W. R., Jr. 1963. *Landscaping Your Home*. Circular No. 858. Cooperative Extension Service, University of Illinois, Urbana.

Robinette, G. 1967. *The Design Characteristics of Plant Materials: Plant Form Studies*. College Printing and Publishing Company, Madison, Wis.

Wyman, D. 1969. *Shrubs and Vines for American Gardens*. Macmillan Publishing Co., Inc., New York.

Wyman, D. 1972. *Trees for American Gardens,* 2nd ed. Macmillan Publishing Co., Inc., New York.

Zube, E. H. (ed.). 1977. *Changing Rural Landscapes.* University of Massachusetts Press, Amherst, Mass.

TERMS TO KNOW

Accent	Groundcover	Screen
Canopy	Perennial	Soften
Easement	Rock Garden	Transition
Enframement		

STUDY QUESTIONS

1. List the principles of design as they apply to landscaping and give a brief explanation of each.

2. List the three parts of a typical residential landscape and explain how these should be designed to fit the personality of the family.

3. List the four methods by which plants may create an accent and give examples of each.

4. Explain how shrub borders provide a more interesting screen than do fences. Under what circumstances are fences the best choice for screening a landscape?

5. List the five functions provided by trees in a landscape and give a brief description of each.

SUGGESTED ACTIVITIES

1. Practice the use of drawing equipment, such as architectural scales, and become skilled at freehand printing for making landscape plans.

2. Make a scale model of a residential landscape using cardboard, paper, and other easily obtained materials.

3. Draw landscape plans of a few residences in your area as a class project.

4. Provide a landscape plan for beautifying your school property. Select plants that are easiest to care for and which will provide the best function while school is in session.

5. Collect catalogs from nurseries in your area and make a list of the plants best used for landscaping. You might include in this list a description of their best features and the sizes they will attain.

CHAPTER FIFTEEN

Turf Grass Maintenance

Americans are enjoying more leisure time than ever before in history. Rising family incomes, shorter work weeks, and an increase in home ownership have created a greater demand for leisure-time activities. As the working population receives more opportunities for free-time activities, the use of public recreational areas is increased. Skilled horticulturists are required for establishing and maintaining these private and public recreation areas. The professional horticulturist will find a greater need for his services in the future in the maintenance of public parks, public and private golf courses, public highway right-of-ways, athletic fields, industrial and office grounds, and residential landscapes.

As the population expands and is given more opportunities for recreational pleasures, less time will be spent by homeowners in the maintenance of their own home lawns. These duties will be relinquished to professional home grounds maintenance services. These horticultural firms will perform many of the duties of home lawn maintenance for those homeowners who can afford this service. Whether the trained horticulturist is employed by a public organization or a private firm, the skills required in turfgrass maintenance are similar. Each horticultural student must become skilled in the various facets of turfgrass maintenance, which include establishment and maintenance of lawns, maintenance of golf courses, and the control of lawn grass pests.

ESTABLISHING AND MAINTAINING LAWNS

The lawn surrounding a home adds beauty to the landscape. Grass creates a foreground for the home or building and is indispensable in the creation of a natural setting for city parks, cemeteries, and highway right-of-ways. Grass also provides a soft carpet for the landscape and aids in cooling the air around a home. One of the greatest values of lawn areas is the prevention of soil erosion from landscaped grounds.

The landscape horticulturist should have a thorough knowledge of lawn establishment and maintenance practices. These practices include (1) construction of the lawn area, (2) selection of lawn grasses, and (3) maintenance of the established lawn.

Preparing the Lawn for Planting

A lawn may be established from seeds, plugs, sprigs, or sod (depending on the type of grass to be grown). Regardless of the type of grass species to be planted, the soil must be properly prepared to ensure adequate growth of the lawn for many years. The failure of lawns to become well established and to remain healthy is often caused by poor subsoil preparation prior to planting.

Most new lawns are established around newly constructed homes or buildings. The building contractor should remove the topsoil from the surface of the lawn and pile it in an area well away from the construction site. This practice will prevent compaction of the topsoil and will allow suitable subsoil grading during the preparation of the lawn. The first step in the construction of any lawn area is the removal of all building materials, such as bricks, boards, roofing, and debris which have been scattered over the soil. This debris should not be buried in the soil to restrict root growth of the lawn grass and landscape plants.

Correct drainage of the lawn determines, to a great extent, the ability of the grass plants to become established. The soil for the lawn should drain away from drives, walks, and buildings. The slope of the lawn from the building should have approximately 2½–4 inches (6–10 centimeters) of fall for each 10 feet (3 meters) of length. The slope should not be greater than necessary to maintain adequate drainage in heavy rainfall. Steep slopes cause a problem in mowing and do not allow grass roots to absorb adequate water during dry soil conditions. The center of the lawn should be slightly higher than the perimeters, while maintaining adequate drainage away from the house or buildings. Retaining walls, although difficult to maintain, may be required to avoid overly steep lawn surfaces.

When topsoil is added to the lawn over the subsoil surface, the walks and drives should be approximately 1–2 inches (2–5 centimeters) higher than the soil. This prevents the water from flowing to these surfaces and standing in pools. As the topsoil is graded and leveled, all depressions should be eliminated. These depressions allow water to stand around the grass roots and may cause the lawn to die in patches, especially during winter. Where the subsoil will not drain adequately, clay tile drainage lines may be needed to allow rapid drainage from depressions in the lawn.

During the grading of new lawns, the original slope of the soil may require alteration. When large existing trees are present in the lawn area, special precautions must be taken to protect the trees. Soil may be removed from around the roots of these trees as long as the large buttress roots are not exposed and only a few of the surface feeder roots are removed. When the grading of the soil surface must cut deeply around these trees, it is best to allow a mound of soil to remain around the tree and taper the cut gradually away from the trunk. Less damage to the tree occurs when the feeder roots are removed from the soil nearest the drip line (outside edges of the branches) of the trees (Figure 15–1).

Occasionally, the new grade for the lawn requires the addition of soil to low areas around existing trees. The larger shade trees are valuable assets to a home landscape and would require many years to

FIGURE 15–1. When the grade around an existing tree must be lowered during construction, a mound of soil should remain around the main feeder roots and the soil tapered gradually away from the tree trunk.

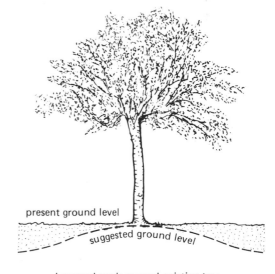

present ground level

suggested ground level

Lowered grade around existing tree

replace, so they should be protected from harm during the grading work. Trees will not survive for long after soil is piled against their trunks at depths of over 4 inches (10 centimeters). When the soil surface must be raised, a tree planting well is recommended to allow air movement around the tree trunk. A rock or tile wall may be constructed around the tree trunk at a distance of no less than 18 inches (45 centimeters) from the tree. Tile drainage lines may be installed at the base of the tree well to remove excess water and allow the proper aeration of the soil around the tree roots (Figure 15–2).

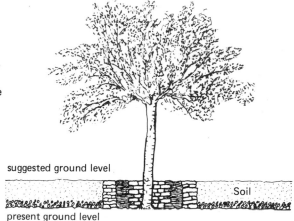

FIGURE 15–2. When the grade of the soil must be raised above the existing level around an established tree, a tree planting well can be constructed to prevent damage to the tree.

suggested ground level

Soil

present ground level

Raised grade around existing tree

Preparation of the Seedbed

Whether the new lawn grass is to be planted from seeds, plugs, stolons, or sod, the topsoil should be prepared in a similar manner. The rapid establishment and maintenance of a healthy lawn depends on the proper preparation of the topsoil prior to planting. Topsoil may vary drastically from one region of the country to another or even one section of a city to another. The soil texture of the new lawn will greatly influence the successful growth of the grass plants.

Highly sandy soils may be found in some areas. These soils drain rainwater rapidly but lack the capacity for storing water and nutrients. Grass plants are fairly shallow-rooted, so the soil surface must contain adequate organic matter to support their growth. Light, sandy soils may be improved by the addition of organic matter, such as well-rotted manure, peat, or compost. These are mixed into the top

6–8 inches (15–20 centimeters) of the topsoil at a rate of 1–2½ inches (2.5–6.3 centimeters) of organic material spread evenly on the soil surface and tilled in.

Heavy soils contain a high proportion of clay, with little sand being present. These soils have a high capacity for water and nutrient storage, but water does not penetrate the soil and drain properly. These soils usually remain dry for most of the season or, when heavily watered, remain wet for long periods of time. Water remains standing around the roots of grass plants in heavy clay soils and may be killed from lack of air. Clay soils are amended by the addition of sand to allow more rapid water penetration and drainage. These clay soils may also benefit from the addition of some organic matter as the sand is being incorporated. A dark-colored clay-loam soil contains ample sand and organic matter and is generally ideal for the growth of lawn grasses.

Agricultural lime is added to the topsoil, when required, to make the fertilizer elements more readily available to the grass plants. A slightly acid soil reaction is desired for best growth of most plants. A soil acidity test should be made of the lawn area before lime is added, however, since an overly limed soil will damage the grass plants. A complete soil test for soil acidity and fertilizer needs may be obtained from any state soil testing laboratory. Soil samples are collected from each area of a lawn and are sent to the soil testing laboratory. The report returned from the laboratory will indicate the soil pH, present amounts of fertilizer elements available, and may also indicate the amount of these minerals that should be added to the soil at the time the lawn is planted.

Soil acidity reaction is determined from a pH scale, where a value of 7.0 indicates a neutral soil reaction (it is neither acid nor alkaline). Grasses and most other landscape plants grow best in soil having a soil reaction between 6.0 and 6.5 (slightly acid soils). When required, lime should be applied to a soil prior to the final disking or tilling. Lime is added at the rate of about 1 pound per 100 square feet of area (49 grams per square meter) to raise the soil 0.1 of a pH unit. The amount of agricultural limestone required for adjusting various soil reactions is given in Table 15–1.

TABLE 15–1:   *Limestone Required for Adjusting Soil Reaction*

PRESENT pH	DESIRED pH	GRAMS PER METER²	POUNDS PER 1000 FEET²
5.0	6.5	490–735	100–150
5.5	6.5	245–490	50–100
6.0	6.5	Up to 245	Up to 50

Lime should be applied to the soil and the surface given a final tilling about 3–6 weeks prior to planting of the grass. Additional liming should not be required again for up to 6 years once the soil reaction has been adjusted. In addition to the lime requirements of new lawns, certain fertilizer elements are required for the proper establishment of the grass plants or seedlings. These include nitrogen, phosphorus, and potassium (potash). The high requirement by grass plants for calcium is met by the addition of lime to the soil.

An average soil will require the addition of nitrogen, phosphorus, and potash prior to the planting of grass. These fertilizer elements should be applied in small amounts before seeding to prevent the burning of roots on new seedling plants. Generally a fertilizer formulation of 6–12–12 or 5–10–10 is applied to the soil surface and worked into the top 6 inches (15 centimeters) before planting. The amount and type of fertilizer to be applied prior to planting is determined by the soil test report received from the testing laboratory.

A firm seedbed is required when grass seed is to be used for planting. When lawns are to be established from vegetative planting, the soil surface is planted in a similar manner. The subsoil is first loosened by plowing, disking, or spading before it is graded to form the proper slope for drainage. Once the topsoil has been replaced and smoothed, the surface is worked with a tine harrow. This harrow has flexible tines that aid in leveling the soil as clods are broken. A rotary tiller may also be used to prepare small lawns for planting (Figure 15–3). The operator must be careful to avoid overtillage of the soil, however, as the rotary tiller will "float" the finer soil particles to the surface. If the soil structure is worked until it is this fine, poor seed-to-soil contact will result, and the soil will compact easily.

A properly prepared seedbed will contain some exposed clods the size of a quarter or less in diameter. The soil surface may be firmed by use of a lawn roller, followed by a light raking by a fine-

FIGURE 15–3. The lawn seed bed is prepared with a tiller to provide the proper soil structure.

toothed harrow or by hand. The lime and fertilizer are applied to the planting surface prior to this final raking.

Selection of Grasses for Lawns

Several different types of lawn grasses are grown in various regions of the country. Certain types of grass are adapted to particular climatic regions, while others may be grown over broadly differing climatic zones of the United States. These lawn grasses may be divided into two classes: cool season and warm season.

Cool season grasses. *Cool season grasses* remain in active growth during the cool, wet seasons of autumn, early winter, and through the spring or early summer months. These grasses generally become semidormant or cease active growth during the dry, hot periods of summer. These grasses are generally planted as seeds or from sod to establish lawns. The most commonly used cool season grasses for lawns are Kentucky bluegrass, tall fescue, ryegrass, and bentgrass.

For best establishment of a lawn, the cool season grasses should be planted from seed in early autumn. The soil temperature is higher at that time, so seed germination is faster. This aids in getting the desired grass seedlings established before annual weeds become a problem during the spring months. Although these cool season grasses may be planted in late winter or early spring, the seeds germinate slowly and winter annual weeds compete heavily with the germinated seedlings. Cool season lawn grasses may be established from sod at any time of the year provided that adequate water is available. The best time for establishing a cool season grass lawn by either method is from mid-August until mid-October, depending on the climate. When started from seed, the grass plants will require at least 6 weeks from sowing until the first hard freeze of winter.

Bluegrass is the most commonly grown cool season lawn grass in the United States. Many different varieties of Kentucky bluegrass are marketed in this country. These include Merion, Newport, Park, Windsor, and Nu Dwarf. These varieties of Kentucky bluegrass are relatively fine-bladed, survive well in the shade of large trees, and spread by underground rhizomes. Bluegrass varieties tend to become dormant during the months of July and August in most areas of the country unless adequate moisture is applied at weekly intervals. Bluegrass may be grown in open sun areas in northern states, but should be used in shaded lawn areas in southern climatic regions. Common Kentucky bluegrass is generally kept mowed at a height of 2–4 inches (5–10 centimeters) and may be damaged at cutting heights of less than 1½ inches (3.8 centimeters).

Merion bluegrass is a variety similar in appearance to common Kentucky bluegrass. This variety is reported to be superior in resistance to leaf spot disease, can tolerate more drouthy conditions, and can survive cutting heights as low as ½ inch (1.3 centimeters). It is susceptible to rust disease in some parts of the country, however. Bluegrass seed is sown at the rate of 1–2 pounds per 1000 square feet (488–976 grams per 100 square meters).

Tall fescue is adapted for use as a lawn grass in the more northern regions of the country and the Midwest. Two varieties of tall fescue are sold for lawn grass use ("Alta" and "Kentucky 31"), with no apparent difference between them. These tall fescue varieties form a rather coarse-bladed, bunched turf unless heavily seeded. This grass provides a rather deep-rooted, drought-tolerant lawn when fertilized and watered properly. When these grasses are seeded heavily at rates of 5–8 pounds per 1000 square feet (2.4–3.9 kilograms per 100 square meters), they will form a tight sod which resists the encroachment of weeds. Frequent mowing at a height of 3–4 inches (7.6–10 centimeters) is required to form a dense, spreading turf of fall fescue. Generally, these fescue varieties grow equally well in sun or shade, except in the southern extremes of their hardiness range, where they are best planted in shaded locations.

Ryegrasses are generally included in seed mixtures with other cool season grasses to provide a quick green lawn. These are called "nurse" grasses because they provide a dense turf which aids the establishment of the more desirable grasses. Use of these ryegrasses in seed mixtures for fall planting is not recommended, because they actually inhibit establishment of the desired lawn grass. Italian ryegrass is an annual grass that survives for only one growing season. Perennial ryegrass will survive for several years, but will not provide a durable turf as well as bluegrass or tall fescue. Their only value is in providing a quick, temporary green turf or for overseeding in a warm season turf to provide a green lawn during the winter months.

Bentgrass is used primarily for providing a dense, fine-bladed surface on golf course putting greens. These lawns are not recommended for home or commercial landscapes because they require considerable care during the growing season. A successful bentgrass lawn would require from 2 to 5 mowings each week at a height of ½–1 inch (1.3–2.5 centimeters). Bentgrass is more susceptible to diseases and cannot tolerate moisture stress because the plants are very shallow-rooted. For this reason, bentgrass seed should not be used in any lawn grass mixtures.

Warm season grasses. *Warm season grasses* grow best during the warmer, dryer months of the year. These grasses generally become brown or dormant during the cold months, from the first killing frost in the fall until midspring. These grasses are best adapted to the southern half of the United States, with certain species better adapted to particular regions.

Warm season grass lawns are established from seed, vegetative plugs, stolons, or sod. Most of these grass seed varieties germinate rather slowly and are sometimes difficult to establish into a tight, self-healing turf. For this reason, vegetative planting is more desirable. Both seeding and vegetative planting are done during the warmer months of late spring and early summer. The warm season grasses most used for lawns in this country are bermudagrass, zoysiagrass, centipede grass, buffalograss, and St. Augustine grass.

Bermudagrasses are more drought-tolerant than other lawn grasses. They produce a tough, fast-growing turf, but may become a weed when used around gardens or other plantings. Bermudagrass survives best when grown in full sun and will withstand heavy foot traffic when maintained properly. Bermudagrass and its improved varieties produce a low-growing, spreading habit of growth. These should be mowed often at a height of 1¼ inches or slightly less, to maintain the spreading habit and to prevent the accumulation of thatch (dead remains of the grass plants).

Common bermudagrass is often used for lawns. It is less expensive to purchase and will provide a suitable turf if properly managed. Hybrids between common bermudagrass and African bermudagrass (also called velvetgrass) have formed greatly improved varieties for lawn use. These include: U-3, Tifgreen, Tiffine, Tiflawn, Tifway, Sunturf, and others. Some bermudagrass varieties are available only as sod or vegetative stolons.

Zoysiagrass lawns are often more durable and attractive in southern regions, when properly established and maintained. This grass is slower in growth than is bermudagrass, but provides drought resistance and the ability to withstand heavy foot traffic when grown in the open sun. Three types of zoysiagrasses are used for lawn and recreation area turf. These are Japanese lawngrass, Mascarenegrass, Manilagrass, and the hybrids Meyer and Emerald zoysiagrass.

Zoysiagrass spreads slowly by creeping stems to form a dense mat which resists the invasion of other grass species and weeds. For this reason the practice of overseeding with a cool season grass species (such as bluegrass) to provide winter color is not practical. The zoysia hybrids Meyer and Emerald are improvements of the parent species. These are more desirable as home lawn grasses, as they provide green color for a longer period each growing season. These zoysiagrasses are rather fine textured and are planted from vegetative plugs, stolons, or sod. Zoysiagrass is kept clipped at a height of 1–1½ inches (2.5–3.8 centimeters), although they will tolerate closer mowing heights.

Centipede grass is a more shade-tolerant grass than is bermudagrass. It is most adapted to the southern tier of states and western areas of the country. This grass is highly desirable where it is adapted, because it requires less watering, fertilizing, and mowing than other warm season lawn grasses. Centipede grass may be established from seed or from vegetative planting. Although this grass species forms a dense, vigorous turf which resists the encroachment of weeds, it should not be used for lawns around farm homes. The grass may escape into pastures, where it quickly crowds out desirable grasses. Centipede grass is a poor pasture grass with little nutritional value to cattle.

Buffalograss is used as a lawn grass in the arid areas of the Great Plains region of the country. This grass is highly drought-tolerant, is adapted to full sunlight areas, and can tolerate close mowing heights. Fertilization, watering, and mowing practices should be adapted to benefit the buffalograss lawn rather than the weeds. High mowing and frequent fertilization and watering will cause weeds to shade buffalograss plants, resulting in the gradual loss of this turf grass from the lawn. Buffalograss lawns may be established from seed or from vegetative planting.

St. Augustine grass is used as a lawn grass in the most southerly states, especially along the coastal areas. This turf grass species is tolerant of saltwater spray and can withstand shaded locations. The growth of St. Augustine grass is by long runners which form a dense, tight turf. It grows best when provided with adequate water and fertilization (primarily nitrogen in sandy soils). This grass is subject to damage from chinch bugs and from brown patch fungus. Rooted runners (sprigs) are used to establish a new planting of St. Augustine grass, because seeds are not available.

Lawn grass mixtures.   Lawn grasses are sometimes combined to form a turf that will have the desirable characteristics of each type. The most common of these combination turfs is the mixture of cool season and warm season grass varieties. These grass types are combined to provide a green turf for most months of the year. The warm season grass grows best during the dry, hot months of the summer. When these grasses become brown after the first killing frost of winter, the cool season grasses are still growing actively. When mixtures such as this are to be used, the grass species should be selected for similar qualities (such as blade fineness) and their compatibility with each other. The most popular lawn grass combination is Merion bluegrass and Meyer zoysiagrass, although such mixtures are not practical in some areas of the country.

Commercially packaged lawn seed is often sold in mixtures of similar grasses. Bluegrass seed is generally mixed with 5–20 percent ryegrass seed. The ryegrass is added to the lawn seed mixture to provide fast germination and to act as a nurse grass for the bluegrass. The Kentucky bluegrass, however, must compete with the ryegrass for light, water, and fertilizer. Even with ideal management, it may require several years for the bluegrass to form a suitable lawn after the ryegrass has died out. The most attractive lawns for homes and recreation areas are established from pure seed sources of the desired grass species rather than from seed mixtures. Seed mixtures of several varieties of the same grass species are more practical, since each variety will provide specific qualities of drought tolerance and disease resistance.

ESTABLISHING TURF GRASS LAWNS FROM SEED

Cool season grass species are planted in well-prepared soils during the cooler months of the year. The most favorable conditions for seed germination and growth of the grass seedlings with the least competition from weeds is from seed sowings made in early autumn. Early spring planting will also provide favorable conditions for seedling growth, although germination and seedling growth will be slower.

Warm season grass species that can be planted from seed require continuous warm temperatures for proper germination and seedling growth. These are sown from early May to mid-July to allow adequate development of the grass plants before cold weather occurs.

Grass seeds may be planted by several methods. The best lawn establishment results when the seed is planted by a seed drill, similar to the type used for planting small grain crops. The planter may be adjusted to allow even the small-seeded grasses to be planted approximately ½ inch apart and at a similar depth in the topsoil.

Since these seed drills are designed for use on large areas, most lawns are seeded by hand or with small spreaders. The seeds may be broadcast onto the soil by use of a cyclone spreader, which applies the seed evenly, at the desired rate recommended. A drop spreader may be used also for this purpose. Both machines should be calibrated before sowing the seed to make certain that the seeds are sown at the recommended rate. The grass seed to be planted should be weighed to determine the amount required for seeding the entire lawn. The seed may then be divided into two equal quantities and each seed lot broadcast onto the soil in different directions. The first half of the seed is applied in an east-to-west direction. Then the planting operation is completed, with the final seed lot applied in a north-to-south direction.

Extremely small-seeded grass species (such as centipede grass) should be mixed thoroughly with an equal amount of fine, dry sand to allow the seeds to be distributed evenly. When completed, the seeds of the lawn grass should be spaced evenly over the soil, but the ground should not be completely covered with seed. The seed is covered lightly by hand raking or by pulling a cloth mat over the soil surface. Small seeds are covered to a depth of ¼ inch (0.6 centimeter), while large seeds may be covered to a depth of ½ inch (1.3 centimeters) (Figure 15–4).

FIGURE 15–4. Lawns that have been seeded are covered with a light layer of straw to prevent rapid drying of the soil during germination.

ESTABLISHING LAWNS BY VEGETATIVE METHODS

Seed is not produced in sufficient quantities or is not true to the variety type with some grass species. In some cases the use of spot plugs of grass sod may be the most practical method for establishing a new lawn. Most warm season grass lawns must be established from plugs, sprigs (stolons), or sod. All cool season lawn grasses may be established from sod, when a quick groundcover is desired.

The subsoil and topsoil is prepared in the same manner for both seeding or grass planting by vegetative methods. Small sod plugs may be planted with a specially designed tool. The sod plugger is used to cut round plugs from a turf sod. Each plug is cut at the same diameter and to the same depth by the plugger. A similar-sized plug is removed from the top soil of the area to be planted. Since the hole and the sod plug are the same size, the plug may be easily pressed into the planting hole. The plugs are generally planted in rows and spaced about 12 inches (30 centimeters) apart in each direction. Slow-growing grasses (such as hybrid zoysiagrasses) are planted at 6-inch (15 centimeters) spacings. The sod plugs are firmed by pressing them into the soil slightly with the foot or by a roller. The soil surface should be given a thorough soaking with water each week to encourage rapid spreading of the plugged grass. Frequent mowing also encourages the establishment of a dense lawn from sod plugs (Figure 15–5).

Some grasses may be planted from sprigs. These are individual plants, runners, or stolons which root when placed in contact with the soil. A turf sod is shredded into individual stems approximately 6 inches in length. These sprigs are broadcast onto the soil surface and pressed into the soil in straight furrows made by a disk. The disk presses the sprigs into the ground so that they will root readily when irrigated regularly. Bermudagrass, zoysiagrass, St. Augustine grass, and centipede grass may be established from sod sprigs.

A lawn may be established most quickly when sod is used. However, unless a high-quality sod is available and seeding or another method for vegetative planting is not practical, the greater expense of a sodded lawn may not be warranted. Seeding may be up to five times less expensive than sodding a lawn, but sodding is easier and more likely to prove a successful lawn. Sodding is also the most practical method for establishing grass on a steep slope, where grass seed may easily be washed away by rainwater.

Sod is grown in commercial sod nurseries. It is cut into strips to a depth of 1 inch (2.5 centimeters). These strips are 18 inches wide and up to 6 feet in length. The sod strips are cut with a sod-cutting

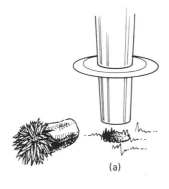

(a)

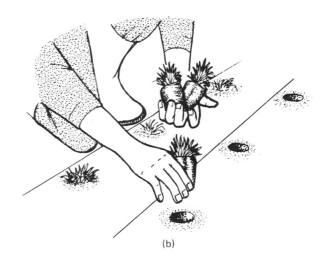

(b)

FIGURE 15–5.   Planting a lawn with sod plugs: (a) Plugs are removed from a sodded area. (b) The plugs are planted in holes prepared in the lawn.

machine, rolled, and placed on a truck (Figure 15–6). A lawn may best be sodded in the early fall or spring for cool season grasses or in late spring after growth begins for warm season grasses. New roots, which aid in anchoring the sod, are being formed at this time.

(a)   Sod strips are cut from a field using a sod cutter machine.

(b)   Large rolls of sod strips are loaded into trucks at a sod nursery.

FIGURE 15–6.   Preparation of sod strips.

FIGURE 15–7.   The sod strips are installed in the lawn on carefully prepared soil. The strips are placed snugly together, topdressed with soil, and rolled to provide good root-to-soil contact.

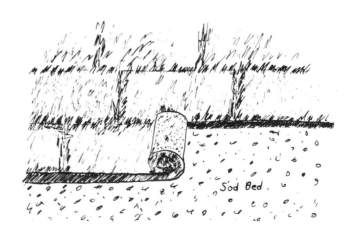

The strips of sod are laid snugly together and the ends cut to fit the lawn perimeters. The firmed topsoil surface is spread under the rolls of sod to even out depressions. Sod pieces are then placed in spots where the sod strips do not match. A top dressing of topsoil is spread over the cracks between the sod strips to aid in a uniform coverage of the turf (Figure 15–7). The newly laid sod is then rolled to firm the roots and soil into close contact. On steep slopes or grades, the sod is in danger of being washed away in heavy rains. This may be prevented by placing wooden stakes at right angles to the slope and to a depth flush with the sod surface. The newly sodded lawn should be watered liberally and kept moist until new root growth is established.

MAINTAINING AN ESTABLISHED LAWN

The maintenance practices of an established lawn determine the length of time the lawn will remain healthy, attractive, and free of serious problems requiring renovation. A lawn requires the proper application of water and fertilizer for growth. If these are not applied at the correct time or at the proper rate, the grass plants may be damaged. Mowing practices of the lawn grass should also be altered as the grass plants grow.

Maintenance practices for lawn grasses should be timed according to the growth characteristics of the grass plants. Cool season grasses begin root growth in early spring after the ground thaws and soil temperatures reach 40–45°F (4–8°C). Growth of the grass blades begins shortly thereafter. These cool season grasses grow most rapidly at temperatures near 60°F (15°C), but become dormant when temperatures remain near 80°F (27°C). Warm season grasses begin active root and shoot growth when soil temperatures reach 60°F (15°C) and remain healthy during the warmer months. The fertilization and watering of these grasses should be timed to favor growth for the particular type of grass.

Lawn Irrigation

An actively growing grass plant contains nearly 80 percent water. Rainfall during the growing season is generally not adequate for maintenance of healthy turf grass lawns. Frequent water applications are required, whether by irrigation or from natural rainfall, to promote the deep rooting of the plants. Grass is highly efficient in extracting water from the soil when it is available. Irrigation of lawns should be timed so water will penetrate to a depth of 1 foot (30 centimeters) and the soil allowed to dry somewhat before additional water is applied. This generally requires the application of 1 inch (2.5 centimeters) of water each week during periods of active growth. This irrigation must be applied before the grass plants reach the point where irreversible wilting occurs.

Grass plants require watering just before wilting is visible. The lawn will take on a slightly blue-green coloration as wilting of the grass plants begins. If irrigation water is not applied immediately upon the onset of wilting, the grass plants may not recover and will die. The grass may be tested routinely to determine the water status of the plants by the "footprint" method. A worker can walk across a lawn and then observe the speed of the grass blades in returning to their natural positions. Grass plants having sufficient water will return quickly to their normal stature and the foot prints will not be visible. Plants that are at or near the wilting point will not resume an upright position, so the footprints will remain visible for some time.

Frequent, shallow irrigation stimulates root growth near the upper surface of the soil. During the dry season of midsummer these shallow-rooted grass plants are unable to reach the moisture at the lower depths. For this reason less frequent irrigations of adequate water will stimulate deep rooting of the grass. The roots of these plants are more capable of surviving the hot, dry periods. When watering is begun during a drought period, it must be continued periodically as long as the dry weather persists. If the lawn is allowed to go dormant during a drought period, addition of water by irrigation will generally promote the growth of weeds (such as crabgrass) rather than aiding the growth of the lawn grass.

Timing of water application to lawns is especially important in disease control of turf grasses. Irrigation of a lawn should be done during the morning or early afternoon of a warm day so that water may be evaporated from the leaf surfaces. Irrigation at periods when free water is permitted to remain on the leaf for long periods allows disease pathogens to develop. This free water is required for germination of the pathogen and the invasion of the grass plant.

Fertilizing Lawn Grasses

All lawn grass species respond to proper fertilization at those times when active growth is most pronounced. Addition of fertilizer can improve a poor-quality lawn or maintain a healthy, dense and beautiful lawn. Grass plants require certain chemical elements for proper growth. The essential elements required by plants for growth are listed in *Table 15–2*.

TABLE 15–2:   *Essential Elements Required for Plant Growth*

MAJOR ELEMENTS		MINOR ELEMENTS
FROM AIR		
AND WATER	FROM SOIL	FROM SOIL
Carbon	Nitrogen	Copper
Hydrogen	Phosphorus	Manganese
Oxygen	Potassium	Iron
	Calcium	Molybdenum
	Magnesium	Boron
	Sulfur	Chlorine
		Zinc

Of the *16 essential elements*, carbon, hydrogen, and oxygen are supplied by the air and water to the grass plants. Calcium (and sometimes magnesium) is supplied to the plant from lime which is added to the soil. The minor nutrients are required by the plants in very small quantities. These elements are usually available in the soil in sufficient amounts. The major elements are generally added to the soil and are called the *fertilizer elements*. The primary fertilizer elements which are required in the greatest amount are nitrogen (N), phosphorus (P), and potassium (K).

Nitrogen is the element required in the most abundance for the proper growth of lawn grasses. Nitrogen promotes the deep green color and continued top growth of grass. Nitrogen is water-soluble and moves freely in the soil. This element is generally applied frequently and in large amounts to the lawn as a granular fertilizer. The addition of water to the lawn carries the nitrogen into the soil and in contact with the plant roots.

Phosphorus is required by plants for the maintenance of healthy, attractive growth. Phosphorus is necessary for the development of a healthy root system, flowers, and seeds on plants. The element is not readily transported through the soil by water, so should be incorporated into the top 4–6 inches (10–15 centimeters) of the soil during seedbed preparation. Some soils contain sufficient amounts of phosphorus for the growth of lawn grass. However, when a soil test indicates that phosphorus is needed, this element may be applied to an established lawn. Since the phosphorus does not move into the soil readily, the turf should be aerified or otherwise opened to place the phosphorus fertilizer into the root zone of the grass plants.

Potassium is an important element for improving the disease resistance and overwintering capabilities of lawn grasses. Because the addition of potassium to a lawn increases root growth, the plants are more tolerant of dry conditions and wilt less easily. Most of the soil potassium is bound by the soil particles and is not readily available to the grass roots. However, an abundance of potassium in the soil often occurs, making exchangeable and water-soluble forms of this element available for use by the plants. Potassium may be either applied to established turf as a top dressing or incorporated into the soil at the time of planting.

Fertilizers that contain these three major elements are called *complete fertilizers*. Some complete fertilizers are formulated to include sulfur, calcium, magnesium, and the minor elements. The fertilizer formulations most commonly recommended for application to lawns will have approximate N–P–K ratios of 3–2–1 or 2–1–1. Some examples of these fertilizer ratios are those having an analysis of 15–10–5, 10–5–5, or 12–6–6. In locations where phosphorus and potassium are present in the soil in adequate quantities (as determined by a soil test), the addition of nitrogen alone may be all that is necessary. Some fertilizers that contain nitrogen (and sometimes another elemental carrier) are ammonium nitrate (33–0–0), ammonium sulfate (21–0–0), and urea (45–0–0). Extreme care must be excercised when applying high-analysis fertilizers to lawns, to prevent damage to the grass plants.

Fertilizer rates are usually expressed in pounds of actual nitrogen (N) per 1000 square feet. If a soil test recommends the addition of a 10–5–5 fertilizer, 10 pounds of the 10–5–5 formulation is required to equal 1 pound of actual nitrogen per 1000 square feet. Nitrogen is generally applied at no higher rate than 2 pounds of actual nitrogen at each application. The metric equivalent to this recommended rate is 976 grams per 100 square meters. When higher rates are required, the fertilizer applications are spaced over the growing season according to the type of grass being grown.

Cool season grasses should be fertilized in early spring and again in autumn when the plants are growing most actively. The first application of fertilizer can be applied as the grass is beginning to grow. A second fertilizer application may be made again in mid-spring, but not after warm, dry weather has begun. Fertilizer should not normally be applied during the summer unless the lawn is kept irrigated regularly and the grass plants maintained in active growth. Under these conditions a very light fertilizer application could be made in midsummer. The autumn fertilizer application is made during the month of September. This fertilizer allows the grass plants to grow actively during the cool fall months and accumulate food reserves for overwintering.

The normally recommended rate for fertilizing a cool season lawn is at the rate of 1 pound per 1000 square feet (488 grams/100 square meters) at each application. Bluegrass lawns should not be fertilized with more than 5 pounds per 1000 square feet (2440 grams/100 square meters) of actual nitrogen during a single growing season. Generally, half this amount is satisfactory.

Warm season grasses are fertilized as soon as they resume active growth in mid or late spring. Fertilization during early spring before these grasses begin active growth may only stimulate the establishment of weeds in the lawn. The warm season grasses may be fertilized as often as once every 4–6 weeks during the summer months. Fertilizing this frequently stimulates a dense turf which resists invasion from weeds and tolerates heavy foot traffic. The last application of fertilizer to these lawns is made no later than mid-August, so the grass will slow in growth before cold weather occurs. Fertilization later in the season may stimulate a soft, luscious growth of the grass that would be damaged during winter.

Bermuda grass is fertilized at the rate of up to 6 pounds of actual nitrogen per 1000 square feet (2928 grams/100 square meters) each season in most areas where it is grown. Greater quantities of fertilizer are required in the most southern regions of the country, where this grass is growing actively all year. Zoysiagrass generally requires less fertilization than does bermudagrass. However, during the first year of its establishment, the same rate of fertilization may be used to stimulate a rapid filling in of the lawn. The native buffalograss of the plains areas grows in relatively poor soil. Heavy fertilization and watering of this grass species generally results in the stimulation of weeds and other grasses in the lawn. An occasional light application of nitrogen fertilizer will maintain good green color in buffalograss lawns.

Mowing Lawns

The development and maintenance of a healthy and attractive lawn requires frequent mowing at the proper height. Mowing also slows the development of weeds in a lawn, since grass plant growth occurs near the base of the plant while most broadleaved weeds grow from the shoot tips. The practice of mowing removes a portion of the leaf area of the grass plant, so mowing should be done frequently to prevent serious damage to the grass. When grass is allowed to grow unusually tall between mowings, the base of the plant is shaded. Removal of the top growth exposes the lower portions of the plant and they are "scalded" by the sun. If the grass clippings are not removed after mowing this tall grass, the plants may be smothered and diseases will become established. For this reason, no more than one-third of the new growth of the grass plant should be removed at a single mowing. The lawn should be mowed regularly; and the shorter the height of the cut, the more often the lawn must be mowed to prevent scalding.

Mowing heights.   Cool season grasses are mowed at a height of no less than 1½ inches (3.8 centimeters) during the spring and autumn months. During the warmer months the cutting height may be raised to 3 inches (8 centimeters) to reduce drought damage and to retard crabgrass growth. Crabgrass plants may be shaded by the higher cutting height of the grass during the months of May through September. The cool season grasses may then be cut at the lower height again in the autumn. Close mowing of most cool season grasses during the warmer months will weaken and may kill the lawn.

Warm season grasses are kept clipped at lower heights than are most cool season grasses. The normal height of cut for these grasses is 1–1¼ inches (2.5–3 centimeters) for the best appearance of the lawn and slowest accumulation of thatch. Bermudagrass hybrids may withstand a cutting height as low as ⅝ inch (1.6 centimeters) if mowed frequently.

Removal of clippings.   As long as the amount of grass clippings from a mowing are not excessive, they may be left on the lawn. These clippings contain valuable nutrients and provide organic matter for the soil. The clippings should be removed, however, if they do not readily sift down into the soil during mowing. This condition may be avoided by more frequent mowing. Large quantities of grass clippings may smother the plants and pose a disease problem for the lawn. Warm season grasses have a tendency to form thatch (an impenetrable barrier of dead and living plant parts). Removal of the grass clippings from these lawns following mowing aids in the reduction of thatch accumulation and the spread of diseases.

Mowing equipment.   Two basic types of mowers are used for lawn maintenance: the reel and the rotary mower. *Reel mowers* are used more often on public grounds and golf courses for maintenance of large turf areas. These mowers use a scissor action to cut the grass blades as they pass across the mower blade. When properly sharpened and adjusted, the reel mower provides the highest-quality cut for a lawn.

Rotary mowers are more popular for home lawn use because they are generally easier to handle and less expensive to purchase than reel mowers. They are also more dangerous to operate and should be fitted with various safety features during operation. The high-speed rotation of the rotary blade may pick up and project debris from the lawn for long distances. Either a grass catcher or a metal deflector should be in place over the discharge chute of the rotary mower. All loose debris and small stones should be removed from a lawn to prevent accidents during mowing. The operator must be particularly careful to avoid placing either the hands or feet under the mower while it is in operation.

Regardless of whether a reel or rotary mower is used, the cutting blades must be kept sharp at all times. A blade may be dulled by hitting small stones or debris in the yard. The grass plants themselves will also dull the cutting edges of a mower. Tall fescue, for example, dulls a mower blade rapidly, making it necessary to sharpen the blades once each month of regular use. When the mower blades have become dulled, the newly clipped lawn will appear slightly browned a few days later. Close inspection of the grass plants will show that the grass blades have been ripped rather than finely cut. Jagged, ripped blade tips turn brown and give a ragged appearance to the lawn.

The height of the mowing cut is determined by placing the mower on a solid, flat surface. The height to the cutting blade on a rotary mower or to the bed knife on a reel mower is then determined. The adjustment of the height of cut is made according to the directions in the owner's manual. Generally, a screw on each side of the bed knife of a reel mower may be loosened and the cutting height readjusted. Rotary mowers may be raised or lowered by adjusting the clips found at each wheel.

RENOVATION OF LAWNS

Sometimes an established lawn will deteriorate to the point where it cannot be maintained in an attractive and healthy condition. If the lawn does not require complete reconstruction, it may be renovated by first correcting the causes for lawn failure and then replanting new grass in the old lawn. Factors that normally cause failure of a lawn and may be corrected by renovation include the lack of plant nutrients, soil compaction, acid soils, heavy shade, and the encroachment of undesirable grasses or weeds.

Lawn renovation begins with an assessment of the cause for failure of the grass. A soil sample should be taken, particularly in the most damaged areas, to determine the nutrient status of the soil. This should be done at least a month before the actual work is to be done so that the results will be available. The actual renovation should be done at the same time normally selected for planting the grass. Weeds may be eliminated by spraying the lawn with selective herbicides.

Dead plants and thatch are removed either by giving the lawn a thorough raking or by use of a renovating machine. A power rake is often used for this purpose. A vertical cutting machine having blades that cut into the soil slightly between the grass plants will remove much of the thatch layer of the turf. A power vacuum is then used to remove the plant debris from the lawn (Figure 15–8).

When compaction of the soil is a cause for lawn failure, the top-soil may be opened by use of an aerifying machine. This compaction may be caused by rainfall, artificial watering, and foot traffic which compresses the soil particles and makes water, air, and root penetration difficult. The aerifying machine loosens and aerifies the soil under the turf without disturbing the grass plants (Figure 15–9). Frequent aerification of most lawns will prevent the need for complete renovation. The aerifier opens the soil by actually removing cores of soil. Hollow spoons are fitted on the revolving shaft of the machine. These spoons cut holes in the turf and scatter the soil plugs in the lawn. The soil is then raked over the lawn to loosen the soil and fill the holes. This loosened soil in the holes allows more rapid air, water, and root penetration of the lawn grass.

Lawns that have been renovated by thatch-removing machinery will require a fine layer of topdressing to provide a seedbed or to place soil around the existing plant roots. Even healthy lawns will benefit from an application of topdressing once each year. Common

(a)

FIGURE 15–8.   Thatch removal in a bermuda-grass lawn: (a) A vertical cutting machine removes dead plant material from the lawn. (b) A power vacuum may be used to remove the thatch left after the vertical cutting machine has been used.

(b)

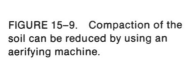

FIGURE 15-9. Compaction of the soil can be reduced by using an aerifying machine.

materials used for this purpose are topsoil mixed liberally with sand and organic matter (such as leaf mold or peat). The topdressing is thorougly mixed and applied to the soil by a drop spreader. A heavy canvas or burlap mat is pulled over the lawn to even the distribution of the topdressing and to aid in filling depressions. Application of lime and fertilizers must be done before planting, if required. The lawn is now ready for seeding or vegetative planting, as was described earlier.

Protecting Lawns from Winter Damage

A healthy lawn will withstand cold winter weather more readily than plants that are already in a weakened condition. A fall application of fertilizer will stimulate strong root and shoot growth of cool season grasses. Warm season grasses should be allowed to slow in growth during autumn by making the final fertilizer application of the year no later than mid-August (in all but the most southern regions of the United States). A properly fertilized cool season lawn is more likely to resist diseases, weed encroachment, and repair traffic damage that occurs in late fall through winter.

Soil compaction may be alleviated by use of an aerifying machine to open the soil before winter. Most regions of the country receive plentiful moisture during the winter months. If the soil is badly compacted, however, this moisture will not be absorbed by the soil. The lawn should be watered heavily during the fall months to make certain that the soil is moist to a depth of up to 2 feet before the ground freezes.

The grass should be allowed to grow slightly taller as it goes into winter. The taller grass cover will be less damaged by alternate freezing and thawing and will also aid in preventing the heaving of the soil during periods of fluctuating temperatures. Large accumulations of leaves from trees in the lawn should be removed. The leaves will shade the grass and allow diseases to become established in the lawn. The leaves should be removed and placed in a compost pile. This compost may then be used later as a topdressing for the lawn.

LAWN PEST CONTROL

The proper maintenance of any lawn grass area must include the control of weeds, insects, and diseases. A healthy lawn that is kept properly mowed, fertilized, and irrigated will be more capable of resisting pests than will weakened grass plants. Attention to these practices will improve the chances for reducing lawn pest problems.

Weed Control in Lawns

Weeds do not generally become established in a dense healthy lawn. When the grass plants become weakened by insect infestations, diseases, or through poor management practices, weeds will quickly crowd out the grass. Any measures to control lawn weeds must be accompanied by a suitable management practice that will improve the grass stand; otherwise, weeds will again become established.

Many annual weeds may be controlled by use of *preemergence herbicides* applied before the seeds germinate in the spring. These herbicides form a chemical barrier to seedling growth at the soil surface. A preemergence herbicide applied before forsythia blooms is effective in controlling the earliest emergence of crabgrass and other spring weeds. The chemical barrier is broken when the lawn is walked upon or mowed, so use of a preemergence herbicide provides only temporary control of weeds in the lawn.

Many *post-emergence herbicides* are available and must be carefully selected for control of the weeds present in the lawn. Most broadleaf weeds are killed or injured by foliar applications of 2,4-D. Dandelions treated with this chemical in the autumn and again to newly germinated seedlings in the spring are readily controlled. Some broadleaf weeds (henbit, chickweed, knotweed, and nutsedge) will not be killed by 2,4-D, but are more easily controlled when a combination of 2,4-D and 2,4,5-T (Silvex) are used. These herbicides may be applied most successfully during the spring or fall months to cool season lawns. If a fertilizer application is made before the herbicide is applied, the lawn grass may fill in the areas left when the weeds are killed. Crabgrass may be controlled after the young seedlings are present in the spring by use of various methylarsonate chemicals. The most widely used chemical for this purpose is DSMA (disodium methanearsonate), but other chemicals are also available for postemergence crabgrass control.

Occasionally areas of a lawn must be eradicated of all plant growth to allow the establishment of a more desirable lawn grass: for example, when a cool season lawn is being overtaken by a warm season grass or a persistent weed species. A nonselective systemic herbicide may be used to kill all grasses in the area to be sprayed. Generally, several applications at 1- to 2-week intervals are required to completely kill all the plants. Some of these nonselective herbicides persist in the soil for 6 weeks or longer, so a suitable period must be allowed before reseeding of the lawn may be done.

Nonvolatile formulations of all herbicides should be used around home lawns. The chemicals should be applied at low pressure with coarse nozzles on calm (no wind) days to avoid the drift of the herbicide onto desirable landscape plants. Whenever any chemical is used, it should be mixed and applied according to the directions on the label. Some of these herbicides may be harmful to the applicator unless protective clothing is worn. The chemicals should never be mixed at rates greater than those recommended on the label. Rates that are too strong or applications made in hot weather may kill only the tissues with which the chemical comes in contact, without being absorbed by the plant. This results in only temporary damage to the plant. The directions on the label should be followed exactly and the chemical applied during the cooler hours of the day. Applications should be made on days when no rain is expected for at least 24 hours.

Herbicide residues are difficult to remove from sprayer tanks, hoses, and nozzles. These residues may persist for a long time after the equipment is cleaned and rinsed. For this reason spray equipment used for herbicidal treatments should never be used for applications of insecticides or fungicides on garden or landscape plants.

Insect Control in Lawns

The insect pests most damaging to lawns in the United States are white grubs of the May and June beetles, sod webworms, Japanese beetles, cutworms, chinch bugs, and other insects that may occasionally injure grass plants. White grubs live upon the decaying organic matter, but the grass roots are also eaten just below the soil line. When a heavy infestation of these grubs occurs, the grass will turn brown and can be pulled up easily. A soil insecticide should be thoroughly soaked into the ground in mid to late spring to control these pests.

Sod webworms may be very injurious to bluegrass varieties and other cool season grasses. These caterpillers feed on the grass blades rather than on the roots. This feeding generally occurs during late summer. The lawn will appear with irregular brown areas and new grass blades will die suddenly. An insecticide sprayed on the foliage after the presence of this insect is confirmed is the best control measure. No artificial irrigation or rainfall should occur for several days if the insecticide treatment is to be effective.

Chinch bugs may be particularly damaging to thick turf, particularly bent grass, in various regions of the country. These insects are small ($1/10$ inch) and may be either red or black in color. They feed on the aerial portions of the grass plant, but are mostly found at the soil line during the day. When the turf is wetted heavily, the chinch bugs may be seen crawling rapidly to the surface. Chemical insecticide sprays applied to the grass foliage are effective in controlling chinch bug populations in established turf.

Controlling Lawn Grass Diseases

Diseases may occur in any lawn when conditions favor their development. A disease may spread rapidly in a lawn comprised of a pure stand of grass, since each plant is alike and equally susceptible. Disease organisms may become established more quickly when the lawn remains wet for extended periods. Enough water should be applied at one time to thoroughly soak the soil to a depth of 6 inches (15 centimeters). Water should be applied early enough in the day to allow evaporation from the grass blades before nightfall. The fungal spores require this free water on the foliage for germination and establishment. Most major lawn diseases become established during the wet, rainy months.

The accumulation of grass clippings and tree leaves on the lawn provides wet areas for disease establishment. Frequent mowing and removal of grass clippings is desirable for preventing diseases. Fungicide applications to the turf during wet periods may also be necessary when cultural practices alone are not adequate in controlling disease outbreaks in the lawn. Some common turf diseases are: damping-off, leafspot, brown patch, dollar spot, melting out, rust, and mildew. Some causes for lawn failure may appear to be caused by diseases, but are actually the result of physical damage to the grass.

Damping-off is found most often in new grass plantings. The young seedlings are generally killed as they emerge from the soil, giving the lawn a spotted appearance. This problem may be reduced by avoiding the use of too much seed when planting. Also, keeping the soil surface too wet during seed germination will favor damping-off of the seedlings. Fungicides formulated with copper are available for control of this disease.

Various *leaf-spotting diseases* may develop whenever the lawn remains wet for extended periods. Leaf spots appear on the foliage and sometimes affect the crowns of the grass. Excessive moisture on the foliage and piles of grass clippings may cause these disease symptoms to develop. Grass species most damaged by leaf spot diseases are Kentucky bluegrass, creeping red fescue, and U-3 bermudagrass. Regular use of lawn fungicides will aid in preventing these diseases.

Brown patch, dollar spot, and *melting-out disease* are somewhat similar in pattern. Brown patch and dollar spot create circular patterns of varying sizes, while melting out appears as irregularly shaped dead patches in the lawn. Bentgrasses are more susceptible to brown patch and dollar spot, while bluegrass and fescue are more susceptible to melting out. Foliar fungicides may be applied as a preventive control measure for these diseases. Urine from dogs (mostly females) will create browned-out spots that resemble brown patch. Injured grass areas are round and may be several inches in diameter.

Rust is very common on most ryegrass and bluegrass varieties each season. Rust is most easily controlled by keeping the grass growing actively by fertilizing and watering properly and by removal of clippings after each mowing. *Mildew* is most predominant in shaded areas where the grass remains moist during the day. Blue grass is most susceptible, but any grass can be attacked by this disease under the proper conditions. Fungicides that are effective in mildew control are used in the prevention of this problem.

GOLF COURSE MAINTENANCE

Maintenance of golf courses has become a more popular area of employment for young people who enjoy working outdoors in the environment of public and private clubs. The increased popularity of golfing as a hobby has created many new facilities and a corresponding demand for trained maintenance employees.

The maintenance of golf course grasses and landscape plants requires more training and experience than is required for home lawns.

Often more sophisticated equipment is used and an intensive program of maintenance is required. The heavy foot or cart traffic, intensive use of the facilities, and the watchful scrutiny of the club members requires that the playing surfaces be kept in excellent condition at all times. Since the golf courses are in heavy use during the daytime hours, most of these maintenance chores must be completed while the course is in use.

The maintenance of the golf course grounds is generally under the direct supervision of the golf course superintendent. He is responsible for crews involved in maintenance of the grass playing surfaces, buildings, roads, landscape materials, and equipment. He is also responsible for obtaining all supplies and equipment used in the operation of the course.

The grounds crews are responsible for maintaining all playing surfaces of the golf course. They must be skilled in performing the various duties required for routine maintenance of the trees, greens, fairways, and bunkers. Specialized equipment and practices are required for each of these areas. Some routine management practices for golf course maintenance will be discussed.

Tees.   Most golf tee areas are long (about 100 feet), so tee markers may be rotated frequently. As grass is worn along tee lines, the markers must be moved to allow healing of the worn areas. This is done as frequently as necessary, generally every 2–3 days. The tees are mowed daily with a tee mower to provide a tough, thick grass surface. The mowing height is dependent upon the type of grass used, generally at the height of ½–1 inch (1.3–2.5) centimeters.

Tees generally receive considerable foot traffic in a small area, so compaction of the soil is frequent. The tee areas are aerified and topdressed as often as required. Occasionally, the tees become so badly worn that renovation of the turf is required to provide a suitable playing surface.

Fairways.   The fairway surfaces are maintained in much the same manner as large recreational lawns (Figure 15–10). The fairways seldom receive the high-intensity management required of greens and tees. Mowing is practiced on fairways about twice a week with tractor-pulled gang mowers. Cutting heights of these lawn areas varies with the species of grass grown, but they are generally clipped at the heights recommended for home lawns. Repairs must be made to badly worn turf areas caused by compaction and divots. The fairways may be aerified weekly or monthly, depending on their use.

Bunkers.   The sand traps around greens and along fairways require constant attention from maintenance crews. The sand is smoothed and kept playable daily with power sand rakes (Figure 15–11). The hand rake used by players to repair the sand traps must be replaced in a position where it will not interfere with normal play. Grasses and weeds that encroach on the sand trap are routinely destroyed with herbicides to keep the trap uniform in outline.

Greens. Golf greens receive more intensive management than any other area of a course. This is mostly because the grass used on a green is kept clipped extremely short and as smooth as a carpet to enhance accurate putting. The grass on the greens is mowed daily, alternating the direction of the cut with each mowing. Specially designed greens mowers are used for this purpose to ensure an even cutting height and to remove the grass clippings. Aerification of the greens is practiced routinely, since the greens are subject to high foot traffic. Watering and fertilization of the greens is intensified to maintain a thick, full turf.

The rapid growth of the grass on greens causes a rapid accumulation of thatch and makes the plants more prone to diseases and insect infestations. The greens must be constantly maintained to re-

FIGURE 15–10. Fertilizer application on a golf course fairway using a cyclone spreader.

FIGURE 15–11. The sand in the golf course bunkers is kept smoothed and playable with power sand rakes.

move this thatch accumulation and to relieve soil compaction. The greens are sliced with a vertical mowing machine to remove the thatch layer and vacuumed. The surface is then topdressed with a soil mixture to encourage thatch decomposition and rapid grass growth into the sliced areas.

The cups and flags are moved rountinely to allow heavily played areas to mend. A cup-hole cutter is used to remove a plug of sod and soil from the area where the cup is to be placed. A cup remover is used to pull the cup from the green, while the sod plug is replaced in the hole. A cup setter then firmly places the cup in its new position. In addition to the maintenance of the playing surfaces, the facilities near the greens and tees must be routinely maintained. Ball washers require filling with washing solutions and clean towels replaced. Trash from receptacles must be emptied daily. Club washers are also emptied and refilled with a cleaning solution.

SELECTED REFERENCES

Beard, J. B. 1975. *How to Have a Beautiful Lawn.* Intertec Publishing Corporation, Kansas City, Mo.

Beard, J. B. 1973. *Turfgrass Science and Culture.* Prentice-Hall, Inc., Englewood Cliffs, N. J.

Conover, H. S. 1977. *Grounds Maintenance Handbook,* 3rd ed. McGraw-Hill Book Company, New York.

Crockett, J. U. 1971. *Lawns and Ground Covers.* Time-Life Books, New York.

Denisen, E. L. 1978. *Principles of Horticulture,* 2nd ed. Macmillan Company of Canada Ltd., Toronto.

Griffen, J. M. 1970. *Landscape Management.* California Landscape Contractors Association, 234 Marshall Street, Redwood City, Calif.

Hanson, A. A., and F. V. Juska (eds.). 1969. *Turfgrass Science.* American Society of Agronomy, Madison, Wis.

Madison, J. H. 1971. *Practical Turfgrass Management.* Van Nostrand Reinhold Company, New York.

Madison, J. H. 1971. *Principles of Turfgrass Culture.* Van Nostrand Reinhold Company, New York.

Musser, H. B. 1962. *Turf Management.* McGraw-Hill Book Company, New York.

Oravetz, J., Sr. 1975. *Gardening and Landscaping.* Theodore Audel and Company, 4300 West 62nd Street, Indianapolis, Ind.

Shurtleff, M. C. 1974. *How to Control Lawn Diseases and Pests.* Intertec Publishing Corporation, Kansas City, Mo.

Sprague, H. B. 1976 *Turf Management Handbook.* The Interstate Printers and Publishers, Inc., Danville, Ill.

TERMS TO KNOW

Aerification
Bunkers
Compaction
Cool Season Grass
Fairways
Fertilizer Elements
Greens
Lime Elements
Nonselective Herbicide
pH
Postemergence Herbicide
Preemergence Herbicide
Reel Mower
Renovation
Rotary Mower
Sod
Sod Plugs
Stolons
Tees
Thatch
Vertical Mowing
Warm Season Grass

STUDY QUESTIONS

1. Discuss the major advantages and disadvantages of warm season grasses and cool season grasses in a home lawn.
2. List the major cool season grasses used for lawns in the United States.
3. List the warm season grasses used for lawns in the United States.
4. Outline the major steps required for establishing a cool season lawn from seed.
5. Describe how a lawn may be established from sod plugs.
6. List the major and minor elements required by plants for best growth.
7. Explain why the clippings should be removed each time a lawn is mowed.
8. Explain why a verticle mowing machine is used on a lawn.
9. How is soil compaction corrected in a lawn?
10. List some of the common diseases that may infect lawn grasses and give a brief description of each.

SUGGESTED ACTIVITIES

1. Make a maintenance schedule for both cool season and warm season grass lawns in your area. This schedule should include dates for planting and fertilization of these lawns.
2. Prepare small areas of ground near the school classroom for planting lawn grasses by seed, sod plugs, sprigs, and sod. Compare these methods for their rates of establishment.
3. Visit a local golf course and have the superintendent explain how each of the golf playing areas is maintained.
4. Collect insect pest samples from lawns and make a display for class use.
5. Make a display of turf grass diseases from pictures or grass samples.
6. Have the students demonstrate proper maintenance of a lawnmower.

CHAPTER SIXTEEN

Landscape Maintenance

The horticulture student who finds employment in any of the various businesses associated with the nursery industry will require a basic knowledge of the establishment and maintenance of landscape plantings. Nursery workers and employees of retail garden centers must be capable of handling the nursery stock before it is sold so that the plants will remain healthy and will transplant well when placed in the landscape. A basic knowledge of each plant's growth characteristics, requirements for establishment, and potential pest problems is necessary whenever an employee is involved in the sale of nursery stock. A fundamental understanding of the principles involved in landscape design also aids employees in the selection of plant material for landscapes.

Many businesses within the nursery industry provide the services of establishing, maintaining, or renovating landscape plantings. Retail nurseries, garden centers, and landscaping service businesses plant that nursery stock which they sell. In addition to planting, these firms may also provide basic maintenance of the plantings for the customers. Horticulturists employed by city governments or park commissions must constantly care for the landscapes on public property. The golf course maintenance crews must, in addition to caring for the turf playing surfaces, maintain the landscape plantings along fairways and buildings. The landscape maintenance service businesses are very lucrative for any horticulture student skilled in the fundamentals of plant care. The student should be trained in the following areas of landscape maintenance: planting trees and shrubs, pruning and training of trees and shrubs for various purposes, fertilizing and watering landscape plantings, and diagnosis and control of various pests commonly found on landscape plants.

PLANTING ORNAMENTAL TREES AND SHRUBS

The objective of any well-planned landscape is to provide a minimum of maintenance for the homeowner. The proper selection of plants for their intended uses is fundamental for establishing a minimum maintenance landscape planting. Certain maintenance problems may be expected, such as fall leaf drop from deciduous trees. However, when plants are selected for their ultimate expected height, spread, rate of growth, and freedom from damaging pests, few problems that require constant attention will result. One of the primary causes for failure of landscape nursery stock to become established is improper planting. Special attention must be given to the planting of woody ornamental trees and shrubs if they are to be expected to grow and remain attractive for many years.

Time of Planting

Trees and shrubs are best planted while dormant. This is during the period from late fall until early spring, after the leaves have fallen from deciduous plants. At this time most plants have slowed or ceased active shoot growth and will be least harmed by the disturbance of the root system. Planting may be done at any time during the dormant season as long as the ground is not frozen. In the southern states, where the ground does not freeze to a depth greater than 6 inches (15 centimeters) and daytime temperatures may reach 50°F (10°C) or higher, the best time for planting nursery stock is in the autumn. Balled-and-burlapped plants may be planted later in the spring than may bare-root stock; however, this should not be done after new leaves are formed on deciduous plants. It is most important that the newly planted stock be watered adequately to prevent the root system from drying, no matter what season it is planted.

Woody plants transplant best when in a dormant condition (without leaves or not actively growing). Fall is the best time for planting in areas where winters are not too severe. The tree roots continue to grow during days when the ground is not frozen. Trees allowed to form roots in the soil from fall plantings will be better able to support new flower and leaf development in the spring. In areas where the ground remains frozen for several feet all winter, trees may better be planted in early spring and pruned severely to balance the reduced root system with the new shoot growth.

The size of tree to be planted is of some consideration. Large trees are often used in a landscape to provide an immediate effect.

These trees must be balled-and-burlapped to be moved successfully (Figure 16–1). Some tree species that do not recover easily from digging and disturbing of their roots must be moved in this fashion at any size (magnolias, flowering dogwood, and tuliptree are examples). In contrast, some easily transplanted tree species (such as elms, maples, and ashes) may be moved while young plants as bare-root stock. These young trees may easily become established, with less shock to growth, and eventually outgrow similar large balled trees in the landscape.

FIGURE 16–1. Balled-and-burlapped trees require careful handling when planting to prevent damage to the root ball.

Soil Preparation

One of the most common causes for failure of tree and shrub growth is poor soil preparation prior to planting. The dark color and crumbly texture of the topsoil does not indicate the amount of organic matter or nutrient status of the soil. A soil test should be made several weeks before the soil is to be prepared. The soil test report will determine the amount of fertilizer and lime (soil pH) required for mixing in the planting holes. Nitrogen-containing fertilizers are not added to the soil around newly transplanted trees or shrubs during the first growing season.

Organic matter is an important ingredient for topsoil to be used in planting trees and shrubs. The added organic matter will provide soil aeration, moisture regulation, and allow deep root penetration for the landscape plants. The soil may be mixed with as much as 5 percent organic matter. This may be done by spreading up to 3 inches

(7.6 centimeters) of peat, compost, or rotted manure over the surface of the soil. The organic matter is mixed into the planting area by deep tilling or by turning the top soil with a shovel. Soils that contain high amounts of clay and drain poorly will also be improved by the addition of sand. The soil removed from the planting hole may be amended with sand and organic matter to improve the aeration and drainage characteristics throughout the root zone of the newly planted stock.

Adjusting the Soil pH

The soil reaction (pH) has an important influence on plant growth. Some plants such as azaleas and rhododendron, grow best in an acid soil; while many others grow best at soil pH levels near neutral (pH 7.0). Most landscape plants are grown in soils having a pH value between 6.0 and 6.5. Soils that have a soil reaction more acid than this will benefit from the addition of lime. Soils can be made more alkaline (less acid) by applying hydrated lime, agricultural limestone, dolomitic limestone, bonemeal, or wood ashes. Generally, less lime is required than will be recommended by the soil test report. The recommended rate may be cut to 25 percent of the lime requirement for newly installed landscape plantings. (For example, if 10 tons per acre (11 kilograms per hectare) of lime is recommended, only 2.5 tons per acre (2.8 kilograms per hectare) will be sufficient for adequate adjustment of the soil reaction for tree and shrub plantings. This is equivalent to 23 pounds per 100 cubic feet (3.45 kilograms per cubic meter) of soil for the planting holes. Lime is easily leached from the soil in heavy rains or irrigation, so additional lime may be required after several years.

Some plants (such as azaleas, rhododendron, and laurel) require a more acid soil reaction for proper growth in the landscape. An alkaline (high pH) soil may be made more acid by incorporating aluminum sulfate, iron sulfate, or sulfur (powdered) into the topsoil. The acid-soil-loving plants will actually tolerate a wide pH range (4.0–7.0). However, optimum growth occurs between pH values of 5.0 and 5.5. The soil may be made acidic most rapidly by the incorporation of powdered sulfur into the soil. The water in the soil reacts with the sulfur to form sulfuric acid, which lowers the soil pH. The various amounts of lime, sulfur, or aluminum sulfate that may be incorporated into the topsoil to change the soil pH are listed in Table 16–1.

The lowering of the soil pH by the addition of chemicals is only temporary and may damage the plants if applied too often. The soil

TABLE 16–1: *Approximate Amounts of Chemicals Required to Alter the Soil Reaction of a Cubic Meter (35 cubic feet) of Topsoil*

pH CHANGE	SULFUR		ALUMINUM SULFATE		LIMESTONE	
	POUNDS	KILO-GRAMS	POUNDS	KILO-GRAMS	POUNDS	KILO-GRAMS
8.0 to 6.5	4.2	1.9	8.4	3.8	–	–
8.0 to 6.0	5.6	2.5	11.2	5.1	–	–
8.0 to 5.5	7.7	3.5	15.4	7.0	–	–
8.0 to 5.0	9.8	4.4	19.6	8.9	–	–
7.5 to 6.5	2.8	1.3	5.6	2.5	–	–
7.5 to 6.0	4.9	2.2	9.8	4.4	–	–
7.5 to 5.5	7.0	3.2	14.0	6.4	–	–
7.5 to 5.0	9.1	4.1	18.2	8.3	–	–
7.0 to 6.5	2.1	1.0	4.2	1.9	–	–
7.0 to 6.0	2.8	1.3	5.6	2.5	–	–
7.0 to 5.5	4.9	2.2	9.8	4.4	–	–
7.0 to 5.0	7.0	3.2	14.0	6.4	–	–
6.5 to 6.0	2.1	1.0	4.2	1.9	–	–
6.5 to 5.5	3.5	1.6	7.0	3.2	–	–
6.5 to 5.0	5.6	2.5	11.2	5.1	–	–
6.0 to 5.5	2.1	1.0	4.2	1.9	–	–
6.0 to 5.0	4.2	1.9	8.4	3.8	–	–
5.0 to 6.5	–	–	–	–	9.8	4.4
5.0 to 6.0	–	–	–	–	7.0	3.2
5.5 to 6.5	–	–	–	–	7.0	3.2
5.5 to 6.0	–	–	–	–	4.2	1.9

may be kept acid for those plants requiring such conditions by the addition of an acid-forming mulch around the bases of newly planted trees and shrubs. The mulch is spread over the ground to a depth of 2–4 inches (5–10 centimeters), especially around the area covering the plant roots. Some common acid-forming mulch materials used in landscaping are fir bark, corncobs, and pine needles. These mulch materials do not create an acid soil condition until they have become partially decomposed, which does not occur until the second year after their application.

Planting Bare-Root Stock

Bare-root plants have been dug from the nursery row without soil being left on the roots. These plant roots dry very quickly when exposed to the sun and air. Roots that are allowed to dry excessively will soon die and the plants will not survive after planting. The planting holes for the bare-root trees and shrubs should be dug before the plants are brought out into the sun.

Newly purchased (or dug) bare-root plants should be planted as soon as possible and the roots protected. These plants are usually wrapped in burlap or paper filled with damp shingletow (wood slivers) to keep the roots from drying. The package should be opened to inspect the plants for damage. At this time the shingletow may be dampened slightly to be certain it continues to protect the plant roots. When only a few shrubs or trees are to be planted, the roots may be plunged in buckets of water while the planting holes are being prepared. Whenever the bare-root stock cannot be planted immediately, they should be heeled into a shallow trench, with the roots covered with soil and in a shaded, protected location.

One of the more common causes for failure of plants following transplanting is from digging too small a hole. When the soil is highly compacted or drains poorly, the plant roots may not be capable of growing out of the original planting hole. A spacious hole should be prepared for the bare-root plants so that the roots may be arranged and spread. The hole diameter should be wider than the root span by at least 4–6 inches (10–15 centimeters) to allow proper placement of the roots in the planting hole. The sides of the hole should be roughened with a shovel to prevent formation of an impermeable soil layer. The hole should also be deep enough to allow the plant to be set slightly deeper than it was growing at the nursery. The base of the planting hole is loosened and amended topsoil added to form a thin layer for easy root penetration. This topsoil is mounded slightly to the center of the hole (Figure 16–2).

Roots are carefully inspected as each plant is placed into the individual planting holes. Broken and dead roots are trimmed with pruning shears to reduce the spread of diseases. The roots are then arranged by spreading them uniformly around the base of the hole with the center placed over the mound of topsoil at the bottom. This root placement allows development of the roots in several directions around the plant for better anchorage.

Plants having graft unions are positioned with the graft slightly below the surface of the soil. Plants that require this treatment are specially grafted flowering shade trees, shrubs, roses, and fruit trees. Some dwarf fruit trees that are grafted with an intermediate stock must be planted with only the rootstock placed below the soil surface. By placing the graft union below the soil, the suckering (shoot growth) from the root stock is reduced. This is especially a problem with hybrid tea roses, which often sucker freely.

The planting hole is filled by first adding a fine layer of the amended topsoil over the roots. This layer of topsoil is firmed with the hands to create good soil-to-root contact. Topsoil is then added until the planting hole is half filled. The soil is again firmed slightly

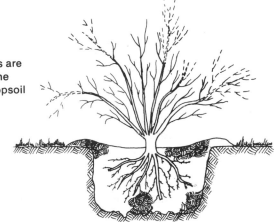

FIGURE 16–2. Bare-root plants are planted in spacious holes and the roots spread before amended topsoil is placed over them.

with the hands. A light application of phosphate fertilizer may then be applied. The newly planted stock will generally require the addition of phosphate to the amended topsoil, since it is not easily leached. High nitrogen or rapidly absorbed fertilizer formulations should be avoided or the roots may be burned from overfertilization. The hole is then filled with water to settle the soil around the roots. Once the water has drained from the soil surface, the remaining volume is filled with topsoil.

A basin is constructed from the topsoil at the top of the planting hole. The excess topsoil is formed to provide a shallow depression at the center of the hole, with a circular mound around its perimeter. This planting basin aids in catching and holding irrigation water until it can drain into the planting hole. Some landscape contractors prefer to add the fertilizer to this basin and then water it into the soil. This treatment reduces the possibility for root burning by contact with the fertilizer before root growth occurs. Once the trees or shrubs have been planted, they may require pruning and staking. These procedures will be described in detail later.

Planting Balled-and-Burlapped and Containerized Nursery Stock

Larger trees and evergreen plants are sold principally as either balled-and-burlapped or containerized nursery stock. These will be more expensive than plants from storage, but will usually become better established during the first year and will grow more rapidly later. The most suitable shade tree sizes for planting in a home landscape are those having a caliper measurement of 1–1½ inches (2.5–3.8 centimeters) taken 6 inches (15 centimeters) above the ground level. For many shade trees, this means they are 8–10 feet (2.5–3 meters) in height. Many tree and shrub species, such as hard maple, some oaks, magnolia, flowering dogwood, and rhododendron, are very difficult to transplant as bare-root stock. It is not recommended that these difficult-to-transplant species be purchased unless they are balled-and-burlapped or have been grown in containers. These plants are usually more expensive because of the added years required to achieve a satisfactory height, spread, or trunk diameter.

The principal advantage of the use of balled-and-burlapped or container nursery stock in landscape plantings is the relatively low amount of disturbance to the root systems. Evergreen plants retain their foliage throughout the year, so transpiration is continuous. These plants have a high water requirement, even during the winter months when warm days occur. The soil ball will retain moisture around the roots of these plants; therefore, less damage to the plant will occur during the time between digging and transplanting. Since little root disturbance is likely with this nursery stock, it is possible to plant even deciduous plants after they have leafed out in the spring. These trees or shrubs may be planted during the summer months, however, only if they are provided with some protection from wind and given adequate soil moisture.

The nursery stock should be planted as soon as it is received from the nursery. While planting, the extra plants should be placed in a shaded area and kept damp. The soil balls on balled plants may be dampened by spraying the burlap frequently with water from a garden hose.

Container-grown plants should be handled carefully so that the roots are not pulled loose from the soil. If the soil ball does not dislodge from the container easily, the container should be cut along the sides with a can-cutting tool and peeled away from the soil ball. The soil ball should be inspected for large masses of roots which circle around the perimeter, indicating that the plant was badly pot-bound. If these roots have grown together or appear to girdle each other, they should be clipped and thinned out to stimulate new fibrous root growth.

Balled-and-burlapped plants should be handled carefully to prevent the soil from being loosened around the roots. These plants should be carried on a large, flat piece of heavy burlap sheeting rather than by the trunk or stems. Large plants have heavy soil balls and the roots will be broken or the soil loosened when they are carried improperly. When balled-and-burlapped nursery stock is placed in a planting hole with the burlap remaining around the soil ball, the nails and twine may be removed and the burlap loosened around the trunk or stems before the topsoil is added. The burlap will deteriorate in a short time, so root penetration into the planting space will not be impaired. If plastic twine was used to tie the branches, it should be removed to prevent girdling of the stems.

Both container-grown and balled-and-burlapped nursery stock are planted at the level at which they were grown previously. The planting hole should provide ample space for the soil ball, with additional amended topsoil around the ball for new root growth. Make the planting hole about 50 percent wider than the ball of soil on the plant. The soil should be loosened at the bottom of the planting hole to a depth of 6 inches (15 centimeters) to aid in drainage of irrigation water. Plants that are to be grown in very heavy clay soils may require additional drainage. A circular hole may be cut with a posthole digger to a depth of at least 15 centimeters below the bottom of the planting hole. This may then be filled with sand or loose gravel to provide added drainage from the planting hole and the plant root zone. Topsoil is added, settled, and irrigation provided in the same manner as was described for planting bare-root stock (Figure 16–3).

Large trees may be moved if adequate equipment and sufficient care is taken. Trees up to 6 inches (15 centimeters) in trunk diameter may be moved by specially designed tree-digging machines. These machines cut a large soil ball around the roots of the trees and transport the tree and soil to the planting site. The planting hole must be dug by the machine previous to digging the tree. This operation is

FIGURE 16–3.   Balled-and-burlapped plants are planted in spacious holes having sides that have been roughened to eliminate the glazed, packed marks from the shovel. The twine is removed and the burlap loosened from the soil ball before the hole is backfilled with amended topsoil.

costly, but the larger shade trees may be successfully transplanted if special attention is given to the care of the tree afterward. The trees are best transplanted during their dormant period, late fall through late winter. Approximately one-third of the small branches should be pruned to compensate for the reduced root system. In addition, some of the lower scaffold branches may require removal to allow the machine to get close enough to the trunk to cut the soil ball.

Planting Azaleas and Rhododendron

Azaleas and rhododendron are difficult plants to grow in areas where their requirement for an acid soil limits their use. These plants can be grown, however, in soils having a pH greater than optimum by amending the soil and constant lowering of the soil reaction.

Most broadleaved evergreens grow best in a soil that is predominantly acid in reaction. Such soils will be found most commonly in the eastern and southern parts of the United States. These plants may be grown in some other parts of the country if properly planted and maintained. However, the limestone soils of some regions makes this very difficult.

A specially prepared soil mixture is necessary in limestone areas to provide a soil reaction acceptable for growth of azaleas and rhododendron. The preferred range for these plants is between pH 4.5 and 6.0, with 5.0 representing an ideal soil condition. At this pH range the necessary iron in the soil is more readily available to the plants. Limestone soils are too alkaline and will have to be modified to make them more acid. The most commonly used material for making the soil more acid is aluminum sulfate. However, this must be applied cautiously, because its continued use may cause an aluminum toxicity which is harmful to the roots of most plants. Other methods for artificially increasing soil acidity is by use of ordinary sulfur (applied at the rate not exceeding 50 milligrams for each square meter or 1 pound per 100 square feet of application). A tannic acid solution consisting of 1 part commercial tannic acid to 50 parts of water will also adequately lower the soil pH.

The acid soil condition may be maintained around conifers and other acid-loving plants by use of acid-forming soil amendments. A good natural soil mixture might consist of equal parts of woodland soil (such as topsoil from oak woods), neutral washed sand, decomposed oak leaf mold, and peat moss. The rotted wood from oak stumps or fallen logs, pine needles, or decomposed oak sawdust may also be substituted for peat.

The existing soil is removed to a depth of 2–2½ feet (60–70 centimeters) from the area where the azaleas or rhododendron are to be planted.

This hole is lined with cinders or small rocks before soil mixture is added. Lime from the surrounding soil will leach into this area unless the planting beds are raised above the existing soil grade. During drought periods the natural upward movement of ground moisture will seep lime into the planting bed. When this occurs the soil mixture will again require treatment with an acidifying chemical.

Care of Azaleas and Rhododendron

Poor drainage, sun, and wind are quite damaging to azaleas and rhododendron. High overhead shade is beneficial, so in home ground plantings a shaded north or east exposure is a desirable planting location. Plants to be placed on west or south exposures will require greater protection. This will reduce the drying effects of hot summer winds and lessen "winter burn" as well. Winter burn results from excessive evaporation and transpiration during sunny winter days when the frozen soil around the roots prevents normal replacement of the moisture lost through the leaves. Covering the soil beneath the plants with evergreen branches or a heavy application of mulch material will reduce this evaporation and subsequent damage to the plants.

Artificial irrigation is necessary at regular intervals unless rainfall is ample. Water should be applied regularly and thoroughly, but the plants should not be allowed to stand in water. These plants require regular applications of the standard acid rhododendron and azalea fertilizers in the spring and again in early autumn. If iron chlorosis should appear in the form of yellowing foliage, a foliar spray of *chelated iron* may be applied to the plants to correct the imbalance of this nutrient. The soil should also be treated with one of the acidifying chemicals at this time to maintain the proper acid pH.

In addition to their need for acid soil, these plants have a shallow root system which will require a light, porous soil mixture to admit air and drain excess water rapidly. Baled peat provides a satisfactory soil conditioner for maintaining this porosity. The faded blooms should be removed from both azaleas and rhododendron to allow the vegetative growth to form new flower buds for the following year.

Additional Care at Planting Time

Tree and shrub plantings should be kept clear of grass and weed growth for several seasons. For trees, a circular area should be cleared of grass in the lawn around the trunk at planting time. This strip should be 3–4 inches (7.5–10 centimeters) deep and may either be left to bare soil or be filled with a mulch material. Cedar or fir bark, dark crushed rock, peat or other mulch materials will help conserve soil moisture and aid in the control of weeds around the tree. The planting circle will also prevent damage to the trunk when mowing (Figure 16–4).

FIGURE 16–4. A planting circle will prevent damage to tree trunks during mowing and provides an attractive space for groundcover plantings.

Mulch materials may also be used around shrub borders. Metal or brick edging strips should be placed around the perimeter of the planting areas to make mowing more convenient and reduce the tillering or other spreading of the lawn grasses into the mulched planting area.

As a further precaution during the first few years, a tree trunk wrap should be used to protect young trees from sun scald during the winter months (Figure 16–5). This is especially important for smooth-barked trees (such as hard maples, oaks, and sweetgum). Tree wrap is applied neatly by starting at the ground level and working up the trunk to the larger branches. Each spiral turn is over-

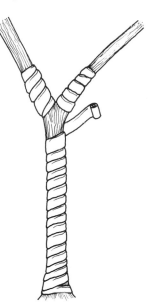

FIGURE 16–5. The trunks of young trees are wrapped with a specially treated paper to protect them from sun scald and other types of physical damage.

lapped one-half the width of the strip so that a double thickness is applied to the trunk. The final turn of tree wrap may be secured with plastic electrical tape. The tree wrap will provide a barrier to rabbit damage and boring insects. To reinforce the wrapping, bind it with a cord or string, starting at the base of the trunk and winding upward in the opposite direction of the wrap.

Rapid water losses through the foliage on hot windy summer days often results in injury to newly planted trees and shrubs. Some protection should be provided for trees located in exposed sites. This protection may consist of a cloth stretched around stakes in an L shape at the south and west exposures. Also, simply placing several cedar or redwood shingles in the ground around small shrubs or seedling trees will divert the drying winds around the plants. Frequent syringing of the tops of newly planted shrubs or providing some type of shade protection will greatly reduce excessive transpiration.

Staking and Guying Newly Planted Trees

Trees having a trunk diameter greater than 2 inches, (5 centimeters) will require additional support until new root growth can provide strong anchorage. This support may be provided by either stakes or guy wires attached to the tree. Three guy wires are attached to the tree by running each length of wire through a piece of garden hose before encircling the tree trunk at a crotch. Anchor stakes should be placed in the ground at a distance from the trunk equal to the height at the crotch.

Wooden stakes made of 2- × 2-inch strips may also be used if the guy wires will interfere with traffic around the tree. These wooden stakes should be set in the planting hole before the topsoil is added. The length of the stake should reach to at least the first strong branch on the trunk of the new tree. Nylon strapping with eyelets is available commercially for tying the stakes to the tree trunk (Figure 16–6). The stakes should remain in place for 2–3 years, or until the tree has formed strong anchor roots. The wires on both the staked and guyed supports may be kept at the proper tension by placing turnbuckles in each support wire and periodically tightened.

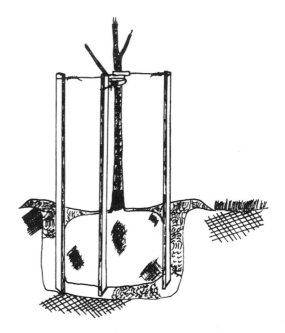

FIGURE 16–6. Newly planted trees will require some type of staking or guying until new roots provide adequate anchorage.

Mulching Materials for Landscaping

Various types of mulching materials may be used around landscape plantings to provide improved plant growth and aesthetic appeal. Some of the beneficial effects which these mulching materials have on plants include soil moisture conservation, soil temperature control, weed reduction, increased organic matter and soil structural quality, and a more pleasing appearance.

Organic mulching materials eventually rot and are mixed with the soil around the roots of the landscape plants. This mulch enriches the soil with nutrients as it provides aeration for water penetration and air movement around the plant roots. Water quickly penetrates through the mulch layer, so runoff is greatly reduced. Mulched soils are less affected by air temperature fluctuations, so the soil remains more favorable for plant growth. These materials are derived from plant parts or waste products. As they decompose, they favorably affect the waterholding capacity, nutrient availability, soil aeration and structure, which influence plant growth.

Organic mulch materials may have a drastic affect on the chemical nature of soils. A fertilizer deficiency, particularly nitrogen, may be induced by the addition of mulches having a low nitrogen content. The available nitrogen is used by the soil bacteria, which aid in the breakdown of the organic materials. A nitrogen deficiency will result unless sufficient nitrogen-containing fertilizer is applied to the mulched soil for plant growth.

The soil reaction (pH) may also be influenced by the decomposition of some organic mulching products. Some mulches (such as pine needles, sawdust, and oak leaves) create an acid soil reaction. These may be suitably used around plantings of azaleas, rhododendron, and other acid soil plants. Use of these mulches reduces or eliminates the necessity for addition of acidifying chemicals to soils around these plants. Some hardwood barks, on the other hand, have a tendency to raise the soil pH (make it more alkaline). Acid-soil-loving plants should not be mulched with materials that create an alkaline reaction. At higher soil pH levels, these plants are deprived of iron. Lack of iron causes a chlorotic symptom on the foliage with dark green veins and yellowed tissue between the veins.

Various organic mulching materials are available on the commercial market for use in landscape planting areas. In addition, certain types of organic materials may be derived from around the landscape or home. Some of these organic mulching materials are listed below.

Bark and wood chip products. Bark and wood chips are by-products of the lumber industry and may be obtained easily in most parts of the United States. These materials are the most decorative of the organic mulching materials and are ideal for reducing weed competition, conserving soil moisture, and modifying soil temperatures. These mulching products may be purchased in variously sized bags or in bulk, when located near a lumber industry center.

Wood chips are derived from the shredding of tree trimmings by utility companies and city cleaning crews. The chips contain both bark and wood material and may be used immediately after shredding, but this product provides a better mulch if composted first. When composted, the wood chips do not tie up the nitrogen in the soil needed for landscape plant growth. The wood chips may be used more successfully for mulching on steep banks, since they do not wash away with rain as readily as will bark products. The bark mulches are sold in several convenient particle sizes:

Chunks.   Bark pieces 1½–3½ inches (3.8–8.9 centimeters) in diameter. These are very attractive in the lawn and deteriorate very slowly (Figure 16–7).

Nuggets.   Bark pieces that are smaller but similar in effect to the chunks. These are graded into sizes ranging from ¾ to 1½ inches (1.9–3.8 centimeters).

Bark Fines.   These are bark particles screened to sizes ranging from ¼ to ¾ inches (0.6–1.9 centimeters) in diameter. These bark particles may be blown easily by wind.

Shredded Bark.   This category of bark material includes the very small bark pieces, together with larger bark strips. This material makes an excellent soil conditioning mulch because the smaller particle sizes are broken into organic matter rather rapidly.

FIGURE 16–7. Bark chunks provide an excellent mulch and are attractive in the landscape.

When wet, the coniferous bark products (such as redwood, cedar, and fir bark) impart a pleasant odor to the air around the landscape. All bark materials used for landscaping mulches are clean and provide a natural earth-tone color to the planting areas. In addition, the different particle sizes may be combined in various patterns to form contrasting textures and accents in the landscape.

Peat mosses. Peat and sphagnum mosses may be used to increase the organic matter content in the root zone, improve the water-holding capacity of the soil, and improve the soil structure whenever other materials are unavailable. These products are generally more expensive for use as mulching materials and do not provide an attractive ground covering around plantings. Peat moss is easily washed away by rainwater or blown by winds, so it must be raked into the planting area frequently.

Manure. Animal waste products provide an excellent soil additive and surface mulch. These materials should be aged or well rotted before being used in the landscape to avoid objectionable odors and any damage to growing plants. These mulch materials may be most economical in areas where manure is plentiful.

Composts. Composts are mixtures of plant debris or other organic materials with soil. These materials are placed in mounds consisting of alternating layers of organic matter and soil to a depth of 4–6 inches (10–15 centimeters) of each, with successive layers placed above. Generally, the compost pile is constructed in alternate layers of these ingredients to a height of 3–5 feet (90–150 centimeters), since the compost pile must be turned frequently to mix the components. The top layer contains only soil, so any escaping gaseous nutrients (mainly ammonia) may be captured and returned to the compost (Figure 16–8).

After the soil layer is added above each layer of organic matter, fertilizer is applied to stimulate the decomposition of the material into compost. Water is applied frequently to dissolve these nutrients and maintain bacterial action, which breaks down the organic matter to form a fine-textured loam soil. The layers are lightly tamped and shaped with a depression at the center to catch and hold the water and nutrients. Any waste material which provides organic matter is suitable for use in forming compost. These may include the lawn grass clippings removed from each mowing, leaves, plants removed from gardens, or even household trash from meal preparation (such

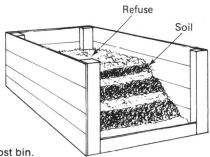

FIGURE 16–8. Diagram of a compost bin.

as eggshells, meat scraps, or other food items). A properly managed compost pile will form enough heat during the decomposition of the organic materials to kill many insect eggs, disease organisms, and weed seeds. The compost forms an excellent material for use as a soil amendment or as a landscape mulch. Table 16–2 lists the amount of each fertilizer to be applied to a single layer of organic matter and soil mixture.

TABLE 16–2   *Fertilizer Required to Decompose*
Each 6-Inch Layer of Compost

FERTILIZER	AMOUNT PER 100 SQUARE FEET
Potash (0–0–50)	2 pounds (0.9 kg)
Superphosphate (0–20–0)	2.5 pounds (1.1 kg)
Limestone	4.2 pounds (1.9 kg)
Ammonium nitrate (33.5–0–0)	3 pounds (1.4 kg)

Inorganic Mulches

Inorganic mulch materials are not formed from living organisms, so they will not add to the organic matter content of the soils. They do, however, reduce soil temperature fluctuations, aid in the control of soil erosion and weed growth, and help in the preservation of the soil structure.

Black polyethylene film.   Black polyethylene film is one of the most commonly used inorganic mulching materials for landscape plantings. A layer of the plastic film over the soil prevents weed growth by blocking out all light. Clear plastic is not effective when used alone for the prevention of weed growth, since the light that penetrates to the soil allows the weeds to remain alive. Excessive soil moisture may accumulate under plastic films, causing serious root-rotting diseases to develop. Generally, these conditions do not occur unless the soil is extremely heavy or compacted. Plants mulched with plastic will require less frequent irrigation, since water evaporation from the soil surface is minimal.

Solid and crushed rock.   Rock is often used in conjunction with black plastic film to provide an attractive and functional mulch around landscape plantings. Most plastic films are destroyed rapidly by the ultraviolet wavelengths from sunlight. These damaging rays are blocked by a layer of material over the surface of the film. Rock materials ranging in size from solid river stones to pea gravel may be used to add a natural appearance to the landscape (Figure 16–9). Organic mulches may be placed above black plastic films; however, the soil-amending qualities of these materials will be lost. One of the major advantages to the use of rock materials over black plastic films as a mulch around plantings is the reduced requirement for replenishing these. Since rock does not disintegrate into finer material, it will not be easily lost from the landscape. Fine sand or light volcanic rock, however, will be easily washed away from the planting area by heavy rainfall.

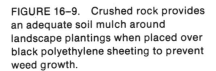

FIGURE 16–9.   Crushed rock provides an adequate soil mulch around landscape plantings when placed over black polyethylene sheeting to prevent weed growth.

FERTILIZING ESTABLISHED TREES AND SHRUB PLANTINGS

Tree and shrub plantings will benefit from annual applications of fertilizer. Fertilizer is best applied to the landscape plantings in late winter and early spring to enhance new root and shoot growth. A fall application is also beneficial provided that the trees have become dormant. Fertilizer applications during the hot summer months may stimulate active growth that will not survive dry weather.

Established trees are fertilized by drilling holes in the soil to a depth of 1–2 feet (30–60 centimeters) with a soil auger or brace-and-bit. Specially designed bits may be obtained to reach these depths. The holes should be placed in close proximity to the feeder roots for maximum benefit to the trees. These holes are located within the area around the tree extending from a distance of 1 foot from the trunk to approximately 1½ feet (45 centimeters) beyond the drip line of the branches (the spread of overhanging limbs). The holes are spaced as close as 12 inches apart nearest the trunk to 1 yard distances outside the drip line on large trees.

Smaller trees and shrub plantings may be fertilized by broadcasting the material over the soil around the planting circle and thoroughly working it in to a depth of about 6 inches. Once the fertilizer is applied to the holes or broadcast over the soil, the ground should be thoroughly soaked. This will place the fertilizer into solution and carry it to the root zone of the plant.

The proportion of the primary nutrients in a fertilizer mixture must be matched to the plants being treated. Certain specialty fertilizers are manufactured in grades usually suitable for specific types of plants. Some formulations, for example, are designed for use on coniferous evergreen plants, while others are to be used on azaleas or other acid soil plants. Most tree and shrub species may be fertilized with those products designed for use on lawn grasses or agriculture fertilizers. These will have a higher proportion of nitrogen than potassium or phosphate in their formulation. They may be applied in any form that is convenient (such as granular, powdered, soluble, or slow-release capsules).

Granular and powdered forms of fertilizer are most convenient for application using a drop or cyclone-type spreader. These materials must then be thoroughly worked into the upper surface of the soil and made soluble through heavy irrigation. Soluble fertilizers are mixed with water and sprayed onto the soil by use of a hose-end proportioner. The soluble fertilizers are also absorbed through the foliage of the plants for rapid release of the nutrients. A more popular method for achieving long-range benefits from any fertilizer application to landscape plantings is by use of slow-release products. These may be either applied to the fertilizer holes drilled around the trees or broadcast onto the soil surface and irrigated frequently. The slow-release fertilizer capsules break down at different rates, releasing small amounts of nutrients for a greater length of time. Another type of slow-release fertilizer product consists of a compressed peg of nutrient material which is simply driven below the soil surface around the plants. As water enters the soil, the fertilizer is slowly placed into solution and carried to the root zone.

Additional nutrients are required for proper plant growth of landscape specimens. Trace elements (micronutrients) are essential to the growth of plants but are needed in only small amounts. The most commonly required trace elements are *iron, manganese, zinc, copper, molybdenum, sodium, boron, sulfur,* and *chlorine*. These elements should not be applied routinely with the agriculture fertilizers, for an overabundance of these minerals in the soil may be damaging to the growth of plants. A light application of fertilizer containing trace elements should be made once each 3–5 years or whenever the soil test report suggests the need for these elements.

When a fertilizer formulated for use on lawn grasses is to be used for the fertilization of other landscape plants, a careful study of the ingredients is required. Many commercial lawn fertilizers contain, in addition to the basic nutrients, certain insecticides or herbicides. Some fertilizer–herbicide combinations injure or kill trees and shrubs when applied to the grass under them. When lawn grasses are being fertilized, the movement of the spreader back and forth around the landscape plants often results in applying much more than the recommended rate of material, and damage results to the plants.

Just as with newly planted ornamental specimens, established trees and shrubs require proper soil acidity, soil structure, and soil moisture. Soil acidity may be adjusted by applying either agriculture lime or an acidifying material to raise or lower the soil pH. Soil structure may be improved or maintained around landscape plantings by working the soil properly or by incorporating organic matter into the soil. Soil moisture may be improved by increasing drainage, irrigating, and the application of mulches.

PRUNING LANDSCAPE PLANTS

Trees and shrubs in the landscape may require periodic pruning or shaping to enhance their beauty and to develop healthy, sturdy growth. The art of pruning plants correctly may be learned easily, yet is an extremely important part of landscape maintenance. One of the most important rules in pruning is: If you do not have a definite reason for pruning a plant or tree, put your tools away without using them. Landscape plants will grow into beautiful specimens having natural shapes if left alone to achieve their normal sizes. When plants are selected for their mature height and spread and located with these dimensions in mind, there is no good reason for them to be sheared and butchered repeatedly during their life in the landscape.

Woody ornamental plants should not be indiscriminantly sheared (unless a highly formal appearance is desired) when pruning. A knowledge of pruning is best gained through the understanding of basic plant growth, experience, and close observation of a professional pruner. The new shoots that will grow after a stem has been cut are derived from buds located along the stem. These located at the base of each leaf stalk in the axil where it joins the stem. These buds are called *axillary buds*. When a stem is cut, the bud or buds nearest the cut end will begin to grow into new shoots. The height and direction of this new shoot growth can be directed somewhat by observing the position of these buds and the direction in which they are growing. The new stems produced by these shoots generally will grow in the same direction the bud was pointing before the shoot emerges from it (Figure 16–10).

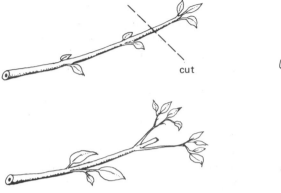

(a)   Plants having opposite buds.   (b)   Plants having alternate buds.

FIGURE 16–10.   Pruning and training of shrubs or trees should be done by carefully observing the direction of subsequent growth after a cut is made.

Severe pruning of a tree or shrub will cause the growth of *adventitious shoots.* Pruning cuts made into larger limbs where no axillary buds may be present stimulates the formation of these buds below the bark surface. Adventitious buds are not observable at the time the cut is made in the branch. The resulting stems are very fast-growing, but weaker than normal shoots. If you understand the fundamentals of plant growth, you can organize the pruning operation to regulate the size, form, and extent of new growth on any ornamental plant.

The purpose of pruning a plant is to remove unwanted growth. The following reasons for pruning trees and shrubs are listed in order of importance:

1. To remove broken, dead, or winter-killed branches (Figure 16–11).
2. To compensate the top and root growth at the time of planting. The loss of roots at digging must be balanced by removal of some of the branches.
3. To clean up a plant that has been seriously injured by insects or diseases.
4. To repair branches that have been damaged by wind or snow.

(a)   The hydrangea stems die to the ground each winter in many parts of the United States.

(b)   The dead stems and blooms may be cut at the soil line.

(c)   The new shoots arise in the spring from below the ground to produce a new plant.

FIGURE 16–11.   Pruning plants that are killed to the ground each winter.

5. To renovate old plants which have sparse blooms or foliage. This is done to rejuvenate new growth, which is more compact and attractive.

6. To maintain a desired size or form of the plant.

7. To maintain a high degree of flower and fruit production.

Pruning Tools for Light Maintenance Jobs

The most basic tools for maintaining landscape plants are high-quality pruning shears. These may be found in sizes ranging from 7 to 9 inches (18–23 centimeters) in length. Pruning shears for commercial work should be of heavy construction and cut with a scissors action. There should be two cutting blades rather than a single blade which cuts against a flat surface. The latter style of pruning shears will crush the stems rather than make a clean cut. The crushed end of the cut will often be killed, leaving a wound that is subject to disease and insect attack. Large-diameter stems may be cut with heavy lopping shears. These cutting shears have long handles which provide adequate leverage to cut stems up to 2 inches (5 centimeters) in thickness.

Smaller tree limbs may be pruned easiest with pruning saws. These are available in various sizes and styles. The most commonly used pruning saw is slightly arched and the teeth are wide-set to prevent binding. Pruning saws cut on the backward stroke, unlike a carpenter's saw. These may be found in lengths ranging from 1 to 3 feet (30–91 centimeters) and may be mounted on short handles or on long poles for large tree work. Large tree limbs are most often trimmed by use of power chain saws. Removal of large tree limbs is best done by a professional arborist.

Pruning Shrubs and Trees After Planting

Woody plants that have been dug "bare-root" have lost much of the root system. To compensate for this loss of feeder roots, some of the top growth should also be removed. Generally, about one-third of the shoots are pruned off. This is done selectively removing the larger shoots while maintaining the shape of the plant. It is not necessary to shear the top of the plant unless a more dense growth habit is desired near the ground.

Shrubs may be pruned to a height of 1–1½ feet (30–45 centimeters) from the ground to stimulate a low branching habit. This is especially important when a dense shrub border is desired. Small bare-root trees having few lateral branches are generally topped at a convenient height to compensate for the reduced root system. Trees and shrubs transplanted from containers or balled with soil will require much less pruning following planting.

At the time the plants are being pruned to balance root and top growth, they should be inspected for damaged or crossing branches. Any stem that is already dead or has been severely damaged should be removed. Stems that rub each other as the plant moves in the wind are called "crossed" stems (Figure 16–12). One of the branches should be selected for removal to allow the other to grow undisturbed. Occasionally a plant that normally grows upright in direction will possess two terminal growing points or leaders. The most centrally located and strongest shoot is kept, while the other leader is pruned back several inches. This pruning method will prevent the tree or shrub from forming a double leader, two tops to the plant. Often these branches form a V-shaped crotch at their juncture which may be easily split when the plant becomes older (Figure 16–13).

Pruning Established Shrubs

Shrubs in a landscape are pruned in various manners and at different times of the year to achieve a specific shape or to enhance blooming and fruit production. When shrubs are given a formal appearance by shearing them into unnatural shapes, more detailed and frequent pruning is required. Modern landscape plantings are allowed to grow into natural shapes which require only an occasional pruning to maintain their beauty. Shrubs are pruned to maintain their size or shape, to influence flowering, or to remove dead and injured branches.

When shrubs have been properly selected for their height and spread in a landscape, pruning to restrict their growth should not be necessary. However, a common practice of planting too many shrubs next the foundation of a house leads to the requirement for excessive pruning to keep the plants in scale with the landscape. Most shrubs, when properly selected for their location, will maintain their beauty and form without the need for much pruning. This results in a more attractive landscape than when the shrubs are given a tightly clipped appearance by frequent shearing. When a shrub is properly pruned, it should not be apparent that any cutting has been done. The exception to this, however, is when a formal landscape is desired. In this situation each shrub or hedge is sheared heavily and frequently

(a)   Original clump birch with
      crossed branches.

(b)   The crossed branches are
      removed and the smallest limbs
      removed to a suitable height on
      the trunk.

(c)   An extra plant arising from a root
      sucker is removed.

(d)   The pruned clump birch.

FIGURE 16–12.   Pruning newly planted trees and shrubs.

FIGURE 16–13.   Pruning a plant having a double leader. This upright holly has developed two terminal shoots. The most centrally located shoot is saved, while the other is cut back to half its length.

to maintain the desired shapes. A formal landscape will require much more maintenance than will an informal planting.

The size of a shrub plant may be reduced by the careful thinning out of the wood in such a manner that the natural shape is maintained. Rather than cutting all branches back to the same height, the longer shoots are cut back to side branches. The height of cut on each branch is varied to avoid a shorn appearance. This type of pruning opens the top growth to light and air movement so that the branches are less crowded and will not die out from heavy shading. The height and spread of the plant may be gradually reduced in this manner by cutting back into older wood or by the removal of the largest stems near the ground. This will force the growth of younger and more sturdy branches near the ground.

The most attractive feature of many ornamental shrubs are their flowers and fruit. It is important to know when the flower buds are produced on these plants so that the blooms may be preserved during pruning. Shrubs that bloom during early through midspring form flower buds on vigorous young wood which has developed the past summer and autumn. If these shrubs are pruned during the dormant season of autumn or winter, most of the blooms will also be removed. Spring-flowering shrubs are generally pruned as soon as the blooms fade and fall from the plant. This allows ample time for new shoots to grow and form new flower buds before winter (Figure 6–14).

Shrubs that are most attractive when they possess a mass of colorful berries or fruit should be pruned during the dormant season. These berries are generally produced late in the growth season and may remain into the winter. The branches that have been pruned will produce new vegetative shoots in the spring. These shoots will form flower buds during this season for later fruit production. Many

(a) Forsythia at the height of its blooming period.

(b) The blooms are nearly gone two weeks later.

(c) The height of the shrub is reduced at this time by removing only the tallest and largest stems from the center of the plant.

FIGURE 16–14.   Spring-flowering shrubs are pruned after the flowers have fallen.

shrubs produce attractive blooms, but the seed pods formed after the blooms fall are unattractive and utilize considerable amounts of the food reserve of the plant. These seed pods should be removed as they form, to allow either new vegetative shoot growth or an extended blooming period for the plant. A listing of suggested pruning seasons is given in Table 16–3.

TABLE 16–3: *When to Prune Some Ornamental Trees and Shrubs*

PRUNE IN WINTER OR EARLY SPRING (DORMANT SEASON)	
SCIENTIFIC NAME	COMMON NAME
Abelia grandifolia	Glossy Abelia
Abies spp.	Fir
Abutilon hybridum	Flowering Maple
Acer spp.	Ornamental Maples
Albezia julibrissin	Mimosa (Silk Tree)
Amorpha fruiticosa	Indigobush Amorpha
Arbutus unedo	Strawberry Tree
Arctostaphylos stanfordiana	Manzanita
Aronia arbutifolia	Chokeberry
Arundinaria spp.	Bamboo
Arundo donax	Giant Reed
Asimina triloba	Pawpaw
Aucuba japonica	Japanese Aucuba
Berberis spp.	Barberry
Bougainvillea spectabilis	Bougainvillea
Broussonetia papyrifera	Paper Mulberry
Buxus spp.	Boxwood
Callicarpa americana	French Mulberry
Callicarpa japonica	Japanese Beautybush
Caryopteris × *clandonensis*	Blue Spirea
Catalpa speciosa	Catalpa
Ceanothus ovatus	Inland Ceanothus
Cephalanthus occidentalis	Button Bush
Citrus spp.	Orange, Grapefruit, etc.
Colutea arborescens	Bladder-senna
Cornus spp.	Dogwoods
Cotoneaster spp.	Cotoneaster
Crataegus spp.	Hawthorn
Euonymus spp.	Euonymus
Euphorbia pulcherrima	Poinsettia
Ficus spp.	Fig
Gleditsia tricoanthos	Thornless Honeylocust
Gymnocladus dioicus	Kentucky Coffee Tree
Hamamelis virginiana	Witch Hazel
Hibiscus syriacus	Rose-of-Sharon
Hydrangea spp.	Hydrangea
Hypericum spp.	St. John's Wort
Ilex spp.	Holly
Kerria japonica	Japanese Kerria
Lagerstroemia indica	Crapemyrtle

Ligustrum spp.	Privit
Lindera benzoin	Spicebush
Lonicera spp.	Berried Honeysuckle
Malus spp.	Flowering Crabapples
Morus spp.	Mulberry
Myrica pennsylvanica	Bayberry
Osmanthus heterophyllus	Holly Olive
Paulownia tomentosa	Empress Tree
Pernettya mucronata	Chilean Pernettya
Photinia villosa	Oriental Photinia
Poncerus trifoliata	Hardy Orange
Punica granatum 'nana'	Dwarf Pomegranite
Pyracantha coccinea	Scarlet Firethorn
Pyrus calleryana 'Bradford'	Bradford Callery Pear
Rhamnus cathartica	Buckthorn
Rhodotypos scandens	Jetbead
Rhus spp.	Sumac
Rosa spp.	Garden Roses
Sambucus canadensis	American Elder
Sophora japonica	Japanese Pagoda Tree
Sorbus aucuparia	European Mountainash
Spiraea × bumalda 'Anthony Waterer'	Anthony Waterer Spirea
Stewartia ovata	Mountain Stewartia
Styrax japonica	Japanese Snowbell
Symphoricarpos albus	Common Snowberry
Taxus spp.	Yew
Thuja spp.	Arborvitae
Tsuga spp.	Hemlock
Viburnum spp.	Berried Viburnums
Vitex agnus-castus	Chaste Tree
Wisteria sinensii	Chinese Wisteria
Zanthoxylum americanum	Prickly Ash

PRUNE AFTER FLOWERING OR DURING GROWING SEASON

SCIENTIFIC NAME	COMMON NAME
Abeliophyllum distichum	Korean Abelia-leaf
Acacia farnesiana	Opopanax
Amelanchier canadensis	Shadblow Serviceberry
Betula pendula	European White Birch
Buddleia spp.	Butterflybush
Calycanthus floridus	Carolina Allspice
Camellia japonica	Camellia
Caragana arborescens	Siberian Peashrub
Carissa grandiflora	Natal Plum
Ceanothus americanus	New Jersey Tea
Cephalotaxus harringtonia	Plum Yew
Cercis canadensis	Redbud
Chaenomeles japonica	Japanese Flowering Quince
Chionanthus retusus	Chinese Fringetree
Chiosya ternata	Mexican Orange
Clethra acuminata	Cinnamon Clethra
Codiaeum variegatum	Croton

Cornus florida	Flowering Dogwood
Cotinus coggygria	Smoketree
Cytisus scoparius	Scotch Broom
Deutzia gracilis	Slender Deutzia
Eleagnus umbellatus	Autumn Olive
Exochorda racemosa	Pearl Bush
Forsythia spp.	Forsythia
Grevillea robusta	Silk Oak
Jasminum nudiflorum	Winter Jasmine
Juniperus spp.	Junipers (Cedars)
Kalmia latifolia	Mountain Laurel
Kerria japonica	Japanese Kerria
Koelreutaria paniculata	Goldenrain Tree
Kolkwitzia amabilis	Beautybush
Laburnum alpinum	Goldenchain Tree
Lonicera fragrantissma	Bush Honeysuckle
Magnolia spp.	Magnolias
Mahonia aquifolium	Oregon Grape Holly
Mahonia bealei	Leatherleaf Mahonia
Melaleuca linariifolia	Snow-in-Summer
Nandina domestica	Nandina
Phellodendron amurense	Amur Corktree
Philadelphus spp.	Mockorange
Physocarpus opulifolius	Ninebark
Picea spp.	Spruce
Pieris japonica	Japanese Andromeda
Pinus spp.	Pine
Pittosporum tobira	Japanese Pittosporum
Potentilla fruiticosa	Cinquefoil
Prunus armeniaca	Flowering Apricot
Prunus cerasifera	Purple-leaf Plum
Prunus glandulosa	Dwarf Flowering Almond
Prunus persica	Flowering Peach
Prunus serrulata	Kwansan Cherry
Prunus triloba	Flowering Plum
Ptelea trifoliata	Hophornbeam
Rhododendron spp.	Azalea and Rhododendron
Salix discolor	Pussy Willow
Sorbaria sorbifolia	Ural False Spirea
Spiraea spp.	Spring-flowering Spirea
Syringa spp.	Lilac
Tamarix spp.	Spring-flowering Tamarix
Viburnum carlesi	Korean Spice Viburnum
Viburnum lantana	Wayfaring Tree
Weigela spp.	Weigela

Pruning shrubs to form a formal hedge requires considerable maintenance, knowledge, and skill. The most important consideration in forming a dense hedge is to maintain adequate foliage over the entire perimeter of the shrubbery. Light should be able to reach all sides of the shrub border, so pruning must be done to shape the plants with the top narrower than the bottom (Figure 16–15). Formal hedges sheared to form a box, with straight sides, often lose foliage on the lower branches. Rapidly growing shrub species (such as privit) may require pruning at least three times a year to maintain their size and shape. Other species which grow slowly will require only a single pruning each year.

When shrubs are planted for use as a dense hedge row, the plants are pruned at a height of 6–8 inches (15–20 centimeters) from the soil. This forces new shoot growth near the ground. The following season another 6–8 inches is removed from the tips of these branches. Continued heavy pruning in this manner until the shrub border has achieved a suitable size will allow a dense, low-branched growth on the plants. Successive pruning is then done merely to maintain the desired size and shape of the hedge.

One of the most important reasons for pruning shrubs is to restore vigor and health. Shrubs that have been heavily damaged by winter conditions may require severe pruning to remove dead and diseased wood. In some cases the only way this can be done is by pruning a shrub back to the ground. New basal shoots should arise from the roots to form a new plant, provided that the roots were not killed also. Broken branches should be removed to reduce insect and disease problems. Shrubs that have grown too thick in their branching should be thinned occasionally. Thinning is done by selectively removing older, larger branches to allow better light and air penetration to the inner foliage.

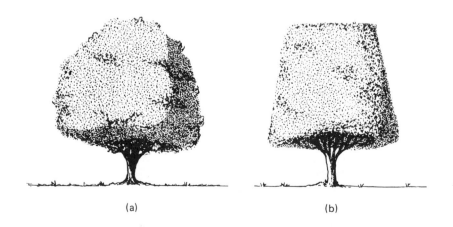

(a)   (b)

FIGURE 16–15.   Plants grown in a hedge row may be (a) left alone to form their own natural shape or (b) sheared so that the top is narrower than the base to provide sunlight to all sides of the plant.

Pruning Evergreen Shrubs

Most narrow-leaved evergreens may be improved in appearance by an occasional pruning. Shrub forms of these evergreen plants will appear most attractive if they are selectively thinned out rather than sheared into unnatural shapes. The growing tips of each branch may be pruned lightly to force the development of lateral branches. This new growth will give the shrub a fuller appearance. Pruning cuts may be safely made on older wood deep in the center of the plant when rejuvenation is required. However, active buds are rarely located on this older wood, so the new shoots must arise from adventitious buds. Regrowth from this severe pruning will require a much longer period of time than will light pruning on the young shoot tips.

Narrow-leaved evergreens having coarser foliage (such as pines) should be pruned in the spring when new succulent growth is present. This new growth is very soft and lighter in color than the mature foliage. This stage of growth is called the "candle" stage because the new shoots are pencil thin and the closely spaced needles resemble a candle. The candles may be pinched at any point to force lateral branching and a fuller plant will result (Figure 16–16). This is most often done on specimen plants, such as mugho pines, to create more compact growth. These narrow-leaved evergreen plants should also not be pruned into older wood below the candles unless severe pruning is absolutely necessary. Few buds are located on this bare wood and the plants may suffer irreparable damage from such pruning.

Narrow-leaved evergreen plants may be observed with yellowing foliage or dropping needles behind the new growth in the fall or spring of the year. This is a natural pruning process of these plants as they shed the old foliage. The new foliage will soon cover the bare wood left after these needles are shed. These branches should not be

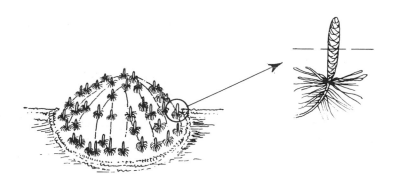

FIGURE 16–16.   Needle-bearing conifers are pruned in the spring when the new "candles" (vegetative shoots) are present.

pruned from fear of disease or insect attack. Extensive dropping of needles in hot weather, however, may indicate a low moisture content in the root zone of the plants.

Broad leaved evergreens will require much less pruning than the narrow leaved plants. Plants of this type (such as azalea, rhododendron, holly, and camellias) may require heavy pruning when they become old and weak-stemmed. They may be cut back to 6–8 inches (15–20 centimeters) from the ground during the dormant season. This will eliminate flowering for one season, but the plant will recover its stature rather rapidly. The old blooms should be removed from flowering shrubs that do not form attractive fruit by early July. This will allow the plants to form new flower buds before the next growing season.

PRUNING ORNAMENTAL AND SHADE TREES

Trees are pruned principally for the purpose of developing strong branches, to preserve their health, and to remove damaged limbs. Generally, no large tree will require heavy pruning by topping as an attempt at rejuvenating its growth. The pruning of large trees is done for the same reasons a shrub is thinned. Large trees should be pruned only by skilled professional arborists. The large limbs are heavy and may fall on property or other workers if not securely tied. Workers involved in the removal of large limbs or trees must exercise extreme caution at all times and be supervised by skilled tree experts.

Pruning Young Trees

When young trees are planted as either bare-root or balled-and-burlapped plants, the only pruning required is to remove an occasional broken stem or crossed branch. The lower branches are left on the trunk during the first season of growth. As the trunk grows in diameter, a few of the lowest branches may be removed by cutting them flush with the trunk. Such small cuts made with a sharp pruning shears will heal quickly and will not require further treatment (Figure 16–17). The presence of these lateral branches aid in the attainment of strength by increasing the trunk diameter. They should be removed only a few at a time from the lower portion of the trunk during the dormant season. Trees that have all the lateral branches removed at the time of planting will require staking or other support for a longer period.

(a) The wound growth around a cut made several months earlier.

(b) The healing process is almost complete one year after these branches were pruned.

FIGURE 16–17. Small pruning cuts made on trees or shrubs will heal quickly without other special treatment.

Further pruning of a young ornamental or shade tree is necessary only to repair damaged limbs or to aid in directing growth. A tree left unpruned will usually form a characteristic shape which is normal for that plant. Pruning will generally not influence this shape unless the tree is repeatedly pruned heavily. Some tree species do, however, produce large numbers of succulent sprouts (water sprouts) near the base of the tree. These should be removed as they form to maintain the overall shape of the tree. Other ornamental trees produce large numbers of small branches which create a dense foliage area in the crown of the tree. These trees may be pruned by selectively thinning the excess branches to allow better air movement and light penetration into the center of the tree. Flowering trees, such as crabapples, may require a light annual thinning to aid in opening the crown.

During the first few years of growth in the landscape, the young trees should be trained to develop a strong framework of branches.

The trunk should be allowed to expand in circumference by gradually removing the lower branches. As the tree gains height, certain branches may be selected to serve as the primary scaffold limbs. The scaffold branches should have wide crotch angles (greater than a 40 ° angle) at the point where they join the trunk. Some trees (especially ornamental flowering and fruit trees) produce naturally narrow crotch angles. These narrow crotches split easily under the weight of snow or ice and may fail in heavy wind storms. As the tree develops, these strongly crotched scaffold branches should be selected uniformly around the trunk at a suitable branching height and all other side branches eventually removed.

Treating Wounds Following Pruning

Following the removal of branches over 1 inch, 2.5 centimeters. in diameter, a protective wound dressing is generally applied to the cut surface. The wound dressing is used for several reasons: (1) to aid the healing process of the tissue, (2) to prevent the invasion of disease organisms or insects, and (3) to reduce water loss from the cut tissue. Larger cuts require several months to years to fully heal, so the wound dressing is applied to protect the plant during this period.

Several types of materials may be used as a wound dressing. Various types of asphalt or latex-base wound-dressing materials are available commercially. Fungicides are often added to these materials to inhibit wood-rotting organisms. When these materials are applied to the wound, the wood should be generously coated at the center, but the bark and cambium should not be treated. The wound-dressing materials will inhibit cambial growth if applied to this region. The wound dressings will crack or otherwise deteriorate in time, so annual reapplications may be necessary until healing is nearly completed.

Cutting Large Limbs

Whenever a branch or limb is removed from a tree, the cut should be made as cleanly and smoothly as possible. Normally, large branches are cut first to a stub, then the stub is removed flush against the trunk of the tree. The purpose for making the cut in two operations is to prevent the heavy limb from stripping the bark on the trunk as it falls.

Stub cutting requires three separate saw cuts (Figure 16–18). The first cut is made from the lower surface of the branch at a distance of

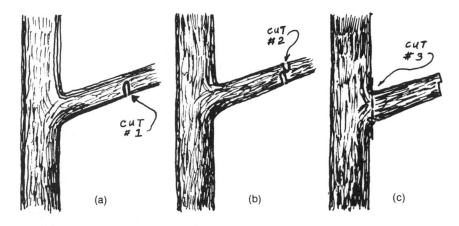

FIGURE 16-18. Larger tree limbs are removed by making three separate cutting operations: (a) The first cut is made from the bottom of the limb to a depth of nearly half the diameter of the limb. (b) The second cut is made nearly above the first and in front of it to a depth that causes the limb to fall. (c) A final cut is made flush with the main trunk to remove the stub remaining after the limb is removed.

1–1½ feet (30–45 centimeters) from the main trunk. The depth of this cut should be approximately 1/3 to 1/2 the diameter of the limb at this point, or until the weight of the limb causes a slight pinching on the saw. The second cut is made above the first, but slightly farther from the trunk. The limb will fall when the cut is sufficiently deep. Any tearing of the bark or splitting of wood will occur on the stub rather than on the tree trunk. The final cut is made along the junction of the stub and tree trunk. Once the stub is removed, the cut surface should be cleaned up with a knife or chisel to remove any splinters or peeled bark. The cut surface may then be treated with a suitable wound compound. When short stubs remain on the tree or branch following pruning, the stub will usually die back to the nearest branch or bud. These stubs, no matter how small, should be removed to prevent disease organisms from infecting the tree. Making a cut at a crotch is called *drop pruning*.

Repairing Damaged Trees

Larger shade trees occasionally require special care to repair damage caused by high winds, lightning, or severe insect attacks. The large broken limbs must be removed by a professional tree service equipped for handling them. When the wood has already begun to decay at a wound, the area must receive further treatment. The decayed tissue must be thoroughly cleaned and removed. The cavity left after the rotted wood is removed is then treated with a fungicide, which contains copper to prevent further decay. Small cavities may be dressed with a wound compound. Larger cavities, however, will weaken the trunk, so more elaborate treatment is required. The cavity may be filled with concrete or braced with cables before the concrete is added. This will form strong support for the heavy limbs once the concrete has dried. The bark and cambium will eventually grow around the treated cavity to support the tree.

Trees may be damaged from mechanical injury to the bark on the lower portion of the trunk. A frequent cause of injury to younger trees is the scraping of the trunk while mowing. Often the bark is torn in several locations around the trunk in a single season. Such severe damage to the trunk may easily kill the tree. Several methods may be used to avoid such damage to trees in the lawn. A tree planting circle will keep the grass far enough away from the tree trunk so mowing will not have to be so close. Tree wrap will prevent minor cuts to the bark as the lawn is being mowed. However, even when care is taken to prevent such damage, some accidents are likely to occur.

When the bark has been split from the trunk by mechanical injury it may be repaired by either cleaning the wound and dressing it, as is done following pruning, or the bark may be replaced and "bandaged." If the bark piece can be found, it should be replaced immediately after it has been removed. If done correctly and before the moisture leaves the tissue, this bark piece may grow back on the tree. The bark piece may either be nailed in place or covered with a strip of tree wrap to hold it in position. Large bark pieces may dry out before healing can take place. These should be nailed in place and covered with damp sphagnum moss. The moss is covered with a strip cut from a plastic bag before the strip of tree wrap is placed over it. This bandage will hold moisture to prevent the tissue from drying during the healing process.

Large surface cuts to the bark should be cleaned with a knife or chisel to form a smooth-walled wound and to remove splinters. These wounds are then covered with wound dressing. The larger wounds should be inspected periodically to make certain that the wound compound is intact. If a break in the seal occurs, serious decay may result.

Trees may be pruned at any time of the year without causing serious damage to the growth. It is often easiest to prune a tree when it is in foliage, since the dead or damaged limbs are most evident at this time. Spring is the most desirable time to prune most trees, because the tissue will heal most quickly following pruning. Trees that "bleed" excessively when pruned (such as birch, willow, or maple) are best pruned in the summer. During this period, these trees do not bleed as profusely and are less likely to become infected by disease.

PRUNING AND CARE OF ROSES

Rose plants require more general maintenance than most other flowering shrubs, but their spectacular beauty surpasses the blooms of other landscape plants. The care of roses begins with proper soil preparation and planting. Routine pruning and removal of blooms is also necessary to maintain strong blooming canes. Disease and insect control is a continuous maintenance duty if the plants are expected to remain healthy and blooming heavily for many years.

Planting Roses

Roses require a well-drained location that is free of standing water. The best soil for roses is high in organic matter and nutrients. The methods outlined earlier for planting bare-root shrubs should be followed for establishing roses. Rose plantings will benefit especially from the application of an organic mulch around their roots. In areas where cold weather may damage roses in winter, an additional layer of straw mulch or even soil may be piled over the crown of the plant. This mulch will insulate the tissue from cold drying winds which often kill exposed rose canes. At the time of planting, the large canes should be pruned to a height of 3–4 inches (7–10 centimeters) above the soil line. This will force the flowering shoots to arise near the ground for a fuller, more compact plant.

Pruning Established Roses

Roses should be pruned in late winter or early spring before new growth starts. At this time all dead or damaged tissue may be removed. This dead growth is generally easily distinguished by its brown color in contrast with the healthy green bark. Pruning is done differently on each of the major types of roses grown in the landscape.

Hybrid tea roses. Hybrid tea roses are pruned to a height of 1 foot (30 centimeters) in the early spring to develop a low, compact plant. These will appear less leggy or open than nonpruned plants. The best blooms result from new growth that arises near the base of the plant. *Grandiflora* roses are pruned in the same manner as hybrid teas (Figure 16–19).

Floribunda roses.   Floribunda roses are grown most often as hedge rows for their nearly continuous blooming. These roses require very little annual pruning, except when rejuvenation of the plants is required. The larger canes may be pruned at the base of the plant to force new shoot growth lower on the stems. These new shoots will then produce more profuse blooming than did the old canes.

Climbing roses.   Climbing roses may be of any type, but differ by producing long canes which are adapted for growth along walls or trellises. These roses will require very little pruning, except to remove dead branches or to alter the direction of their growth. The older canes should be removed periodically to allow new strong shoot growth at the base of the plant. These renewal canes will flower and branch more freely than will older canes (Figure 16–20).

FIGURE 16–19. Hybrid tea roses are pruned heavily in early spring to promote branching near the base of the canes.

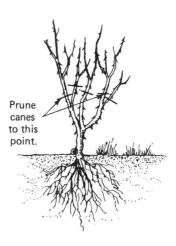

Prune canes to this point.

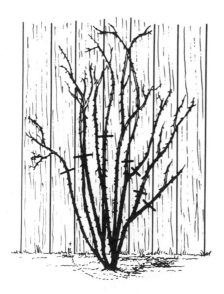

FIGURE 16–20. Climbing roses are pruned in early spring by cutting the larger canes back to one third or one half their length. This causes more branching and a fuller appearance.

Tree roses.   Tree roses are specialty plants produced by budding a profusely flowering cultivar on a hardy rootstock at a height of 2–4 feet (60–122 centimeters). These roses require some protection during the winter months from winds or chilling temperatures (Figure 16–21). In warmer climates the tops may be covered with burlap sacks or filled with straw before covering. In colder climates shallow trenches are cut into the ground next to the tree roses. The plants are then tipped into the trench and covered with soil. In the early spring the roses are again placed upright and pruned of dead or damaged branches. The lower portion of each tree rose consists of a hardy rootstock which does not produce attractive flowers. Any lateral shoots produced on this cane should be removed as they form.

The blooms should be removed as the petals drop to prevent the formation of seed pods. The presence of the seed pods (hips) prevents continued blooming on that flowering stem. When the roses

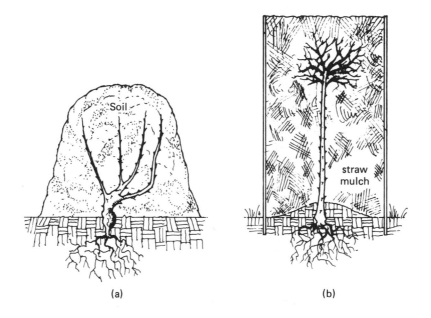

(a)   (b)

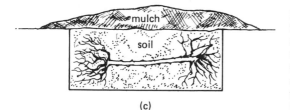

(c)

FIGURE 16–21. Protecting roses during winter: (a) Roses may be mounded with loose soil, straw, or leaves and covered with burlap. (b) Tree roses may be protected by placing a frame around them and filling it with straw or other insulating material. (c) In colder climates, tree roses are best protected by placing them in trenches covered with loose soil and mulch.

are grown for the purpose of cutting the blooms, the stems are cut in such a manner to stimulate further production of strong-stemmed blooms. Each flower stem is cut to a length that will leave at least three leaves on that stem. If the rose has both three-leaflet and five-leaflet leaves, make the cut above the highest five-leaflet leaf that will provide a satisfactory length of the flower stem. If this rule is followed, higher-quality blooms will be available for a longer period during the growing season.

Preventing Pest Problems on Roses

Roses grown in the landscape are subject to the same pest problems associated with greenhouse roses. The most common diseases of roses include black spot and mildew. These diseases infect the foliage and sometimes the blooms during periods of wet weather. They may be controlled by weekly application of suitable fungicides. Under heavy infection, the foliage will be completely lost from the plants, and blooming will cease until new growth appears.

The most common insects found on roses are aphids, spider mites, and thrips, but other garden insects may occasionally damage these plants. Routine inspection and treatment of roses is necessary to control these pests.

SELECTED REFERENCES

Baumgardt, J. P. 1974. *How to Care for Shade and Ornamental Trees.* Intertec Publishing Corporation, Kansas City, Mo.

Baumgardt, J. P. 1968. *How to Prune Almost Everything.* M. Barrows and Company, Inc., New York.

Bruning, W. F. 1970. *Home Garden Magazine's Minimum Maintenance Gardening Handbook.* Harper & Row, Publishers, New York.

Christopher, E. P. 1954. *The Pruning Manual.* Macmillan Publishing Co., Inc., New York.

Conover, H. S. 1977. *Grounds Maintenance Handbook,* 3rd ed. McGraw-Hill Book Company, New York.

Denisen, E. L. 1978. *Principles of Horticulture,* 2nd ed. Macmillan Publishing Co., Inc., New York.

Janick, J. 1972. *Horticulture Science,* 2nd ed. W. H. Freeman and Company, Publishers, San Francisco, Calif.

Oravetz, J., Sr. 1975. *Gardening and Landscaping.* Theodore Audel and Company, 4300 West 42nd street, Indianapolis, Ind.

Pirone, P. P. B. O. Dodge, B. D. Rickett, and H. W. Rickett. *Diseases and Pests of Ornamental Plants,* 4th ed. The Ronald Press Company, New York.

Pirone, P. P. 1972. *Tree Maintenance,* 4th ed. Oxford University Press, New york.

Shurtleff, M.C., and R. Randall. 1975. *How to Control Tree Diseases and Pests.* Intertec Publishing Corporation, Kansas City, Mo.

TERMS TO KNOW

Adventitious Shoots
Axillary Buds
Bare-root Stock
Drop Pruning
Guying

Scaffold Branches
Shingletow
Suckers
Trace Elements

STUDY QUESTIONS

1. Explain how the soil pH may be raised or lowered to provide the best plant growth.
2. Explain why a planting basin is provided around newly planted trees.
3. Describe how a planting area may be prepared for growth of azaleas and rhododendron to ensure proper maintenance of an acid soil.
4. Describe how the various mulching materials are beneficial to a landscape planting.
5. List the reasons for pruning trees and shrubs.
6. Explain how the application of a wound dressing to cut surfaces is beneficial to plants.

SUGGESTED ACTIVITIES

1. Prepare planting holes for bare-root, canned, and balled-and-burlapped nursery stock. Demonstrate the proper planting methods.

2. Demonstrate various methods for staking and guying trees and the use of tree wraps after planting.

3. Make a display of the various mulching materials used in landscaping.

4. Prepare a compost pile for use in making potting soil.

5. Prepare a schedule for landscape maintenance in your area. Include dates for planting, fertilizing, pruning, and pest control in home plantings.

6. Prepare large shade trees for fertilizing through holes drilled in the soil.

Glossary

Abort. Loss of flower parts caused by improper cultural conditions or plant aging.

Abrasion. A wearing away of the surface, as with the removal of a hard seed coat.

Abscission. The dropping of leaves or fruit from a plant by natural separation.

Absorption. A method by which water enters the soil, roots, or plant cells.

Accent. To call attention to a particular location in a landscape or floral design. To create emphasis in a design by use of a focal point.

Accent Plant. A plant that attracts attention by its distinctive color, shape, size, or texture in contrast to plants used with it.

Accented Neutral. A color scheme that has one color predominating with a neutral blended with it. Neutral colors to be used are white, black, gray, brown, or tan.

Accessory. An item used in support of a landscape or floral design, such as a figurine, bird bath, pool, or accent rocks.

Acclimitization. A plant's adjustment to changes in environment, such as to light levels or temperature extremes.

Acid Soil. A soil condition generally considered to be below a pH of 6.0. A soil solution heavy in hydrogen ions.

Acre. A land area measurement of 43,560 square feet. The metric equivalent is the hectare. One acre equals approximately 2.5 hectares (ha).

Adaptability. The ability of a plant to withstand environmental conditions, such as winter cold, heat, light levels, or drying winds.

Adventitious. A bud or shoot produced in an unusual location, other than at a leaf axil or shoot terminal.

Aeration. Horticulturally, the opening up or loosening of soil by the addition of inorganic additives or by mechanical methods.

Aerify. A term used to describe the opening of the soil in turf grass areas by the removal of cores of soil and sod.

Aesthetic. Appearing to be pleasant; appreciative of beauty.

After-Ripening. A process of curing or rest period required by some seeds before germination will occur.

Agriculture. Production of livestock or crops; generally considered to be farming. Horticulture is a division of the agricultural industry.

Agronomy. The science of field crop production.

Air Layering. A method used for propagating some plants by inducing root production along the upper portion of the stem.

Aisle Runner. A white carpet running the length of the aisle of a church sanctuary for wedding ceremonies.

Alkaline Soil. A soil solution having a pH value of 7.0 (neutral) or higher.

Almanac. A book containing weather forecasts and agricultural information based on meteorological and astrological data.

Alpine Garden. A garden comprised of plant species normally found growing at high elevations or in the upper northern climates.

Alternate. An arrangement of stems, leaves, or buds which are spaced singly along the length of a branch, rather than in pairs (opposite).

Analogous. A color scheme that utilizes any three adjacent hues (colors) from the color wheel.

Annual. Horticulturally, a plant that completes its entire life cycle in a single growing season.

Anther. The upper portion of the stamen (male portion of a flower), which bears the pollen.

Anthracnose. A disease that forms sunken, dead areas on leaves and stems and is caused by a fungus organism.

Antidesiccant. A material that inhibits water losses from the foliage of plants.

Apical Dominance. An influence expressed by the terminal bud, which suppresses lateral shoot growth on a plant.

Apical Meristem. The region located at the shoot tip, where upward growth of a stem occurs.

Approach Grafting. A grafting technique in which two plants are attached while on their own root systems. When the grafted area has fully united, the scion plant is severed from its root system, while the upper stem of the stock plant is removed to form a grafted stock and scion. (Also called *inarching*).

Arbor. A structure used for suspending vines. It is open in the center and covered by an open-latticed framework.

Arboretum. An area of land set aside for the growth and display of ornamental trees but which may also include other landscape plants.

Arboriculture. The study of the culture of trees used for shade and ornamental purposes.

Arborist. A person who has studied arboriculture and is qualified to make tree repairs and prune large shade trees.

Area. The two-dimensional measurement of land or other surfaces.

Arid. Regions receiving low amounts of rainfall annually.

Artificial Medium. A growing medium consisting of inorganic and organic constituents, but devoid of topsoil.

Asexual Reproduction. Plant propagation by use of cuttings, grafting, or layering. Production of plants by a method other than from seed.

Asymmetrical. Objects of unequal size or visual weight are placed on opposite sides of a vertical axis, as in an asymmetrical landscape plan.

Atrium. An open-roofed courtyard used for planting which is enclosed on all sides by a building or home.

Auxin. A chemical that can alter plant growth.

Available Nutrient. A nutrient found in a soil solution which may be readily absorbed by the plant roots.

Axil. The area located at the junction of a leaf and stem. A bud generally found in the leaf axil.

Axillary Bud. The bud located at a leaf axil.

Azalea Pot. A container having a height only three-fourths that of its width at the rim.

Backfill. (1) Soil or amendment that is replaced around plant roots or other excavation. (2) The replacement of an excavated area with soil or other amendment.

Backhoe. A machine designed for digging ditches.

Bactericide. A chemical used for the control of bacterial diseases or as a disinfectant.

Balance. A design principle. The placement of objects in strategic locations to create a visual feeling of harmony and proportion in a design.

Balled-and-Burlapped (B&B). The digging of plants for transplanting with a ball of soil around the undisturbed root system. The soil ball is then wrapped with burlap or other fabric to maintain the soil shape.

Bare-Root. Plant material dug from the field and packed without soil remaining around the roots.

Bark. The external covering of a woody plant stem or trunk.

Bark Splitting. Cracks in the external surface of a plant resulting from a sudden lowering of temperature during freezing weather.

Bedding Plants. Herbaceous plants normally grown in outdoor beds to provide colorful blooms or foliage in a landscape.

Bench Grafting. Grafting normally done indoors during the winter months using bare-root stocks and scion pieces which are then healed while in storage.

Bent Neck. A condition found on roses and other cut flowers when water can no longer enter the cut stem, causing the neck to become weak and unable to support the flower.

Berm. A mound of soil used in landscaping for screening an enclosed space.

Biennial. A plant that requires 2 years to complete a life cycle of seed germination to production of seed. A cold period is required before flowers are produced.

Bisexual. A flower possessing both male and female (seed and pollen) organs.

Blind Bud. A bud that ceases to continue growth or fails to produce a flower, but remains vegetative.

Bloom. (1) A term interchangeable with **blossom.** (2) The period of flower production on a plant.

Bonsai. The growing of miniaturized plants by severely restricting the growth of their roots.

Border. Plants located at the perimeters of a landscape to provide a screen.

Botanical Garden. A garden area used for the growth and display of ornamental plants for educational purposes or as a botanical collection.

Bottom Heat. The application of heat to the bottom of a bench surface, generally for enhancing seed germination or root formation on cuttings in a propagation bed.

Boutonniere. A single flower or small assemblagè of flowers worn by a man.

Bract. A leaf that is modified in some manner, such as the colored bracts of a poinsettia flower cluster.

Branch. A shoot arising from the main stem or trunk of a plant.

Break. Lateral shoots that develop following the removal of the terminal bud by pinching.

Bridge Graft. The insertion of a scion piece across an injured area of a plant by forming a graft union above and below the damaged tissue.

Broadcast. To distribute fertilizer or seed on top of the prepared soil rather than by placing it into the soil surface.

Broadleaf Evergreen. Plants that retain their green leaves all year and possess large leaves. Not needle-bearing plants (conifers).

Brushing. A method used to obtain a uniform clipping height of the grass surface on golf greens by the placement of steel brushes in front of the cutting blades of the mower.

Bud Drop. The loss of flower buds caused by stresses in growth or abnormal environmental conditions.

Budding. A grafting method employing a single bud as the scion. The bud is inserted into the stock plant stem to form a new plant having the growth characteristics of the scion.

Budstick. The stem providing the buds used in bud grafting (budding).

Bulb. A reproductive structure found on some plants, containing a reduced stem surrounded by many fleshy leaves. A bulb is generally found underground and serves as an overwintering food-storage organ.

Bulblet. A bulb structure produced either above the ground level in the axil of a leaf or underground from a parent bulb.

Bulb Nose. The top surface of a bulb from which the stem emerges. Generally, the bulb nose is pointed in shape.

Bulb Pot. A growing container that is half as tall as it is wide at the top.

Bulb Scale. A single fleshy leaf with a

portion of the stem from a bulb attached which will form a bulblet and a new plant when removed from the parent bulb.

Bullhead. A term applied to rose buds that are misshapen and either fail to open or produce an abnormally shaped flower.

Bunker. A depression filled with sand located at the perimeters of golf course fairways or greens to act as traps for balls.

Bypass Growth. Vegetative shoots arising around a flower bud, such as with azaleas.

Cactus. A plant having a fleshy stem, thickened epidermis, and leaves reduced to spines.

Caliper. A term used to describe the method by which trees are measured. Trees up to 4 inches in diameter are measured with a caliper at a height 6 inches above the ground. Larger trees are measured at a height of 12 inches.

Callus. A tissue mass which forms at a cut surface (as in a graft union or at the base of a cutting) as a natural healing process in plants.

Calyx. A collective term for the sepals (flower bud leaves) of a flower. The calyx is generally green or leaflike but may be colored or fused to form a cone around the petals.

Cambium. The layer of meristematic (actively dividing) cells producing new shoots, roots, and enlargement of plants.

Candelabra. A fixture designed to support one or more candles for lighting. Used for special lighting effects for ceremonies such as weddings.

Candles. Horticulturally, the new, soft shoot growth on pine trees.

Cane. A major structural branch arising close to the ground, as with rose plants.

Canned Nursery Stock. See **Container Stock.**

Canopy. The overhead portion of a tree which provides shade and screen.

Cash on Delivery (C.O.D.). A demand for payment of goods upon delivery to the buyer.

Center of Interest. A term synonymous with **accent** or **emphasis** in design. Any object that attracts attention or creates a focal point in a design composition.

Certification. An inspection certificate stating that plant material is free of diseases and insects.

Chipper. A machine designed to create small chips from branches.

Chlorophyll. The pigment found in growing plants which imparts a green color and is important in photosynthesis.

Chlorosis. A lack of chlorophyll, which causes a yellow coloration on the plant foliage.

Cleft Graft. A method of grafting where one or more scion pieces are placed in a split (cleft) formed on a cut stock plant.

Clonal Propagation. The reproduction of a group of plants from a single plant by vegetative propagation.

Clone. A group of plants having been vegetatively propagated from a single plant.

Cold Frame. An unheated structure covered with a transparent glazing material and used to protect plants from frost.

Color. A term synonymous with **hue.**

Color Wheel. A diagram illustrating the various primary, secondary, and tertiary colors, with their respective tints and shades as they are applied to design.

Columnar. Plant shape having a tall, narrow habit of growth.

Common Name. The name by which a plant is generally known rather than by its scientific name.

Compaction. The compression of the soil surface so that air and water penetration is inhibited.

Compatible. In grafting terms, a scion and rootstock which will form a successful union and continue to grow as one plant.

Complementary Colors. Any two colors (hues) located opposite each other on the color wheel when used together in a design.

Composite Flower. A flower actually consisting of many smaller flowers, having both ray and disc florets (as in chrysanthemums).

Compost. A soil conditioner composed of soil and decomposed organic matter.

Compote. A stemmed container used for arranging flowers.

Condition. (1) Quality of a flower or plant. (2) To amend the soil by addition of organic matter, sand, or other material. (3) To prepare cut flowers by placing them in a refrigerated cooler in a floral preservative.

Conifer. Plants bearing needles rather than flattened leaves. Most, but not all, are evergreens.

Container Stock. Nursery plants grown entirely in containers rather than being dug from a field.

Convection Tube. A polyethylene tube extending the length of a greenhouse for the purpose of circulating air in conjunction with a fan jet system.

Cool Colors. Colors (hues) composed basically of blue and green tones.

Cool Season Grasses. Grasses adapted to best growth during cooler seasons or climates. Their best growth is during the autumn through the summer months in most of the United States.

Corm. An underground storage organ consisting of a highly compressed stem, buds, and roots.

Corolla. A collective term for flower petals.

Corsage. Any grouping of flowers, net, bow, or other accessories to be worn by a woman.

Cotyledonary Leaves. The first leaflike structures emerging from a seed. These are replaced by true leaves as the seedling begins to produce food by photosynthesis.

Crescent. A design form having the shape of a quarter moon.

Crop Rotation. A schedule that alternates crops according to demand and holidays.

Crown. The canopy or top portion of a tree or plant.

Crown Buds. Vegetative buds or shoots forming around flower buds following unfavorable environmental growth conditions (as with chrysanthemums).

Cultivar. Vegetatively propagated progeny from a single plant.

Cutting. The portion of a plant that is severed for the purpose of forming a new root system and plant.

Cutting Back. A reduction in the size of

a plant to rejuvenate new cane growth.

Damping-off. A rotting of seedlings at the soil surface. Caused by one of several fungus diseases.

Deciduous. Plants that lose their foliage during the winter months.

Defoliant. A chemical that causes plant foliage to drop prematurely.

Dibble. Any pointed object used to press holes in the soil for insertion of seedlings.

Disbudding. The removal of unwanted buds from the stem or terminal of a flowering shoot.

Disbuds. A term synonymous with **standard** flowers. Stems having all lateral flower buds removed.

Division. Propagation by cutting clumps of plants into groups, each with a complete root system and shoots.

Dolomitic Lime. A material used to raise the soil pH and to provide plants with both calcium and magnesium.

Dormancy. A period when plant parts are no longer actively growing because of unfavorable environmental conditions.

Dusting. The application of a fine-powdered chemical to plant foliage to prevent pest damage.

Dwarfing Understock. Rootstocks that cause slower growth or reduced stature in scion cultivars grafted to them.

Easement. A legal right to use an area of land owned by another party.

Embryo. The miniaturized plant within a seed.

Emphasis. A design principle. Creation of a visual accent in a design.

Endosperm. The food-storage portion of a seed.

Enframement. The framing of a landscape or building with plants.

E.P.A. Abbreviation used for the Environmental Protection Agency.

Epinasty. The downward bending of leaves in response to disease or environmental pollutants.

Epiphyte. Plants normally found growing within the branches of trees.

Ericaceous Plants. Plants related to the rhododendron, which grow best in an acid soil.

Espalier. Plants trained to grow flat

against a wall.

Etiolated. Elongated and weak tissue lacking chlorophyll, resulting from plant growth in darkness.

Evaporative Pad Cooling. Cooling structures by use of water passing through excelsior pads. Air is cooled as it passes through the moistened pad and is drawn through a greenhouse by an exhaust fan.

Fairway. The grass area of golf courses lying between the tees and greens.

Feathering. The separation of a flower to form several small florets for use in corsages.

Fertilizer Burn. Dead or damaged foliage and roots of plants caused by over-fertilization or high soluble-salt accumulation in the soil.

Fertilizer Elements. The nutrients of nitrogen, phosphorus, and potassium found in most commercial fertilizers.

Fertilizer Injector. A device that proportions a concentrated fertilizer mixture into the irrigation water.

Field Stock. Plant material grown in field rows rather than in containers.

Flat. A shallow container having the approximate dimensions of 20 × 15 × 3 inches (51 × 38 × 8 centimeters).

Floral Foam. A highly porous block containing a floral preservative used to support flower stems in an arrangement.

Floral Preservative. A chemical mixture added to cut-flower-holding solutions to extend the life of the flowers.

Floret. (1) A single flower in a flower head. (2) A smaller flower created by dividing a larger bloom into groups of petals.

Floriculture. Production of foliage or flowering ornamental plants in fields or greenhouses for commercial sales.

Focal Point. The location within a design that attracts the greatest attention.

Foliage Plant. Any plant grown for its foliage characteristics.

Foliar Disease. A disease organism principally affecting the plant leaves.

Foreman. A supervisor of workers.

Form. A design term synonymous with **shape** or **outline.**

Foundation Planting. A planting of trees, shrubs, or groundcovers in front of a building foundation.

Frond. The leaflike structures of fern plants.

Fumigation. Treatment of an area by use of a chemical which forms a gaseous liquid for elimination of pests.

Fungicide. A chemical used to destroy fungal organisms on plants.

Gallon Can. A container holding approximately 1 gallon of soil, with dimensions of 7 inches in width and 7.5 inches in height (18 × 19 centimeters).

Garden Center. A retail business specializing in the sale of nursery plants and supplies.

Genetics. The science of heredity and the creation of new plant types.

Germination Starting seeds into growth.

Gibberellin. A growth-promoting substance causing longer stems or larger flowers and fruit. Also known as **gibberellic acid.**

Girdling. Roots growing in tightly packed masses in sufficient numbers that some are killed.

Glamellia. A traditional corsage, ordinarily constructed from gladiolus flowers.

Glazing. The transparent or translucent external covering of a greenhouse.

Glazing Bars. The framework that supports the glazing material on a greenhouse roof.

Graft. The joining of a scion with a rootstock to form a new plant.

Green. A highly maintained turf area for use as a putting surface in golf.

Greenhouse. A structure covered with a light-admitting surface for use in the growing of plants.

Greenhouse Range. A group of greenhouse structures.

Greenwood Cutting. Cuttings made after the first flush of active growth has ceased in midsummer.

Goundcover. Plants grown for their low, spreading habit to prevent the growth of weeds.

Growth Retardant. Chemicals applied to plants to restrict growth and form a more compact shape.

Guying. Supporting a tree by use of cables, rope, or wires until root growth is established.

Ha Ha. A sunken ditch that serves as a fence.

Hanging Basket. A container suspended from a ceiling or wall.

Hard Pinch. The removal of a shoot at the point where lignified (hard) tissue is present. A hard pinch will remove two or more of the lateral buds as well as the terminal shoot.

Hardening-off. The gradual reduction of environmental conditions used in production to prepare plants for withstanding conditions found in the landscape.

Hardiness. The adaptation of a plant species to environmental conditions found in a specific locality, such as an ability to withstand low temperatures, cold or dry winds, and drought.

Harmony. An intangible design principle. The successful blending of a design so that all components become a part of the whole.

Head House. A building located adjacent to a greenhouse for use as a storage facility and work area.

Heading Back. The removal of the shoot tips of branches to induce fuller growth of lateral shoots closer to the ground.

Heating Cable. An insulated electrical device used to produce heat under seed germination flats and propagation beds.

Hectare. The measure of area in metric equivalents. A hectare is equal to 100 square meters (2.47 acres).

Hedge. Plants grown closely together to form a row of massed foliage. These plants may be either severly pruned or left to grow in a natural form.

Heeling-in. Placing plants in the ground for temporary storage to keep the roots in contact with soil moisture.

Herb. Plants grown for their flavor, aroma, or medicinal value.

Herbaceous. A plant that does not produce extensive areas of lignified tissue but remains soft and succulent.

Herbicide. A material used to kill unwanted plants (weeds).

Hogarthian Curve. A design having the shape of a modified S form, the upper portion using two-thirds of the figure and creating a free-flowing motion which leads the lines of the design to the focal point.

Hook. A portion of a stem where a flower has been cut near the main cane (as in roses), leaving a portion of the lateral stem.

Hormone. A chemical produced in one portion of a plant and creating a growth reaction in some other location.

Horticulture. The science of producing and using ornamental plants, fruit, and vegetables.

Hotbed. Outdoor growing structures heated by an external source, such as steam, hot water, electricity, or manure.

Houseplant. A plant adapted to environmental conditions found in homes or offices, where light levels and humidity are low and temperatures are high.

Hue. A design term having the same meaning as **color.**

Humidity. The amount of moisture or water vapor found in the air.

Humus. Partially decomposed organic materials mixed with soil.

Hydroponics. The production of plants in a liquid solution or gravel medium supplemented with all required nutrients for proper growth.

Incompatibility. In grafting, when two plant species will not form a successful graft union.

Infection. The penetration of a disease organism into the tissue of a plant.

Infestation. The feeding of an insect population on the surface of a plant.

Injector. A mechanical device designed to add measured amounts of a liquid into the water line.

Insecticide. Any material designed to kill or inhibit insect populations.

Intensity. The visual quality of brightness or dullness in color created by the addition of gray.

Internode. The area of a stem between two nodes or lateral shoots.

Interstock. A stem or root piece placed between two others in a double-grafted plant.

Inventory. Merchandise in stock in a store.

Landscaping. The design and alteration of a portion of land by use of plant material and land reconstruction.

Lateral Buds. Buds arising from nodes located below a terminal shoot.

Lathhouse. A structure shaded by lath or other material for the purpose of storing plant material.

Lawn. Any area covered by grass and kept clipped by regular mowing.

Layering. A method of propagation whereby stems are bent over and covered with soil to form new roots while still attached to the parent plant.

Leach. Drenching of a soil medium with water to remove excess fertilizer salts.

Leader. The major supporting stem or trunk of a tree or shrub.

Leaf-bud Cutting. A portion of a plant which includes a leaf and a portion of the stem with an attached axial bud. The cutting may be rooted and will form a new plant.

Leaf Spot. A symptom caused by certain diseases, where discolored spots are formed on the leaves of affected plants.

Leggy. A plant having few leaves at the base, thereby exposing a major portion of the stem.

Line. A design element which defines the use of material in forming an outline or skeleton of a design or leads the eye to a focal area.

Liner. Nursery stock of small size having been planted at close spacings in a nursery bed to form roots and is now ready for transplanting.

Loam. Soil having a mixture of sand, clay, and silt particles in nearly equal quantities.

Long Day. A period of time having a continuous dark period no longer than 7 hours.

Lopper. A cutting device on a long handle used in trimming trees and shrubs.

Mass. (1) A closely spaced grouping of plants or flowers in a design or arrangement. (2) A flower form consisting of large petals or closely packed florets.

Mat Watering. Growth of potted plants on an absorbant mat which provides a constant supply of water to the base of the plants.

Matting. The act of working top-dressing materials through a turf grass zone by raking or pulling a mat over the surface.

Maze Garden. A formal landscape planting surrounded by hedges or other barrier plantings designed in the form of a maze or labyrinth.

Meristem. The portion of a shoot, bud, or roots where cells are in active division. The cellular layer of a plant, which forms new cells.

Mildew. A white-colored growth on plants infected by certain fungi.

Miniatures. Cut flowers consisting of several flowers supported on short, branched stems. Also known as **nondisbuds.**

Minor Elements. Those nutrients required by plants in only trace amounts.

Mist Propagation. The application of a fine mist at periodic intervals to the foliage of cuttings during propagation.

Monochromatic. A color scheme consisting of one color (hue) with its tints and shades.

Mother Block. The parent plants from which cuttings are taken at regular periods.

Mulch. A material applied to the ground to prevent excessive drying of the soil surface, to prevent rapid changes in soil temperature, as a soil amendment, for decorative purposes, or to prevent weed growth.

Narrow-Leaved Evergreen. A conifer. An evergreen plant having needlelike leaves.

Natural Form. A plant grown without crowding, pruning, or other growth-restricting influences.

Needlepoint Holder. A stem-anchoring device used in flower arranging, consisting of many sharp-pointed spikes or nails. Also called a "frog" or pinholder.

Nematocide. A chemical designed to kill or inhibit the growth of nematode populations.

Neutral. A pH value of 7.0 in soil reaction.

Node. The location on a stem where leaves, buds, or flower stalks are attached.

Nosegay. A grouping of closely spaced flowers, net, and accessories in a hand-held floral arrangement.

Nurse Grass. A vigorously growing, fast-germinating grass species often added to lawn seedings to obtain a more rapid establishment of a lawn.

Nursery. A business that specializes in the growth of plants used in landscaping.

Nursery Stock. Trees, shrubs, vines, and

other plants grown in a nursery.

Organic Matter. The refuse derived from decaying material which was once a part of living organisms.

Ornamental Horticulture. The branch of horticulture specializing in the areas of floriculture, turf grass management, nursery stock production, and landscaping.

Own Root. Plants grown on their own root systems rather than having been grafted to roots of another plant species.

Pallet. A low, portable bench used to grow and transport bedding and potted plants.

Parts per Million (PPM). A unit of measurement commonly used for expressing the amount of material in a solution or other mixture. In metric units, 1 part per million is equal to 1 milligram of material in 1 liter of the mixture or solution.

Peanut Flower. A rose flower having a nearly normal shape, but distinctly smaller and possessing fewer petals than a normal flower for the cultivar.

Peat Moss. The partially decomposed organic remains of sphagnum moss, used as a soil additive to improve the water retention, aeration, and porosity of a planting medium.

Peat Pot. A small pot comprised of compressed peat fibers used for bedding plants that are to be transplanted without removal from their containers.

Pedicel. The stem that supports a flower.

Peeling. Removal of outer petals from flowers to improve their appearance.

Perennial. Any plant that survives for several years, regrowing each spring to form flowers and seeds.

Perlite. An inorganic soil conditioner derived from volcanic rock.

Permanent Grasses. Any turf grass species that persists from one year to the next in an area where it is adapted.

Persistent. Remaining for long periods of time, as with chemicals in the soil.

Pesticide. Any material used in the control of weeds, insects, disease organisms, rodents, or nematodes.

Petiole. An individual leaf stem.

Petty Cash. Money held back to pay for the incidental expenses of a business.

pH. A measure of alkalinity or acidity, as with soil solutions. Measured as the negative logarithm of the hydrogen ion concentration of a solution.

Photoperiod. The influence of daylight or night length on plant growth.

Photosynthesis. The manufacture of sugar by leaves in the presence of light, carbon dioxide, and water.

Pinching. The removal of a shoot terminal to induce lateral branching of plants.

Planter. Any container designed to provide suitable conditions for plant growth.

Planting Basin. A circular, saucer-shaped depression surrounding a newly planted tree or shrub which provides a catch basin for water.

Planting Strip. An area located between a street and a sidewalk.

Plat. A legal chart of a subdivision, city, or township, showing the boundries of each separate lot or property.

Plot Plan. A map showing a property with all utilities, vegetation, and buildings drawn to scale.

Plugging. The propagation of a turf grass by means of vegetative sod pieces (plugs). Sod plugs consist of the grass plants, roots, and soil and are approximately 5 centimeters (2 inches) in diameter.

Polychromatic. A color scheme consisting of three or more unrelated hues.

Pompon. A spray-type chrysanthemum having a flower shaped like a ball.

Porous. Any object having small holes which allow the passage of liquids or gases.

Post-emergent. A chemical herbicide used to control weed growth after they have germinated.

Post-High School. Educational institutions providing training beyond the high school level.

Precooling. A process of storing plants at low temperatures for several weeks to allow flower bud formation and to break bud dormancy.

Pre-emergent. A chemical herbicide used to control weeds before they emerge from the soil.

Primary Colors. The hues of red, yellow, and blue. Colors from which all other hues are derived.

Procumbent. Plants having branches

that remain flattened against the ground but do not root into the soil.

Propagation. Any sexual or vegetative method used to increase plants, such as by seeds, cuttings, grafting, division, or layerage.

Proportion. A design principle. A pleasing relationship in size of the parts of a design to the whole composition.

Pruning. The removal of selected branches of a plant to allow more vigorous growth and flowering.

Pyramidal. Plants having a pointed shape and tapered sides.

Quarantine. The isolation of plants to determine whether diseases or insect populations are present or to prevent their spread to healthy plants.

Quonset-style Greenhouse. A greenhouse structure having an arch shape and covered with polyethylene or rigid fiberglass plastic.

Radiating. A floral design having a fan shape and generally constructed of spike flowers.

Ray Floret. The outer petals of a composite flower, such as chrysanthemums, bearing only anthers.

Reel Mower. A grass-cutting machine having a series of blades that clip the grass blades by a scissors action.

Rejuvenate. Renewal of growth and vigor of plants by selective pruning, fertilization, soil aeration, or pest control.

Relative Humidity. A measure of the water vapor content of air expressed as the percentage required to reach saturation at a specific temperature.

Renovation. The drastic making over of a turf grass surface, including cultivation and replanting without complete reconstruction.

Residual. Having a long-lasting effect or influence.

Resistant. Able to withstand continued contact with chemicals, adverse conditions, or certain pests.

Respiration. The utilization of carbohydrate food reserves by plants, with a release of carbon dioxide and water.

Rest Period. A period of inactivity in plant buds during adverse growing conditions or while buds are maturing for later active growth.

Retail. A business involved in selling products directly to the public.

Retaining Wall. Any structure used to hold an embankment or raised ground surface.

Rhythm. A design principle. The creation of visual movement in a design.

Rock Garden. A garden constructed of rock outcroppings in a manner to appear as a naturalized design of wild flowers.

Rolling. A method used for firming and smoothing a ground surface following seeding of a lawn.

Root Ball. The root mass and soil in a container or the root mass and soil that remains after digging a plant.

Root-Bound. The closely packed mass of roots on a plant that has grown too large for its container. Also called **pot-bound.**

Root Pruning. The removal of a portion of the roots on a plant by passing a cutting blade under the plants in the field or pruning of roots in a container to stimulate root branching.

Rooting Hormone. A chemical applied to the cut basal portion of a cutting to stimulate root formation.

Rootstock. The lower portion of a grafted plant, which includes the roots and becomes the support for a grafted scion.

Rose Pot. A pot having a height equal to 1½ times its width at the rim. Used for deeply rooted plants, such as roses and other woody plants.

Rotary Mower. A grass-cutting machine having a single, rapidly spinning blade.

Runner. A trailing stem located above the ground surface, which roots into the soil at various locations and forms a new plant.

Rust. A foliar symptom observed on plants infected by various parasitic fungi.

Saucer. (1) The depression formed around a newly planted tree or shrub for holding irrigation water. (2) The water tight container placed under pots to protect furniture.

Scaffold Branches. The main support branches of a tree arising nearest to the ground.

Scald. Damage caused to the bark of trees by freezing or drying of the sun. (2) The dying of turf grass when subjected to extended periods of waterlogged soil or standing water.

Scalping. Mowing a lawn at such a close height that the major portion of the foliage is removed.

Scarification. The opening of a hard seed coat to water by wearing away a portion of the surface with an abrasive substance.

Scion. The portion of a graft that will become the top portion of the newly formed plant. A scion may consist of a shoot or merely a single bud.

Scorch. The browning and drying of leaf margins caused by excessive heat or wilting.

Screen. To conceal or mask an area from view or winds.

Secondary Colors. Hues created by mixing equal portions of any two primary colors. Secondary colors are green, orange, and violet.

Seedbed. A soil medium suitable for seed germination.

Selective Herbicide. A chemical used to kill only certain types of plants, such as broadleaved weeds in lawns.

Sepal. An individual leaf of the calyx found at the base of a flower.

Service Area. An area of a landscape set aside for necessary family use; such as garbage storage, wood storage, or clothes drying.

Setback. A line located inside the property lines, designating the extent to which buildings may be constructed.

Shade. In design, the result of adding black to a color.

Shade House. A structure covered with a material designed to reduce the amount of sunlight.

Shifting. Transplanting or repotting of plants into another container.

Shingletow. Wood chips and shavings placed around the base of nursery stock to prevent drying of the roots.

Short-Day Plant. A plant that forms flowers only under conditions of short days or long night periods, such as chrysanthemums.

Sleepiness. A condition of a flower causing the petals to curl toward the center; the entire flower appears cup-shaped or partially closed.

Slip. A soft wood (herbaceous) cutting.

Sod. The combination of grass foliage, stems, roots, and accompanying soil used for establishing completed turf grass areas.

Soft Pinch. The removal of only a portion of a shoot terminal at a point where the tissue has not as yet become woody.

Soften. Use of plant material placed strategically to break the harsh lines of a design or building.

Softwood Cuttings. Cuttings taken in early morning when stems are filled with water and the air is cool.

Soluble. A material that is easily formed into a solution when mixed with water or another liquid.

Soluble Salts. The accumulation of various mineral and fertilizer salts in the soil around plant roots.

Spike Flower. Flower stems having a series of short-stemmed flowers progressing in maturity from the tip.

Spiking. The perforation of the soil surface to allow better air and water penetration to grass plant roots in a lawn.

Splice Graft. A method used for rapidly joining two plant parts in a graft.

Split Calyx. The tearing of the cup-shaped calyx of carnations.

Split Complementary. A color scheme consisting of three colors on the color wheel; one hue used with each of those located on either side of its complement.

Sport. A variation in growth habit found on a single stem or flower of a plant.

Spray. A cluster of flowers on a stem, as with pom-pon chrysanthemums.

Spreader–Sticker. A material mixed with chemical sprays to cause them to adhere to plant foliage for uniform coverage.

Sprigging. Propagation of turf grasses by rhizomes or stolons placed in shallow furrows or small holes.

Standard Pot. A container having nearly equal height and width at the rim of the pot.

Started Eye. A grafted rose stock having the scion bud started into growth before it is dug from the field.

Starter Solution. A fertilizer mixture applied to newly planted plants to aid in their establishment.

Sterilization. Killing of weeds and other pests in soil with heat or chemicals.

Stock Blocks. A field or greenhouse containing plants used for obtaining cuttings or scions.

Stolon. A stem growing along the ground or under the surface which produces roots and new plants at its nodes.

Stratification. The storage of seeds between moist layers of material at cold temperatures to break seed dormancy requirements.

Stunted. Any plant that has experienced unfavorable growth causing it to be smaller than normal.

Succulent. A plant or plant part consisting of soft tissue mostly filled with water; may be easily broken or cut.

Sucker. The tender, fast-growing side shoots arising from lateral buds of plants.

Sun Scald. The drying and cracking of limbs, trunks, or branches caused by the sun warming the surfaces of frozen plants.

Symmetry. Having the same elements or parts duplicated on each side of a vertical axis.

Syringing. The spraying of plants with water to add moisture to the surrounding air or to wash the foliage of dirt and pests.

Systemic. (1) A disease that extends throughout the interior of a plant and infects all parts. (2) Any chemical having no harmful affects on a plant, but when absorbed by the roots is translocated throughout the plant and is poisonous to pests.

Tee. A turf grass area of a golf course used for driving golf balls toward a green.

Temporary Grasses. (1) Grasses used to provide a quick ground cover while permanent lawn grasses become established. Also called **nurse grasses.** (2) Any grass covering used for a few months to provide a ground cover.

Terminal Bud. The bud located at the shoot tip arising from the main stem of a plant.

Terrarium. A transparent enclosure used for growing plants.

Tertiary Colors. Hues obtained by mixing adjacent primary and secondary colors.

Texture. Visual coarseness or softness of objects used in design.

Thatch. A layer of accumulated undecomposed organic matter, comprised of dead and dying portions of grass plants, found above the soil.

Thermal Transmittance. The amount of heat that may pass through a surface.

Tint. The result of adding white to a color.

Tipburn. The burning or browning of tissue at the leaf tips or margins.

Top Dressing. (1) A mixture of soil, amendments, fertilizer, or other materials applied over the top of grass plants and worked around the roots and stems. (2) Any material applied over the soil surface after plants have become established in growth.

Topiary. The severe clipping of plants into unusual shapes, such as animals and balls.

Topsoil. The fertile surface soil, having abundant organic matter.

Toxic. A substance that is injurious to humans or plants.

Trace Elements. The minor nutrients required by plants in very small amounts to maintain growth.

Transite. A product manufactured from asbestos and cement. Used to form benches and cover walls in some greenhouses.

Transition Area. An area between the accent and corner plantings. Used to pull a design together into a single composition.

Transpiration. The loss of water from plants through the stomata.

Transplant. (1) The moving of a plant to a larger pot or container. (2) The young seedling ready for placement in an individual pot.

Turf. Any grassy area maintained by frequent mowing, fertilization, and watering used for lawns, roadsides, or playing fields.

Ultraviolet. Light of longer wavelengths than can be seen by the human eye.

Undercut. (1) The removal of a hook on rose canes while pruning. (2) The practice of severing the roots of nursery plants in the field to stimulate a more concentrated and dense root system prior to digging.

Understock. The basal (bottom) portion of a grafted plant, which possesses the roots.

Union. The area where the scion and understock are joined to form a single plant in grafting.

Unity. A design principle. The blending of all parts of a design into a pleasing composition. Created by the repetition of the same flowers, colors, or textures throughout the design.

Value. The degree of black or white added to a color to form its tint or shade.

Vaneer Graft. Grafting of a scion piece to the side of the understock plant.

Variegated Foliage. Leaves having various patterns of different colors other than green.

Variety. A group of plants within a species having similar growth characteristics.

Vascular Disease. A disease organism that has invaded the vascular (food-and water-conducting) tissues of a plant and has spread to all parts.

Vegetative Buds. The buds that give rise to new shoots and branches, but not flowers.

Vegetative Growth. The accumulated leaves, stems, and roots, but not flowers.

Vegetative Propagation. The increase of plant populations by any method other than from seed, such as cuttings, grafting, division, or layering.

Vermiculite. The product derived from expanded mica used as a soil conditioner.

Vernalization. Exposure of plants to cold temperatures to allow the formation of flower buds.

Vertical Mowing. The cutting of lawn grass areas with blades that cut into the soil surface in a plane parallel to the grass plants. Used for the removal of thatch in lawns.

Viable Seed. Seeds that are alive and capable of germination.

Warm Color. Colors (hues) composed basically of yellow and red tones.

Warm Season Grasses. Grasses best adapted to semiarid growing conditions or which remain in active growth during the warmer, dryer months of the year.

Weeping. A plant having a pendulous (drooping) habit of growth.

Wettable Powder. A material that forms a suspension of chemical particles when mixed with water.

Wetting Agent. A material added to spray mixtures to increase the ability of the liquid to remain on the plant foliage.

Whip. A young, unbranched tree grown from a grafted understock in a nursery.

Whip and Tongue. A method of grafting a scion and understock for maximum cambial contact.

Wholesale. Firms selling their products to retail businesses rather than to the general public.

Wilting. The drooping of leaves caused by insufficient water in the cells.

Windbreak. A group of plants placed in locations where they might screen out winds and snow drifting around a home or landscape.

Windburn. The drying effect of hot winds on plants, which causes them to lose water from tissues more rapidly than the water can be replenished from the soil.

Winter Injury. Any damage to plants which might be attributed to winter conditions, such as freezing of tissue, bark splitting, and killing of branches.

Xylem. The water-conducting cells of a plant.

Appendices

APPENDIX A

The figures found in Table A–1 will aid you in determining the amount of fertilizer required for any stock solution mixture. Several things must be known before you may determine the proper amount of fertilizer to use. You must know:

1. *The dilution ratio of the fertilizer proportioner.* Most injectors have this ratio stamped on them someplace or the figure may be found in the user's manual that accompanies the proportioner.

2. *The parts per million of fertilizer desired for the crop.* This figure will differ among crops. The greenhouse manager usually determines the rates to be used based on the growth and type of crop being fertilized.

3. *The analysis of the fertilizer to be mixed.* The percentages of nitrogen, phosphoric acid, and potash are found in a three-part number on the label of the fertilizer bag. For most mixing situations, you will use the percent of nitrogen (first set of numbers) on the label. The decision as to which fertilizers are to be used is the responsibility of a crop manager. He/she must determine this from the general growth and types of crops being fertilized.

Some examples of typical fertilizer stock solution preparations

Problem

You are instructed to prepare a stock solution of a fertilizer having an analysis of 15–10–30. The dilution ratio of the proportioner is 1:24. The fertilizer rate desired is 100 ppm of nitrogen. How much 15–10–30 should be mixed in each gallon of stock solution?

Solution of Problem

Find 15 percent nitrogen on the left-hand column of Table A–1. *Move to the right two columns (1:24 ratio). The answer to this problem is that you should apply* 2.13 ounces of this fertilizer to each gallon of stock solution.

Problem

You are instructed to prepare a stock solution of a fertilizer having an analysis of 30–10–10. The dilution ratio of the proportioner is 1:128 and you want 150 ppm of nitrogen delivered to the crop. The fertilizer stock container holds 5 gallons. How much fertilizer should be added to this stock solution container?

Solution of Problem

Find 30 percent nitrogen on the left-hand column of Table A–1. Move to the right until you reach the column under the 1:128 dilution ratio. This column shows that 5.69 ounces of 30–10–10 is added *to each gallon of stock solution to deliver* 100 ppm *to the crop.*

However, you desire a mix that contains 150 ppm *of nitrogen. Multiply* 5.69 ounces by 1.5 *to get this figure.*

$$(5.69 \text{ ounces}) (1.5) = 8.54 \text{ ounces/gallon}$$

The stock solution container holds 5 gallons of liquid, so to make 5 gallons of stock solution you must multiply by 5.

$$(8.54 \text{ ounces/gallon}) (5) = 42.7 \text{ ounces/5 gallons}$$

42.7 ounces equals 2 pounds and 10.7 ounces, *since there are 16 ounces in 1 pound.*

TABLE A–1: *100-ppm Chart*
(ounces required per gallon of stock solution)

SOME COMMON PROPORTIONER RATIOS

FERTILIZER ANALYSIS % N, P_2O_5, K_2O	1:16	1:24	1:30	1:50	1:100	1:128	1:150	1:200
1	21.33	32.00	40.00	66.67	133.37	170.67	200.00	266.67
2	10.67	16.00	20.00	33.33	66.68	85.34	100.00	133.34
3	7.11	10.67	13.33	22.22	44.46	56.89	66.67	88.89
4	5.33	8.00	10.00	16.67	33.34	42.67	50.00	66.67
5	4.27	6.40	9.00	13.33	26.67	34.13	40.00	53.33
6	3.56	5.33	6.67	11.11	22.23	28.44	33.33	44.44
7	3.05	4.57	5.71	9.52	19.05	24.38	28.57	38.10
8	2.67	4.00	5.00	8.33	16.67	21.33	25.00	33.33
9	2.37	3.56	4.44	7.41	14.82	18.96	22.22	29.63
10	2.13	3.20	4.00	6.67	13.34	17.07	20.00	26.67
11	1.94	2.91	3.64	6.06	12.12	15.52	18.18	24.24
12	1.78	2.67	3.33	5.56	11.11	14.22	16.67	22.22
13	1.64	2.46	3.08	5.13	10.26	13.13	15.38	20.51
14	1.52	2.29	2.86	4.76	9.53	12.19	14.29	19.05
15	1.42	2.13	2.67	4.44	8.89	11.38	13.33	17.78
16	1.33	1.88	2.50	4.17	8.34	10.67	12.50	16.67
17	1.25	1.88	2.35	3.92	7.85	10.04	11.76	15.69
18	1.19	1.78	2.22	3.70	7.41	9.48	11.11	14.81
19	1.12	1.68	2.11	3.51	7.02	8.98	10.53	14.04
20	1.07	1.60	2.00	3.33	6.67	8.53	10.00	13.33
21	1.02	1.52	1.90	3.17	6.35	8.13	9.52	12.70
22	0.97	1.45	1.82	3.03	6.06	7.76	9.09	12.12
23	0.93	1.39	1.74	2.90	5.80	7.42	8.70	11.59
24	0.89	1.33	1.67	2.78	5.56	7.11	8.33	11.11
25	0.85	1.28	1.60	2.67	5.33	6.83	8.00	10.67
26	0.82	1.23	1.54	2.56	5.13	6.56	7.69	10.26
27	0.79	1.19	1.48	2.47	4.94	6.32	7.41	9.88
28	0.76	1.14	1.43	2.38	4.76	6.10	7.14	9.52
29	0.74	1.10	1.38	2.30	4.60	5.89	6.90	9.20
30	0.71	1.07	1.33	2.22	4.45	5.69	6.67	8.89
31	0.69	1.03	1.29	2.15	4.30	5.51	6.45	8.60
32	0.67	1.00	1.25	2.08	4.17	5.33	6.25	8.33
33	0.65	0.97	1.21	2.02	4.04	5.17	6.06	8.08
33.5	0.64	0.96	1.19	1.99	3.98	5.09	5.97	7.96
34	0.63	0.94	1.18	1.96	3.92	5.02	5.88	7.84
35	0.61	0.91	1.14	1.90	3.81	4.88	5.71	7.62
36	0.59	0.89	1.11	1.85	3.70	4.74	5.56	7.41
37	0.58	0.86	1.08	1.80	3.60	4.61	5.41	7.21
38	0.56	0.84	1.05	1.75	3.51	4.49	5.26	7.02
39	0.55	0.82	1.03	1.71	3.42	4.38	5.13	6.84
40	0.53	0.80	1.00	1.67	3.33	4.27	5.00	6.67
41	0.52	0.78	0.98	1.63	3.25	4.16	4.88	6.50
42	0.51	0.76	0.95	1.59	3.18	4.06	4.76	6.35
43	0.50	0.74	0.93	1.55	3.10	3.97	4.65	6.20

	1:16	1:24	1:30	1:50	1:100	1:128	1:150	1:200
44	0.48	0.73	0.91	1.52	3.03	3.88	4.55	6.06
45	0.47	0.71	0.89	1.48	2.96	3.79	4.44	5.93
46	0.46	0.70	0.87	1.45	2.90	3.71	4.35	5.80
47	0.45	0.68	0.85	1.42	2.84	3.63	4.26	5.67
48	0.44	0.67	0.83	1.39	2.78	3.56	4.17	5.56
49	0.44	0.65	0.82	1.36	2.72	3.48	4.08	5.44
50	0.43	0.64	0.80	1.33	2.67	3.41	4.00	5.33
51	0.42	0.63	0.78	1.31	2.62	3.35	3.92	5.23
52	0.41	0.62	0.77	1.28	2.56	3.28	3.85	5.13
53	0.40	0.60	0.75	1.26	2.52	3.22	3.77	5.03
54	0.40	0.59	0.74	1.23	2.47	3.16	3.70	4.94
55	0.39	0.58	0.73	1.21	2.42	3.10	3.64	4.85
56	0.38	0.57	0.71	1.19	2.38	3.05	3.57	4.76
57	0.37	0.56	0.70	1.17	2.34	2.99	3.51	4.68
58	0.37	0.55	0.69	1.15	2.30	2.94	3.45	4.60
59	0.36	0.54	0.68	1.13	2.26	2.89	3.39	4.52
60	0.36	0.53	0.67	1.11	2.22	2.84	3.33	4.44

APPENDIX B

TABLE A–2: *Sample Potted Chrysanthemum Schedule*

DATE	CULTURAL PRACTICES REQUIRED
January 10	Plant cuttings. Begin lighting at night for 4 hours. Apply 30–10–10 fertilizer at 100 parts per million daily for 2 weeks.
January 24	Pinch each cutting. Stop night lighting and begin short-day treatment. Fertilizer may be increased to 150–200 parts per million at each watering.
February 7	Apply foliar application of growth retardant (not required for short cultivars)
March 7	Disbud the chrysanthemums. Apply second spray of growth retardant to tall cultivars (not required for medium cultivars)
April 4–8	Plants in flower. Water and fertilize well before removing from greenhouse.

For this schedule, a *tall-growing* pot mum cultivar having a *10-week* response was selected. Different flowering response cultivars will require an adjustment in this schedule.

APPENDIX C

TABLE A–3: *Sample Schedules for Poinsettia Production*

DATE	CULTURAL PRACTICES REQUIRED
August 20	Plant rooted cuttings in finishing pots. Use night lighting for long days; keep temperature at 72°F at night.
September 10–15	Use cycocel for height control and start night lighting, if not begun earlier. Night temperature should be lowered to 67°F.
September 17	Pinch the terminals of each cutting. Keep temperature at 67°F at night.
October 4	Begin short-day treatment by turning all night lights off and covering benches nightly with dark cloth from 5:00 P.M. until at least 8:00 A.M. Lower the temperature at night to 62°F.
October 8–15	Apply cycocel again if growth appears to be too fast. Raise night temperature back to 67°F from October 19 until flowers are well formed.
December 10	Scheduled flowering of these poinsettias.

For this sample poinsettia schedule, a 10-week response group variety (Eckespoint C-1) was selected for flowering in mid-December in central Ohio. This schedule is suitable for poinsettia scheduling in most parts of the United States, with only slight variations in timing in the most northern or southern states.

APPENDIX D

Gloxinia production schedule

Propagation

Gloxinia may be grown from tubers, leaf cuttings, or from seeds (Figure A–1). Most propagation of gloxinia is by seeds, since this method is the most economical. Seeds are sown on a fine sterilized soil mix that has been covered with ½ inch of shredded sphagnum peat moss, and germinated at 70°F (21°C) in a humid chamber. The seeds are not covered after being sown on the peat. Seeds germinate in 2–3 weeks.

When seedlings are large enough to handle, transplant them to 2½-inch peat pots. When these plants are well rooted, shift them into

5- to 6-inch pots for flowering. The final soil mix should be equal parts of soil, peat, and sand or other coarse material.

Scheduling flowering

Gloxinia may be flowered in their finishing pots 6–7 months from sowing. The following chart lists various sowing dates and scheduled flowering:

SOWING	FLOWERING
December or early January	June
June 1	December
late July	Mid-February
Early October	Mother's Day (early May)

Plants are grown at 65–70°F (18–21°C) and are given as much light as possible without allowing the leaves to scorch. Watering is critical under high light conditions. Do not overwater gloxinia, or root rotting may occur.

FIGURE A–1. Gloxinia.

APPENDIX E

Calceolaria production

Propagation

Calceolaria or "pocketbook plants" are propagated from seeds which are sown on a fine sterilized medium (such as equal parts of fine peat and vermiculite). The seeds are not covered with any medium. Seeds should be sown thinly and germinated in a humid chamber at 70°F (21°C) under continuous lighting. Seedlings are transplanted to 2½-inch pots as soon as they may be handled safely.

Flower scheduling

Calceolaria will normally flower in May when seeds are sown 6–9 months earlier. The following chart shows a typical schedule for flowering plants in 5- or 6-inch pots.

SOW SEEDS	START LIGHTING	FLOWERING
Early June	November 15	February 15
July	December 20	March 1
Late August	January 20	April 1

Plants are shifted to finishing pots when well rooted. Temperatures are maintained at 50°F (10°C) for flower initiation and flowering. Better flowering occurs when night lighting is applied according to the schedule above. Use 75- to 100-watt bulbs, spaced 4 feet apart and 2 feet above the bench from 10:00 P.M. until 2:00 A.M. each night.

Stem rotting may be a serious problem with calceolaria. This problem may be avoided by not planting them too deeply, providing adequate air circulation, and avoiding overwatering.

APPENDIX F

Cyclamen production

Propagation

Cyclamen normally requires 15 months to flower from sowings made from August until December. This production time may be reduced by application to the growing plant of the growth-regulating chemical gibberellic acid.

Seeds are sown on the surface of neutral peat or other fine seeding medium. The seeds are covered with a light application of sphagnum peat. Germination is done in the dark at 65°F (18°C) for 3–4 weeks. Seedlings are then transplanted to 2½-inch pots and placed in a 55°F (13°C) greenhouse. Seeds are normally sown in April.

Timing of flower production

Plants are grown in small pots during the summer in a cooled, shaded greenhouse. In mid-August the plants are shifted to 5- to

6-inch pots for flowering. Fertilization is applied on a regular basis, using soluble liquid fertilizers with each watering. Water should be applied carefully to avoid root- and crown-rotting diseases.

Gibberellic acid is applied at the rate of 25 parts per million between 60 and 75 days before the plants are scheduled to be in flower. A growth-regulator application sprayed on the foliage on October 1 will provide flowering cyclamen plants for Christmas holiday sales (Figure A–2).

FIGURE A–2. Cyclamen. (Courtesy Ball Seed Company)

APPENDIX G

Use and Maintenance of Equipment

The selection of the proper piece of equipment to perform each task is a basic part of any horticultural job. The type and quality of tools utilized by employees will depend entirely upon the size and nature of the business operation. Production greenhouses or nurseries will use different types of machinery or tools than will a golf course or maintenance business. The maintenance of this equipment is the duty of each employee responsible for their use. A basic understanding of the proper selection, use, and maintenance of the more commonly used equipment is necessary for every employee of an ornamental horticulture business.

Use and Maintenance of Gasoline Engines

The small engines commonly used to power machinery are either two-cycle or four-cycle gasoline engines. With two-cycle engines, the oil is mixed with the gasoline in the fuel tank to provide lubrication of the engine parts. These are called two-stroke engines because a power stroke is permitted with each revolution of the crankshaft. The oil used in these engines is especially designed for mixing with gasoline, and no other grade of oil should ever be used.

All four-cycle engines require the addition of oil through an oil-fill plug. The location of this oil-fill hole will vary with the make of engine. Before any four-cycle engine is started, the oil level should be checked. On some engines an oil-level plug is removed to observe the level of oil in the hole (Figure A–3). The oil should be maintained at the top of the hole to the point of nearly overflowing. The newer engines may be equipped with a dipstick oil-level indicator. The oil level should be maintained at the full mark on the dipstick at all times. A gasoline-powered engine should not be operated even for a short time with less than the recommended amount of oil.

FIGURE A–3. The oil level is checked on this mower by removing the crankcase plug and reading the dipstick oil level indicator.

The oil should be changed on all four-cycle engines at intervals recommended by the operator's manual. New engines will require more frequent oil changes than those which are properly adjusted and broken in. Generally, oil is changed after every 30 hours of operation. However, when the oil is checked, the operator may easily determine if it is dirty or dark in color. This would indicate the need for an oil change.

Procedures for changing oil are specified by the owner's manual, which should be retained for easy reference. Before the oil is to be drained, the engine should be started and allowed to operate for a short time. This will warm the oil to allow it to flow more easily. The agitation of the oil against the moving parts will also gather most contaminants and dirt so they will be removed with the oil. The oil is removed from most engines through a drain plug located at the bottom of the crankcase. The oil should be drained completely into a container before the drain plug is replaced. The spent oil may be disposed of only by placing it in a used oil drum for use in cleaning hand tools or returned to a service station for recycling. The oil should *never* be dumped onto the ground or into a drainage ditch.

New oil is replaced by pouring it into the oil-fill cup or hole (Figure A-4). A funnel is helpful for this since the openings are generally small. The oil should be poured slowly to allow the oil to flow into the engine. The engine crankcase should not be overfilled with oil, or damage may result. The engine will require only 1¼ pints (0.59 liter) of oil. A high-quality detergent oil is the only type that should be used. Acceptable oils for use during summer at temperatures above 40°F (4°C) are marked as SAE 30 API Service SE, MS, SD, or SC. During winter weather use a 10-weight oil.

After the oil has been changed, a routine check of the engine should be made. All nuts, bolts, and screws should be tightened, if necessary, to avoid the loss of parts or malfunction of the engine. The

FIGURE A-4. Oil is replaced by pouring slowly into the crankcase oil-fill cup. A funnel allows easier filling through the narrow neck of the filler tube.

air breather should be cleaned or replaced. The sparkplug should also be removed to check for engine fouling or any indication of a future problem in ignition. Install a new sparkplug if the old one shows signs of burned electrodes. The various recommended sparkplugs used in most engines are AC type GC46, Champion type J-8, or Autolight type A71. The mower or other piece of equipment may be cleaned at this time also. A clean machine will perform much better than one caked with grass, dirt, or grease. One important rule to follow when working on any engine-driven piece of equipment is to *disconnect the sparkplug first*. The sparkplug wire is disconnected from the plug and grounded to the housing of the engine. This will prevent any accidental starting of the engine, which could damage critical engine parts or severely injure the operator.

Winter storage of gasoline-powered equipment. When equipment is used only during the summer months (such as mowers), the engines should be carefully inspected and stored properly. The equipment should first be thoroughly cleaned to remove dirt, grease, and rust. This cleaning will reveal areas where unusual wear is taking place. Rusted areas may be repainted to prevent further damage. The air cleaner is removed and cleaned, relubricated, or changed as specified by the manufacturer (Figure A–5). The gasoline should be drained from the tanks also. Gasoline left in the tanks for extended periods will accumulate water and possibly some dirt, which may damage the engine. Gasoline stored in cans over winter will also become fouled. Only fresh gasoline should be placed in the engine tanks in the spring.

Engines that will be stored in unheated buildings during winter should be drained of their oil in the fall. The crankcase is then filled with a low-viscosity oil (10 weight). The engine should be cranked over periodically during the winter to keep oil around all the engine parts. In the spring when the equipment is to be used again, the oil is again removed and replaced by the recommended type for summer use.

Battery care. Larger machinery having batteries should be stored with a full charge. The battery should be checked periodically to maintain this full charge or it may freeze. If this is not feasible, the battery should be removed from the tractor and stored in a heated building. The battery terminals should be cleaned periodically during normal operation to maintain a good electrical contact. This will ensure that the battery is kept fully charged. Water must be added periodically, since the charging process evaporates water but not the acid. The water level should be checked routinely and kept at the prescribed point on the indicator. Distilled water or at least soft water is best used in batteries. Ordinary hard water or well water con-

FIGURE A–5. Routine maintenance of all power equipment should include the cleaning or replacement of the air filter.

tains large quantities of dissolved minerals, which will reduce the life of the battery.

Battery corrosion may be easily cleaned for better operation. The crusted corrosion is first brushed from around the terminals and the top of the battery. The battery cables are then removed and cleaned with a steel brush to remove most of the corroded particles. A mixture of baking soda and water is used to neutralize the acid and clean the terminals down to the bare metal. Once cleaned and rinsed, the cables are replaced and tightened securely.

Safe Operation of Power Equipment

The safe operation and handling of power equipment is a very important part of employment. Not only is this equipment potentially dangerous to workers and the operator, but it is also very expensive to replace. Every employee who operates power equipment should be familiar with a few of the fundamental principles of their safe operation:

1.   A mower should be started on a level, clean surface. Loose objects (such as sticks, nails, or rocks) may be swept up by the blades and projected for a great distance.

2.   Start a manually operated mower or other machinery by firmly placing one foot on the housing, while the other foot is a safe distance away. This will hold the mower against the ground and prevent the blades from being lifted into the air and possibly coming in contact with the operator's feet.

3.   Always operate a mower with either a grass catcher or a deflection plate in place on the discharge chute. Avoid pointing the discharge at any other person while the mower is in operation.

4. Always dress appropriately for the work to be done. Wear durable shoes and trousers with long legs when operating tillers, dirt shredders, and mowers. Flying debris is sometimes discharged from the machines and may strike the operator.

5. Disconnect the sparkplug wire and ground it securely to the engine housing before working on any part of the machine. This is particularly important when placing the hands anywhere around mower or tiller blades. This equipment is extremely dangerous and should be handled in the same manner as a weapon. The sparkplug can inflict a serious shock when in operation, so for no reason should it be touched unless the machine is turned off.

6. Always fill the gas tank before an engine is operated. If the machine runs out of gas during its operation, move it to a safe location and let it cool before more gas is added to the tank. A hot engine can ignite spilled gas, causing a fire or flashback which could explode the gasoline can while it is in your hands. Never fill a gasoline tank around a flame or source of a spark.

7. When a lawn mower is to be left idle after a period of operation, place it on level pavement or bare ground. The heat from the manifold can accumulate under the mower housing and may kill the grass in this area. Gasoline should be added to the tank in the same locations. Spilled gasoline will burn grass blades and form an unattractive spot in the yard for several weeks.

8. Read the operator's instruction manual before operating any piece of equipment with which you are unfamiliar. Have an experienced operator instruct you in the potential hazards or peculiarities of the machinery. An unexpected problem could require an instantaneous decision and quick action to avoid serious damage to health and property.

APPENDIX H

Metric Conversion Factors

APPROXIMATE CONVERSIONS TO METRIC MEASURES

SYMBOL	WHEN YOU KNOW	MULTIPLY BY	TO FIND	SYMBOL
		LENGTH		
in	inches	2.5	centimeters	cm
ft	feet	30	centimeters	cm
yd	yards	0.9	meters	m
mi	miles	1.6	kilometers	km
		AREA		
in²	square inches	6.5	square centimeters	cm²
ft²	square feet	0.09	square meters	m²
yd²	square yards	0.8	square meters	m²
mi²	square miles	2.6	square kilometers	km²
	acres	0.4	hectares	ha
		MASS (weight)		
oz	ounces	28	grams	g
lb	pounds	0.45	kilograms	kg
	short tons (2000 lb)	0.9	tonnes	t
		VOLUME		
tsp	teaspoons	5	milliliters	ml
Tbsp	tablespoons	15	milliliters	ml
fl oz	fluid ounces	30	milliliters	ml
c	cups	0.24	liters	l
pt	pints	0.47	liters	l
qt	quarts	0.95	liters	l
gal	gallons	3.8	liters	l
ft³	cubic feet	0.03	cubic meters	m³
yd³	cubic yards	0.76	cubic meters	m³
		TEMPERATURE (exact)		
°F	Fahrenheit temperature	5/9 (after subtracting 32)	Celsius temperature	°C

Appendix H Continued

APPROXIMATE CONVERSIONS FROM METRIC MEASURES

SYMBOL	WHEN YOU KNOW	MULTIPLY BY	TO FIND	SYMBOL
		LENGTH		
mm	millimeters	0.04	inches	in
cm	centimeters	0.4	inches	in
m	meters	3.3	feet	ft
m	meters	1.1	yards	yd
km	kilometers	0.6	miles	mi
		AREA		
cm²	square centimeters	0.16	square inches	in²
m²	square meters	1.2	square yards	yd²
km²	square kilometers	0.4	square miles	mi²
ha	hectares (10,000 m²)	2.5	acres	
		MASS (weight)		
g	grams	0.035	ounces	oz
kg	kilograms	2.2	pounds	lb
t	tonnes (1000 kg)	1.1	short tons	
		VOLUME		
ml	milliliters	0.03	fluid ounces	fl oz
l	liters	2.1	pints	pt
l	liters	1.06	quarts	qt
l	liters	0.26	gallons	gal
m³	cubic meters	35	cubic feet	ft³
m³	cubic meters	1.3	cubic yards	yd³
		TEMPERATURE (exact)		
°C	Celsius temperature	9/5 (then add 32)	Fahrenheit temperature	°F

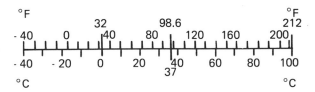

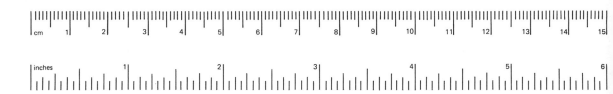

Index

Accent plantings, 238, 383, 385–387, 400
Accented neutral color scheme, 270
Acclimatization, 223–228
Adventitious shoots, 468
Agronomist, 22
Almanac, 6
American Association of Nurserymen (AAN),
    12–14, 54
American Plant Register of Horticulture, 12
Analogous color scheme, 270
Aphids, 141, 234
Appleseed, Johnny, 8–9
Apprentice Training Service, 14
Arboretum, 22
Arboriculture, 22
Arborist, 22, 35
Artificial soil mixes, 120, 125, 127–128, 147–148,
    167, 196–198, 369–370
Asexual propagation, 123–127, 147, 166–167, 170,
    174–175, 347–355
Asymmetry (design), 266, 382
Automatic irrigation, 134, 136–138, 371–372
Axillary buds, 467
Azaleas, 64–65, 165–168, 373, 456–457

Background trees, 395
Balance (design), 86, 266–267, 282, 291, 337, 382,
    397–398
Balled-and-burlapped stock, 52, 364–366, 453–456
Bare-root stock, 52, 364–365, 451–453

Bark amendments, 128, 369, 461–462
Bartrom, John, 6
Bartrom, William, 8
Battery care, 514–515
Bedding plants
    containers, 130; judging, 57, 84–86; propagation,
    119–123, 138, 147, 201
Benches, greenhouse, 101–106
Bench grafting, 352
Bent neck, roses, 80, 82
Bentgrass, 421–422
Bermudagrass, 423
Bluegrass, 421–422
Bosenberg, H. F., 13
Botanical garden, 4, 6, 22
Bouquets, floral, 279–285, 311–312
Boutonniere, 300–301
Bows, 297–299
Breeding, 9–10, 13–14, 15, 22, 30–31
Bromeliads, 206–207
Bud grafting, 181, 351, 354–355
Buffalograss, 424
Bulbs, 169–171
Bullhead, roses, 80–82
Bunker, golf, 442–443
By-pass growth, 64–65

Cacti, 207–208
Calceolaria, 509–510
Carnation, 16, 68–73, 173–179, 256, 299–304

Case-cooled bulbs, 170
Cash-on-delivery (C. O. D.), 250, 326
Centipede grass, 423–424, 427
Chemicals
   height control, 157, 162–163, 172; pesticides,
   143–147; pruning, 373; rooting hormones, 126,
   166–167; sterilants, 358, 375
Chinch bugs, 440
Chlorosis, 229–230
Chrysanthemums, 9, 65–68, 73–77, 152–160, 257,
   507
Circular floral designs, 281
Climbing roses, 13, 486
Coldframe, 115
Color, design, 269–271, 337–338
Colorado Carnation Growers Association, 16
Commission marketing, 8
Complementary color scheme, 270
Compost, 463–464
Conifers, 52, 363, 388–389, 479–480
Conservatory, 11
Containers
   bedding plants, 130; floral design, 271–273;
   greenhouse, 128–130; hanging baskets, 214–215;
   nursery, 369; planters, dish garden, 204;
   terrarium, 191–192
Container stock, 16, 52–53, 346, 368–374, 453–457
Controlled temperature forcing, 170–171
Convection tube ventilation, 110–112
Cooling, greenhouse, 110–114
Cool season grass, 421–422, 432–433
Copper naphthanate, 104
Corsages, 290–311
Crescent floral design, 279, 283
Crop manager, 21, 24–25, 30
Crossbenching, 102–103
Crown buds, 73–74, 76
Crystal Palace, 11
Customer types, 326–329
Cut flowers, 57, 63, 68–82, 153–160, 173–185,
   255–262
Cuttings, 123–127, 153–155, 161–162, 166–167,
   174–175, 347–350
Cyclamen, 510–511
Cyclamen mite, 234

Damping-off disease, 120, 359, 441
Dark cloth treatment, 155–156, 159, 162–163
Deficiency symptoms, 227–231
Deliveries, 21–22, 29, 249–251, 260
Design elements, 268–271

Design principles, 265–268
Detached houses, 95
Diagnosing Plant problems, 228–236
Dilution ratio, 132–134, 504–507
Disbudding, 68, 69–73, 74–77, 157–158, 177, 507
Diseases
   greenhouse plants, 139–140, 162; house plants,
   233–234, 239; lawns, 440–441; roses, 488
Dish gardens, 201–208
Displays, 336–339
Draftsman, 22, 34
Dutch houses, 96–97

Easter lily, 168–173
Elements of design, 268–271
Emphasis (design), 265–266, 291, 382, 385–387, 400
Enframement, 395
Entryway plantings, 385–387, 400
Environmental control, 107–114
Environmental Protection Agency (E.P.A.), 236
Equipment
   maintenance, 511–516; pesticide application,
   144–146; pruning, 470; storage, 514–516
European design, 264–265
Extension Service, 22

Fairchild, Thomas, 9
Fairway, golf, 442
Faneuil, Andrew, 10
Feathered carnation, 302–304
Fences, 7, 393–394
Fertilization (fertilizer)
   analysis, 131–134, 160, 215–216, 222, 372, 432–433,
   504–507; container stock, 371–372; elements,
   131–132, 431–432, 466; formulations, 160, 215–216,
   420; greenhouse plants, 130–134, 160; house
   plants, 215–216, 222, 227–231; injector
   (proportioner), 132–138, 160, 371, 504–507;
   landscape plantings, 453–454, 465–467; lawns,
   430–433; mixing, 131–134, 148, 160, 171, 504–507;
   nursery, 357–358, 371–372; slow release, 227, 371,
   466; soluble liquid, 132–138, 148, 215–216,
   371–372, 466; trees, 465–467; tube watering,
   136–137, 361, 371–372
Fiberglass greenhouses, 98, 100–101
Field stock, 346, 359–368
Florafax,® 334
Floral designer, 21, 26–27, 253–255, 260, 264, 279
Floral designs, 57, 86–87, 279–318
Floral foam, 274–275
Floral picks, 276–277

Floral preservatives, 257
Florist industry, 10–11, 21, 23–29
Florists' Transworld Delivery Association (FTDA), 13, 344
Flower shop manager, 21, 249
Flower types, 273–274
Foliage plants, 57, 82–84, 193–194, 202–204, 209–211
Foliar diseases, 233–234
Form (design), 269
Foundation plantings, 384–388
Franklin, Benjamin, 6
Full-service shop, 247–248
Fumigation, soil, 370
Funeral designs, 312–318
Future Farmers of America (FFA), 42–44, 59

Garden center, 22, 32–33
Gasoline engines, 512–514
Genetics, 9–10
Germination, seed, 119–123, 201, 356–359
Gladiolus, 77–79, 257, 306–307
Glass greenhouse, 10–11, 94–99
Gloxinia, 508–509
Golf course, 22, 35–36, 441–444
Grading stock, 15–16, 178–179, 368, 375
Grafting, 181, 351–355
Grass, 415–444
Greenhouses, 10–11, 93–116, 347, 374
Greenhouse worker, 21, 23–24, 116, 147–148
Greens, golf, 5, 443–444
Greenskeeper, 22, 35–36
Greenwood cuttings, 349
Grew, Nehemiah, 9
Ground benches, 104–105
Groundcovers, 52, 402–408
Groundskeeper, 22, 36–37
Growth retardants, 157, 162–163, 172
Gutter-connected greenhouses, 95–96

Ha ha's, 7
Hanging baskets, 208–216
Harmony, design, 86, 268
Harvesting crops, 178–179, 183–185, 364–366, 374
Haydite (calcine clay), 127
Heating, greenhouse, 107–112
Heikes, W. F., 13
Herbicides, 371, 374, 538–539
Herbs, 4–6
History
    floral design, 264–265; greenhouses, 10–11, 93–94; nursery, 6, 12–14; ornamental horticulture, 3–16

Hogarthian curve design, 285
Homestead Act, 12
Horizontal floral designs, 283–285
Hot bed, 115, 347
House plants, 187–216, 219–242
Hue (color), 269–271
Humidity, 112, 125, 136, 139–140, 192, 221, 226–227

Indole-butyric acid, 166, 350
Insects
    greenhouse plants, 140–143, 234–236; house plants, 234–236; lawns, 439–440; roses, 488
Intensity (color), 269–270
Interior design, 21, 28, 238–239
Interior displays, 336–338
Interior landscape and maintenance, 21, 28, 238–239
International Code of Nomenclature for Cultivated Plants, 17
Irrigation
    greenhouse, 120, 132–138, 148; house plants, 220–223, 231–232; lawns, 429–430; nursery, 362, 371–372

Jamestown, Virginia, 4–5
Jefferson, Thomas, 7–8
Judging
    floral designs, 86–87; floriculture contest, 57–88; horticulture contest, 42–88; landscape design, 54–57; nursery contest, 45–57

Land-Grant colleges, 12
Landscape
    architect, 22, 33; contractor, 22, 34; design, 379–412; designer, 22, 33; judging, 54–57; maintenance, 22, 34, 447–488; principles, 383–402
Lawn
    cool season, 421–422, 432–433; establishment, 416–429; fertilization, 419–421, 430–433; irrigation, 429–430; maintenance, 429–435; mowing, 433–435; pest control, 438–441; renovation, 435–437; seeding, 418–421, 425–426; vegetative propagation, 426–429; warm season, 422–425
Layerage, 87, 350–351
Leaf burn, 230
Lengthwise benching, 102–103
Lewis and Clark Expedition, 7–8
Light acclimatization, 224–226
Light requirements, 240–242
Lime, 131, 171, 361, 419–420, 450–451
Line (design), 269

Line flowers, 273
Liners, 167–168, 359, 361–362
Liquid fertilization, 132–138, 215–216, 371–372, 466, 504–507
Logan, Martha, 6–7
Long-day treatment, 155–156, 159, 162–164, 507–508
Loudon, J. C., 11

Maintenance
  equipment, 511–516; landscape, 22, 34, 447–488;
  lawns, 415–444; terrarium, 200–201
Mass flower, 273
Mathematics, sales, 325–326
Mat irrigation, 105–106
Maze gardens, 5
McMahon, Bernard, 9
Mealy bug, 142, 235
Mendel, Gregor, 9–10
Merchandising, 336–339
Metric conversion table, 517–518
Minor elements, 131, 466
Mist propagation, 125–127, 161, 166–167, 174–175, 347–348
Monochromatic color scheme, 270
Morrill Land-Grant Act, 12
Mound layerage, 350–351
Mower operation, 515–516
Mowing, 433–435
Mulches, 460–465

Naphthalene acetic acid, 166, 350
Narrow-leaved evergreens, 52–53, 363, 388–389, 479–480
National Horticulture Judging Contest, 42–47, 59–60
Natural cooling (bulbs), 170
Needlepoint holder, 275–276
Net (corsage), 295–297
Nosegay (bouquet), 10, 310–311
Nursery
  careers, 21–22, 30–35; history, 6, 12–16;
  production, 345–375
Nutrient deficiency symptoms, 227–231
Nutrients, plant, 131–133, 431–432, 466
Nutritional acclimatization, 227–228

Observer, 45
Occupations, 19–38
Oil, 512–514
Orangeries, 94
Orchids, 304–306
Oriental designs, 265

Pad cooling, 112–115
Paint, 106–107
Pallets, 103, 105
Park superintendent, 22, 36
Parts per million (ppm), 133–134, 504–507
Patents, plant, 13, 15
Paxton, J. C., 11
Peanut flower, 80–82
Peat, 125, 127, 162, 232, 369–370, 463
Peninsular benching, 102–103
Perlite, 125, 127, 162, 370
Pest control
  greenhouse, 138–148, 195, 201; house plants,
  232–238; lawns, 438–441; nursery stock, 374–375;
  roses, 488; trees, 480–485
Pesticides, 143–146, 148, 236–238
Petty cash, 326
pH, 131, 361, 419–420, 450–451, 456–457
Photoperiod, 66, 155–156, 159, 162–164
Pinching
  azaleas, 167; carnations, 175–176;
  chrysanthemums, 156–158, 506; poinsettia, 162,
  507; roses, 182–183
Pine bark, 128, 369, 461–462
Plant breeding, 9–10, 13–15, 22, 30–31
Plant care (houseplants), 219–242
Plant polish, 221
Plant quarantine inspector, 22
Plant store, 187–216, 219–242
Plant Variety Protection Act, 15
Planters, 201–208
Planting
  azalea, 167, 456–457; carnation, 175;
  chrysanthemum, 153–155; container stock,
  453–456; dish garden, 205–206; Easter lily, 171;
  hanging basket, 214–215; nursery stock, 451–457;
  poinsettia, 161–162; rose, 181–182; terrarium,
  195–199
Plot plan, 380
Plugging lawns, 426
Poinsettia, 160–165, 508
Potted plants, 57, 63–68, 152–173, 258, 507–511
Precooling, 168, 170–171
Preservative, 106–107, 257
Primary colors, 269
Prince Nursery, 8
Principles of design, 265–268
Propagation
  asexual, 123–127, 147, 153, 161, 166, 170, 174–175,
  181, 348–355, 426–429; azalea, 166–167; carnation,
  174–175; chrysanthemum, 153; Easter lily, 170;
  grafting, 351–355; lawn grass, 425–429; layerage,

350–351; mist, 125–127, 161, 166, 174–175, 347–348; nursery, 347–359; poinsettia, 161; propagator, 22, 30; rose, 181; seed, 119–123; 147, 356–359; 425–426
Proportion, design, 267, 337, 382
Proportioner (fertilizer), 132–137
Pruning
   azaleas, 167; container stock, 372–373; equipment, 470; landscape plants, 467–488; nursery, 362–364, 372–373; roots, 364; roses, 182, 485–488
Public area, 383

Quarantine inspector, 22
Quonset-style greenhouse, 98–100

Radiating floral designs, 282
Raised benches, 102–103
Recreation area, 383
Reel mower, 433–435
Renovation, lawn, 435–437
Rest period, 356
Retaining walls, 410–411
Rhododendron, 9, 373, 456–457
Rhythm (design), 267–268
Ribbon, 297–299
Rock garden, 408–412
Rock wall, 410–411
Root pruning, 364
Root rots, 232
Rooting hormone, 166, 350
Rootstock, 181, 351–355
Roses, 13, 59, 79–82, 180–185, 256, 485–488
Rotary mower, 433–435
Ryegrass, 422

Sales
   clerk, 21–22, 28, 31–32, 251–253, 321–340; manager, 21–22, 25; techniques, 321–340
Scale (design), 267, 291, 397–398
Scale insect, 142–143, 235
Scarification, seed, 357
Scion, 181, 351–355
Screen, 393–396
Secondary colors, 269
Seed
   germination, 119–123, 201, 356–359; soil mixes, 120, 358
Shade (color), 270
Shade trees, 52, 363–364, 394–395, 398–399
Shadehouse, 116, 347
Shakers, 8

Short-day treatment, 155–156, 159, 162–163, 507–508
Shrubs, 363, 389–394, 448, 451–459, 465–480
Side graft, 352–354
Sidewalks, 381, 387
Simple layerage, 350–351
Sleepiness, 69, 72
Slow-release fertilizer, 227, 371, 466
Society of American Florists (SAF), 12
Sod, 22, 37, 427–429
Sod webworms, 440
Softening (design), 383–385, 400
Soil
   acidity, 131, 361, 419–420, 450–451, 456–457; compaction, 435–436; elements, 131–132, 431–432, 462; sterilization, 120, 138–139, 358, 370; testing, 361, 419
Soil mixes
   greenhouse, 120, 123, 127–128, 147–148, 162, 167, 171, 175, 358; nursery, 369–370; planters, 206; terrarium, 196–198
Soil preparation
   greenhouse, 120, 123, 127–128, 147–148, 167, 171; landscape, 449–450; lawns, 416–421, 427–429; nursery, 357–358, 360–361
Spacing
   flowers, 175–176, 181; landscape plants, 389–391, 401–402
Specialty shop, 248
Spider mite, 141–142, 234
Splice graft, 352–353
Split calyx, 69–73, 179
Split complementary color scheme, 270
Sprays, funeral, 312–316
Sprigging, lawn, 426
St. Augustine grass, 424
Standard grades, 179, 375
Steam sterilization, 120, 138–139, 358, 370
Stem supports, flowers, 277–278, 291–294
Stock blocks, 359
Stock buyer, 21, 29
Stock plants, 124
Stock solution (fertilizer), 132–134, 504–507
Storage
   equipment, 514–516; plants, 173, 368
Stratification, seed, 356
Succulent plants, 207–208
Surveyor's plot plan, 380
Symmetry (design), 266–267, 382

Tall fescue, 422
Tape, corsage, 294–295
Tees, golf, 442

Teleflora Delivery Service (TDS), 334
Telephone sales, 333–335
Terrarium, 188–201
Tertiary colors, 269
Texture (design), 269, 385–386
Thatch, 434, 436
Thermoperiod, 166
Timber Culture Act, 12
Tint (color), 270
Topiary, 5, 7, 373
Toxicity, pesticide, 143–144
Trace elements, 131, 466
Trade Mark Law, 13–14
Transition planting, 383, 388
Transplanting, 122–123, 359
Trees
    fertilization, 465–467; landscaping, 394–402;
    planting, 451–456; pruning, 22, 35, 480–485;
    staking and guying, 363–364, 459–460; wound
    repair, 482–484; wrapping, 458–459
Triangular floral design, 279, 282
Turfgrass management, 22, 35–37, 415–444

Understock, 351–355
United States Department of Agriculture (USDA),
    13, 15
Unity (design), 87, 265, 268, 337
Utility area, 383

Value (color), 270
Vegetative propagation, 123–127, 147, 153, 161, 166,
    170, 174–175, 181, 348–355, 426–429
Veneer graft, 352–354
Ventilation, greenhouse, 110–114

Vermiculite, 127–128, 370
Victory Garden Program, 14
Vines, 406–408
Vocational Education Act, 14–15

Walcott, Henry, Jr., 6
War Manpower Commission, 14
War Production Board, 14
Ward, Dr. Nathaniel, 188–189
Wardian case, 189
Warm colors, 270–271
Warm season grasses, 422–425
Watering
    greenhouse, 120, 132–138, 148; house plants,
    220–223, 231–232; lawn, 429–430; nursery, 362,
    371–372; terrarium, 199
Wedding designs, 311–312
Weed control, 371, 374, 538–539
Whip, tree, 363–364
Whip-and-tongue graft, 352
White fly, 141, 234
Wholesale florist, 21, 28–29, 259–262
Williamsburg, 5
Windbreaks, 396–397
Window display, 336–338
Winter damage, 437–438
Winter protection, 373–374, 437–438, 487–488
Wire services, 13, 344
Wiring flower stems, 277–278, 291–294
Wound repair (trees), 482–484
Wreaths, 317–318
Wrist corsage, 309

Zoysiagrass, 423